The topic of this book is "creation." It breaks down into discussions of two distinct, but interrelated, questions: what does the universe look like, and what is its origin? The opinions about creation considered by Norbert Samuelson come from the Hebrew scriptures, Greek philosophy, Jewish philosophy and contemporary physics. His perspective is Jewish, liberal and philosophical. It is "Jewish" because the foundation of the discussion is biblical texts interpreted in the light of traditional rabbinic texts. It is "philosophical" because the subject matter is important in both past and present philosophical texts, and to Jewish philosophy in particular. Finally, it is "liberal" because the authorities consulted include heterodox as well as orthodox Jewish sources. The ensuing discussion leads to original conclusions about a diversity of topics, in the philosophy of religion, including the limits of human reason and religious faith, the character of religious belief, the relevance of scientific models to religious doctrine, and the nature of the creator/creature relationship between God and the universe.

JUDAISM AND THE DOCTRINE
OF CREATION

JUDAISM AND THE DOCTRINE OF CREATION

NORBERT M. SAMUELSON

Professor of Religion, Temple University, Philadelphia

Published by the Press Syndicate of the University of Cambridge
The Pitt Building, Trumpington Street, Cambridge CB2 1RP
40 West 20th Street, New York, NY 10011-4211, USA
10 Stamford Road, Oakleigh, Melbourne 3166, Australia

© Cambridge University Press 1994

First published 1994

Printed in Great Britain at the University Press, Cambridge

A catalogue record for this book is available from the British Library

Library of Congress cataloguing in publication data

Samuelson, Norbert Max, 1936–
Judaism and the doctrine of creation/Norbert M. Samuelson.
p. cm.
Includes bibliographical references and index.
ISBN 0 521 45214 7 (hardback)
1. Jewish cosmology. 2. Philosophy, Jewish. 3. Creation – History of doctrines.
4. Rabbinical literature – History and criticism.
I. Title.
B157.C65S25 1994
296.3'4–dc20 93-46180 CIP

ISBN 0 521 45214 7 hardback

B
157
.C65
S25
1994

Contents

Preface	page	ix
Introduction		1

PART 1: THE MODERN DOGMA OF CREATION — 23

Chapter 1: Creation in Franz Rosenzweig's *Star of Redemption* — 29

Chapter 2: A critique of Rosenzweig's doctrine: is it Jewish and is it believable? — 68

PART 2: A JEWISH VIEW OF CREATION — 75

Chapter 3: Classical Jewish philosophy — 81

Chapter 4: Classical rabbinic commentaries — 107

PART 3: THE FOUNDATIONS FOR THE JEWISH VIEW OF CREATION — 155

Chapter 5: The account of creation in Genesis — 157

Chapter 6: The account of the origin in Plato's *Timaeus* — 167

PART 4: A BELIEVABLE VIEW OF CREATION — 199

Chapter 7: Creation from the perspective of contemporary physics — 206

Chapter 8:	Creation from the perspective of contemporary philosophy	241

Notes 265
Select bibliography 319
Indices 344

Preface

This book is the result of several radical revisions and close to twenty years of conversations about Judaism, philosophy and creation with friends, colleagues and students. I no longer am sure who contributed what to this final product, and I fear that I will not mention many who deserve recognition. But I will do the best that I can. The two primary forums that have allowed me to exchange ideas about creation were the graduate religious studies program at Temple University and the annual meetings of the Academy for Jewish Philosophy (AJP). I wish first to thank all of the students who have met with me on a regular basis over the past two decades to exchange ideas, read texts and argue (particularly Jacob Staub [Reconstructionist Rabbinical College], Michael Paley [Columbia University], Almut Bruckstein [Jerusalem], Martin Srajek [Champaign, Illinois] and Julius Simon [Philadelphia]), as well as my colleagues in the AJP who shared their own scholarship and ideas with me and offered open, constructive criticism of my own views as I worked them out.

A third avenue for learning available to me over the years has been Temple University's generous program of academic study leaves and faculty exchanges that enabled me to be a visiting scholar at several academic institutions, where I had exceptional opportunities to meet with scholars from all over the world and share ideas with them. In particular, I want to thank my colleagues at the Oxford University Centre for Postgraduate Hebrew Studies in the winter of 1987 (notably Ron Nettler, Kyu Jung, and Daniel and Adini Frank) and at

The Chicago Center for Religion and Science in the spring of 1992 (notably Phil Hefner and Tom Gilbert).

I mentioned above that this book has gone through several metamorphoses. It began as independent research on creation in various disciplines. At the initial level of study I received assistance from many who read first drafts of the chapters related to their expertise. I wish to express my thanks to all of them for making many thoughtful, critical suggestions that always resulted in my developing a richer, more precise, deeper understanding of the texts in question. These colleagues were Robert Wright (Temple University) [on Genesis], Ron Hathaway (Temple University) and Ken Seeskin (Northwestern University) [on the *Timaeus*], Menachem Kellner (Haifa University) [on the medieval commentaries], Gad Freudenthal (Chatenay-Malabry, France) [on Gersonides' philosophy], Don Lichtenberg (Indiana University) [on physics], and Tom Dean (Temple University) and Joseph Cohen (St. John's College) [on philosophy of religion].

The first person to see my entire manuscript was William Scott Green (University of Rochester). Through his careful reading and constructive criticism, I came to realize that I was trying to do too much in a single text. I separated my consideration of how the different Jewish and Western sources have interpreted creation from my discussion of the philosophical meaning inherent in the biblical text itself. My conclusions about the latter, thanks to the generous support of Jacob Neusner (University of South Florida), were published as part of the South Florida studies in the history of Judaism (*The First Seven Days: A Philosophical Commentary on the Creation of Genesis*. Atlanta, Scholars Press, 1992). But I had not yet decided on a proper format to present the historical study of creation that is the subject of this book until I had extensive conversations about my organizational problems with Peter Ochs (Drew University). No one's advice has been more important to me in completing this book than has Peter's.

I wish to dedicate this book to everyone whose personal affection has supported me over the years in this work. Learning and writing are extremely important parts of my life, but without their love all of my work would be empty. In particular, I want to thank my children who produced the most important creativity in my personal world – my *Enkelkinder* Ellie, Johnny, Annie and James. Finally, I want to single out James for thanks – because he specifically asked me to, but also because of his intelligence, inquiring mind, love of nature and mathematics, and (most important of all) because of his love of his SABA and SAVTAH.

Introduction

The topic of this book is "creation." It breaks down into discussions of two distinct, but interrelated, questions: What does the universe look like, and what is its origin? Answers to the first belong to "cosmology," and answers to the second to "cosmogony." Responses to both questions necessarily are speculative, since, in principle, there is no way to base an answer on direct experience. All human beings are part of the universe in both time and space. With respect to time, there always was a time before and there always will be a time after what anyone can experience. Similarly with respect to space, everyone is located within the universe. Hence, there is no vantage point from which anyone can see how the universe began or what the universe as a whole looks like. The best we can do is use what we know to infer answers to both questions about creation. Yet, there is no reason why there should be any legitimacy to this logical move. The universe may have been entirely different at its origin than it became afterwards. Similarly, the characteristics of the universe as a whole may be entirely different from the characteristics of any of its parts. Consequently, no answers to questions about creation can be called "knowledge." At best they are "good stories," i.e., imaginative and/or informed opinions.

The opinions about creation to be considered in this book come from the Hebrew scriptures, Greek philosophy, Jewish philosophy and contemporary physics. As such the perspective in this work is Jewish, liberal and philosophical. It is "Jewish" because the foundation of the discussion is biblical texts that are interpreted in the light of traditional rabbinic texts. It is "philosophy" because the subject matter is important in both past and present philosophical texts in general, and to Jewish philosophy in particular. Finally, it is "liberal" because the authorities consulted include heterodox as well as orthodox Jewish sources.

THE LIMITS OF HUMAN REASON

How the universe began matters to me because it is a fundamental dogma of Judaism, but also because it is in and of itself an important question for two different (but related) reasons. First, to know myself includes knowing how I am a part of the universe, and to understand the universe involves knowledge of its origin as well as its end. To say the same thing in a slightly different way, the origin of the universe is not an event in the past at time zero. As the killing of Lincoln is a single act that at least spans the moment that John Wilkes Booth pulled the trigger to the moment of Lincoln's death, so the creation of the universe is a single act that spans the moment of its origin to its end. Second, to know God includes knowing creation. As a consequence of his radical unity, God has a single act with a single product, and there is no distinction between God as agent, his act, and the product of his activity. In other words, the creator is the act of creation, which in turn is the created world.

While I want to know creation, I also recognize that, from any absolutist perspective, I never will know it. Perfect knowledge always remains a limit-ideal, where to know is pictured as an asymptote. The use of this mathematical characterization is modern, but not the claim about the limits of knowledge itself. Every text examined in this book consistently affirms that creation is more or less but never completely knowable. Timaeus, like Pythagoras, could not avoid positing an element of the irrational in his construction of the physical world. Since "1" is the unit measure, the hypotenuse of his elementary isosceles right triangle has a length of $\sqrt{2}$. Similarly, one adjacent side of his elementary half-equilateral scalene right triangle has a length of $\sqrt{3}$. Both are irrational numbers. These facts in themselves were sufficient motive for him to insist that the universe cannot be only a rational place generated by the mind of the deity. Similarly, while Ibn Daud and Maimonides considered their stated cosmology to be the most reasonable picture of the universe, they did not consider it to be knowledge. Rather, they maintained that in principle human science could attain nothing more certain than probable opinions in this area. Only Gersonides believed that the creation of the universe could be demonstrated. However, no facts were at stake in this issue. The data from astronomy led Ibn Daud and Maimonides to conclude that human beings have no knowledge whatsoever in this area. Gersonides argued that this conclusion is too strong. Certainly there is a difference between human and divine knowledge. What God knows as a cause of the object known in a single act is knowable to human beings only as an effect of the object known through what could be expressible only in an infinite conjunction of distinct propositions. While human knowledge has divine knowledge as its goal, this end can only be approximated. Ibn Daud and Maimonides would not consider "approximate" truth to be knowledge; Gersonides did.

To know what is approximately but not definitively true is to know a story. All of the accounts of creation considered in this book, including that of modern physics, are a story. None of them can purport to be a detailed, literal explanation of the origin of the universe. In the *Timaeus* Plato offered as proof that an ideal state can be made actual, the claim that the universe itself is such a state. The deity formed a rational model of an ideal living creature and materially generated it. That living organism is the universe. The details of this story are Timaeus' account of its creation. The story is a MYTHOS, which Timaeus describes as a kind of "bastard reasoning" where you can give a probable account of something like the space of the universe that sort of exists, but not really. Timaeus' description of creation, like its counterpart in Genesis, is a, but not *merely* a, *story*.

The term "myth" has been employed throughout this book in the way that Plato used it in the *Timaeus*. It is the method of expression appropriate to what exactly is vague and precisely is uncertain. In this context, if both the Genesis account of creation and the *Timaeus* within the context of the other literature of their times are read as movements away from the major role of myth in scientific explanations, the midrash[1] clearly reversed this trend. However, this reversal in direction was only temporary. While subsequent accounts of creation became increasingly elaborate, the role of mythology in them decreased significantly. Still, the medieval Jewish philosophers also told stories about creation rather than gave scientific explanations. Like the author(s) of scripture and Plato, they used the language of art to say that the morality play of creation is more a product of divine will/desire/intention than a chance happening of necessity, where morality and the law have primacy over any other possible kind of value. However, Rosenzweig's new thought of creation explicitly utilized Plato's use in the *Timaeus* of myth. For Rosenzweig, it was the appropriate way to discuss something that on one hand is not something but on the other hand is not nothing at all.

Given the limitations of human language, any theory of creation must be, as it always has been, a myth, viz., a story that pictures all that is in general, that is informed, however inadequately, by everything that we know in particular. It is a mode of thinking as appropriate to the contemporary period as it was to the worlds of the Hebrew scriptures and the classical rabbis.

THE RELEVANCE OF SCIENTIFIC MODELS OF CREATION

The starting point for determining Jewish (or Christian) dogma is the inherited text(s) of the Bible. In general, no biblical text (including those relevant to creation) is so specific that it can determine, by itself, any doctrine. However, the biblical text is not so vague that it can mean anything. On the contrary, an honest reading of the text will limit the range of possible interpretations. In other words, while the Torah sets limits on interpretation, it does not determine the belief. Rather, within its boundaries, total freedom is given to speculation. In fact such speculation is itself a duty of the law. With specific reference to the biblical doctrine of creation, we will see that the boundaries are that there exists something that is nothing, out of which God, through an act of will, creates, throughout eternity, a universe that, in virtue of his intention, has meaning and moral value. However, this is only a general outline of what a belief in creation would look like. It is no more specific about creation than the imperative, "You shall not steal" is specific about laws of private property. To determine in detail which specific possible cosmogony and cosmology within the parameters of the biblical text is true is in itself as much a positive religious duty as it is to legislate concrete civil laws within a state. As in the case of Jewish law we must turn empirically to the data of legal case precedence, so in the case of Jewish philosophy we must turn to the results of empirical studies in science.

Prima facie Rosenzweig's writings present a strong (if not the strongest) argument against my claimed dependence of a belief about creation on scientific studies. He argued that the source of theology's content claims about the origin and general nature of the universe should be revelation. His understanding of the theology that transcends philosophy was in form at least a return to midrash. Nachmanides divided the forms of biblical commentary into four kinds – homiletic or midrashic, linguistic or grammatical, philosophical, and mystical. Rosenzweig had rejected both the philosophical and mystical medieval ways of interpreting holy texts in favor of what he called "grammatical analysis," by which he meant linguistic/grammatical commentary. However, it would be better characterized as a union of both homiletic and grammatical analysis.

In response I want to claim that Rosenzweig's position is a result of the radical separation he made between logic and language, and that separation has no justification either in Rosenzweig's arguments or in fact. While he recognized that thinking about reality would require a language that could encompass depth, i.e., a set of signs that could express more than two dimensions, he mistakenly inferred that such thought must transcend reason and concluded that theology, from the perspective of mathematical thinking, must be a "mystery." However, merely to think multidimensionally in terms of algebra in no sense is mysterious. Rosenzweig drew lines where no separation is called for. In fact the mathematical symbolism of modern physics is more than adequate to picture Rosenzweig's demands for depth.

Our primary source for informed opinions about the origin and nature of the universe are the models developed in particle physics and astrophysics. Yet, to understand them presupposes a level of mathematical sophistication far beyond that presupposed even by the *Timaeus*. At the same time, it is not the case that the mathematics alone can present an adequate tale of creation. The mathematics employed by the physicists are a pure formalism whose interpretation is left for philosophers and theologians. This task can be understood as a kind of (fairly sophisticated) modern midrash. In other words, the consequence of rejecting Rosenzweig's radical separation between science and religion does not diminish the value of studying the Hebrew scriptures and midrash; rather, it increases the value of philosophical commentaries on the biblical text for determining religious dogma.

LIBERAL JEWISH PHILOSOPHY

That my work in philosophy is "Jewish" means that it is not "Christian," "Muslim," 'Buddhist," etc., but it does not mean that it is not "general." The use of religious adjectives to modify the noun "philosophy" both clarifies my sources and places the study within the general category of philosophy of religion. However, there are no such things as religious truths any more than there are philosophical, sociological, scientific, etc. kinds of truths. The adjective specifies the discipline but not any special kind of truth. (I am not an Averröist.) At the same time I do not believe there is some kind of generic brand of philosophy of religion that can simply be called "philosophy of religion." When people in Western civilization say they are doing "philosophy of religion" without the qualification of a specific religious tradition, what most likely they are doing is "Christian philosophy," and in their naivete they are unaware that what they have to say is specifically Christian.

There is a difference between what I have called "factual" and "moral" truths. The former are truths about what was, is and will be the case; the latter are truths about what ought to be the case. However, the meaning of the term "truth" is the same in both cases. What is different are the cases to which these values apply and the rules of evidence for making these judgments. However, this has nothing to do with a distinction between what kinds of claims fall within the respective domains of science and religion.

The most apparent implication of the above paragraph is my operative assumption that reasoning necessarily is contextual. Whatever anyone says is inescapably tied to being a certain person with a specific background bound by both time and place. However, contextual does not mean subjective. The use of reason on this or any topic is intended to determine truth, and truth is not contextual. There is a critical difference between the truth status of a claim and the grounds for making a claim. The claim itself, to the extent that it is clear, is either true or false. However, context inescapably affects the status of the evidence. We reason because we have no more reliable way to judge truth. At the same time we need to recognize that we can never be sure that what we reason to be true ultimately is true. Our reasoning inescapably is limited by our ignorance – ignorance of unimagined relevant facts, hidden assumptions, hidden ambiguities, and hidden alternative ways of formulating our question. These unavoidable limitations mean that all conclusions remain tentative. The best arguments are only that, viz., the most reasonable case possible at this time and place. Certainty necessarily is not attainable.

It might be claimed that what it means to say that all reasoning is contextual is that it is determined by adopting certain premises. This interpretation is not incorrect, but it is misleading. Obviously, inferences in a valid argument follow from the premises. However, the model of forming syllogisms is not the only model for reasoning. In fact, it may be the least applicable one to the questions that most interest Jewish philosophers.[2] On the other hand, if you mean to say that all human reasoning is unavoidably limited by all kinds of ignorance, both of facts and of alternative perspectives, then the answer is yes. Clearly how you ask a question affects your answers.

Clearly different contexts do affect truth judgments. For example, as Spinoza argued, while in general democracy would be a preferable system of government to monarchy, there may be cases where monarchy is preferable. However, all this means is that the context in which a claim is made is itself part of the claim. It does not in any way relativize the truth claim. In a context in which democracy is wrong it is absolutely wrong; in another context it would be absolutely right. A clear distinction has to be made between how you *judge* what is/ought-to-be the case and what is/ought-to-be the case.

It might be objected that by my own admission any answer to questions about creation can have only the status of an opinion as opposed to knowledge. The objection is valid if the terms "opinion" and "knowledge" are used as they functioned in medieval philosophy, where only informative claims that are necessarily true and are recognized as such fall within the domain of knowledge while all other informative claims fall within the domain of opinion. In this context I see no problem in saying that my judgments about creation at best will be opinion. However, they will not be *merely* opinion. In the medieval sense of the term, there simply does not seem to be anything that can qualify for membership in the class of knowledge. Now, given that the term so defined names an empty class whereas in ordinary usage a great many claims seem to qualify for membership, we ought to draw the conclusion that something is wrong with this definition. In fact we can differentiate the members of the class of opinions into relative grades of more and less likely views, and I would apply the term "knowledge" to the more certain members of the class. It is in this sense that I claim that my interest is in knowledge.

The recognition of these limitations is the ultimate basis for preferring historical to mathematical philosophy. The value of the former over the latter approach is that in examining what other philosophers from other ages and civilizations have said about our concern we have the best chance to minimize the role of ignorance in our reasoning. However, this is in no way to say that all judgments are equally valid. Our truth judgments ultimately depend on reasoning not because reasoning is infallible but because every other known alternative is more fallible. Furthermore, it is legitimate to say that some judgments are more reasonable than others even though it is possible that a less reasonable judgment may in fact be true. That what is less reasonable may be true is not a license for irrationality. The point is that a reasonable judgment is more likely to be true than a less reasonable one, and that is sufficient grounds for basing judgments on reason.

The price that must be paid for granting these grounds is that the term "knowledge" cannot intelligibly be restricted to claims that are certain. Rather, what it means to say that we know something is that that something is the more reasonable, conceivable alternative, and for that reason it is more likely to be true than any of the recognized alternatives. In other words, truth is a value that ranges over probability functions. To say that p is true and q is false means that p has more probability than q. In other words, all that it can ever mean to say that something is true is that it probably is the case, and there are objective standards to measure probability.

If my philosophy is qualified by the adjective "Jewish," my Jewish philosophy is qualified by the adjective "liberal." Let me explain what I mean by these terms with the following example: At the first meeting of the Academy for Jewish Philosophy in 1980 considerable time was spent discussing the importance of analyzing classical Jewish texts. I recall one particular discussion between David Bleich and me about the value of reading Gersonides. Both of us agreed that wherever relevant, as certainly is the case in talking about creation, Gersonides is to be consulted not merely because he is Jewish but because in a strictly technical sense he reflects rabbinic philosophy at its best. The fact that Bleich is a traditional Jew and I am liberal made no difference. In general, there ought to be no difference between how a liberal and a traditional Jew will do Jewish philosophy. Both will read the relevant classics of rabbinic tradition with as much care and skill as possible in order to determine what they have to say.

Bleich also said that once he had determined what Gersonides said on a question like creation or divine providence, the issue of truth was settled precisely because Gersonides is the best source in the rabbinic tradition for these questions. It is at this point that the difference between a liberal and a traditional Jew becomes significant. For me the view of Gersonides or any other rabbi is simply that, their judgment. It is a view to be treated with utmost seriousness, but it does not settle the question. In general, both a liberal and a traditional Jew as Jews will consult the voice(s) of rabbinic tradition. That examination is what makes their study Jewish. However, in the end their motives for doing the same thing are quite different. For a traditional Jew the investigation in itself is sufficient to settle truth. For a liberal Jew the reading of traditional texts is only a first step. It answers the question: What is (are) the view(s) of the tradition? To answer that question is a critical part of answering the further question: What is true? But the two questions are not the same. The most fundamental difference between a traditional and a liberal Jew is that the judgment, "This is what the tradition teaches" does not mean "This is true."

The obvious question is, if the tradition does not determine truth, then why consult the tradition at all. In the discussions between Jewish philosophers and contemporary astrophysicists on the origin of the universe at the 1984 meeting of the Academy, this was the most critical question that the astrophysicists posed.

No one denied that many of the statements by rabbis like Maimonides and Gersonides about the cosmos are false. For example, all of the medieval Jewish astronomers believed that there is a sphere of stars that are not subject to generation and corruption, while all of us know, without any reasonable doubt, that these stars are as subject to birth, growth, decline and death as are the material inhabitants of the planet earth.

Everything changes; the sole difference is the unit of time involved in that biography. Why then, the astrophysicists asked, since we clearly know that the medieval astronomers were wrong in their scientific judgments, do we pay any attention to them at all? In more general terms, given that we know that the past was wrong, what difference does it make, if our interest is in knowing truth, what the past had to say? In part, the called-for response is that in all probability any answer at any time will eventually be discovered to be wrong. Modern science is no less vulnerable this way than ancient science. However, that is not the point. While truth is absolute, truth judgments are not. While the goal is to know truth, all that we can ever know is what is most reasonable, and what is most reasonable is itself bound by time and place. We err neither because there is some better way to determine truth than reason nor because we reason poorly. We err because ignorance is unavoidable. We are not only ignorant of facts; we are also ignorant of alternative ways to ask questions. The value of consulting historical texts is precisely that, viz., ideally it opens our mind to alternates that otherwise, because of our limitations within a particular time and place, we would not imagine.

It might be objected that when I say that some factual or scientific statements about the cosmos made by the rabbis are false, I am giving authority to the claims of contemporary science over and above those of Judaism. However, the objectors in this case seem to think that people make truth judgments in the same way that politicians vote in legislatures – by party. They picture a science party with its platform, a Judaism party with its platform, and they want to suggest that truth judgments have something to do with loyalty, where to be loyal means to vote with the party. I think that this model is inappropriate. Consider the judgment of the leaders of Poland to send troops on horseback against armored German tanks. Were those who thought this was a good idea more "loyal" than those who judged it to be foolish? It seems clear to me that what mattered was to determine what in truth would be the consequences of this military strategy irrespective of any consideration of loyalty. In general terms, loyalty to a people and judgments about the truth of what that people believe ought to be independent of each other.

However, that does not answer why Jewish philosophers should examine the particular texts that they do examine. The number of possible testimonies from the past greatly exceeds the number of texts that any of us can confront. Why then choose the ones we choose? In part the answer is, because they are good. We cannot look at all rabbinic texts; we give priority to those that in strictly technical terms are the best ones, i.e., the clearest, most complete, and best reasoned. Hence, for example, Gersonides has priority over other rabbinic sources in discussing creation precisely because of this rabbi's rigor and detail in handling the question. But again, technical skill is not the only reason. Why, for example, does Gersonides' *Wars of the Lord* have precedence over Ibn Rushd's commentaries on the works of Aristotle? The answer at this level is, because Gersonides was a rabbi and Ibn Rushd was not.

I believe that this seemingly irrelevant piece of information to determining truth can be justified as follows: I consult my tradition precisely because it is mine, which means that whatever else I may look at I should at least make explicit my own intellectual heritage precisely because it is mine, and, in all probability, as part of my personal inheritance, it will affect my thinking anyway.

At this point it is important to mention another difference between a liberal and a traditional Jew as a philosopher. The shared Jewish traditions are not identical. As a liberal Jew I grant equal legitimacy to the voices of heterodox Jewish philosophers as I do to orthodox ones, provided they are good philosophers. Hence, while I as a liberal Jew am as interested in the testimony of medieval philosopher rabbis as is my orthodox counterpart, I also include among my advisors giants such as Baruch Spinoza, Hermann Cohen, Martin Buber and Franz Rosenzweig. Again, they are as much a part of my inherited Jewish tradition as are Saadia, Maimonides, Gersonides and Chasdai Crescas.

There is at least one more reason for examining my own

tradition – it reasonably functions as an epistemic authority on both factual and moral grounds. At this point I refer to Barry Kogan's "Reason, Revelation, and Authority in Judaism: A Reconstruction."[3] We listen to traditional rabbis such as Maimonides and Gersonides on questions about the origin of the universe not because they are in authority but because they are authorities. In general, where Jewish tradition has proven to be reliable for advice in the past, it is reasonable to continue to consult it. However, the voices of that tradition lose their status as authorities when, and precisely to the extent that, they fail to be reliable.

While rabbinic tradition is *an* authority, it is not *the* authority. It would be if I believed that it is the word of God. Given my unavoidable ignorance, I certainly would accept the judgment of an infallible authority over my own if I knew it. However, as influenced by God as I believe rabbinic tradition to be, it remains in my judgment in principle a human work, i.e., it is no less subject to human limitations than is any other set of texts. Consequently, in answering any questions, I am open to all reasonable authorities, be they Jewish or not. Specifically in terms of my creation project those sources include Plato's *Timaeus* as well as the best judgments of contemporary astrophysics, even though these "texts" are not "my own personal inheritance" as a Jew. (They are as an American.)

There is any number of ways that my relation to Judaism can be affected by my judgment about creation. For example, how I worship God depends on what is my conception of God, and that conception is significantly dependent on my understanding of creation. It is also true that nothing I will conclude will have the consequence that I will cease to be a Jew. But that does not mean that my judgments are independent of my commitment to Judaism. What is dependent is how I will fulfill this commitment. Support of the State of Israel is a relevant example. While I cannot imagine not being committed to the survival and well-being of the Jewish people, that does not mean that nothing can affect my support of any particular government or policy of a Jewish polity. I believe that to advocate what is either false or immoral ultimately is not in the best interest of my people, and these judgments cannot be dictated by any politicians, be they rabbis or prime ministers.

What was said above about the value of consulting historical texts entails no distinction between any "kinds of truths." Nor do I imply that the dogma of creation is or may be a religious truth (i.e., a truth to be accepted by those who believe in the truth of the Torah) which may also be scientifically false. That many Jews or rabbis believed something in the past is not the same as saying that it is a religious truth. For example, most rabbis and most Jews in the past believed in the legitimacy of the institution of slavery. Should the modern State of Israel legalize some form of this institution? To answer this question I would want to look at the best available evidence. That evidence would include the testimony of Jewish tradition, but that does not mean that there is some kind of religious truth, distinct from other truths, that says slavery is good. Similarly, while the adjectives "scientific" and "philosophical" have meaning when they modify ways of asking and exploring answers to questions, I have no idea what they mean when they modify truth values. For example, I do not understand what it would mean to say that religiously there is a sphere of fixed stars but scientifically there is not. It seems to me that there is no such sphere, although it is true that most medieval Jewish philosophers and scientists believed that there was.

Crescas asserts that heretics are people who form their beliefs independent of the teachings of the Torah. In this sense traditional Jews need not judge liberal Jews to be heretics because at least my kind will consult the tradition. To consult the tradition is to be open to its influence. It might be objected that what makes liberal Jews heretics is that ultimately they rely on reason more than the tradition. However, you cannot compare consulting a tradition with using reason. They are not the same kind of activities. Presumably you should read the texts of the tradition rationally. How else could you read them?

CREATION AS A JEWISH DOGMA

Many contemporary Jews believe philosophy never really was an authentic Jewish enterprise and that dogma no longer has a place within Judaism. Against this assertion I would argue as follows: Jewish tradition has always seen the law that God revealed to Moses to encompass everything. "Everything" includes science. The God of our ancestors, who revealed his Torah through Moses to our people, is the creator of the entire universe. He can only be known through his act(s). Hence, to learn as much of absolutely everything and anything as our brief stay on earth allows is in itself to come to know our God of holiness, justice and truth.

The contemporary emphasis in religious thought on praxis over doxis is an unjustifiable departure from the major conceptual orientation found in the chain of classical rabbinic thinking. The classical rabbis carefully explained fundamental beliefs such as God's unity and the act of creation for the same reason that they elaborately described basic duties such as observing the festivals and resting on the Sabbath. As the revealed law shows, God demands true belief as well as true practice. As the fulfillment of the duties requires a careful, detailed understanding of what is to be done, so acceptance of the doctrines demands a careful, detailed understanding of what is to be believed. Hence, it is no less a religious duty to find proper cosmological schema to interpet what scripture says about creation than it is to determine with precision what does and does not constitute working on the Sabbath or exactly when a festival begins and ends.

By "duties of the heart" (CHOVOT HA-LEVAVOT) the rabbis meant those true doctrines that Jews are obligated to believe. The phrase stands together with the "duties of the limbs" (CHOVOT EVARIM), which designates those correct forms of behavior that Jews are obligated to perform. That premodern Judaism universally maintained that the former no less than the latter is part of Jewish law (HALACHAH) means that traditional Judaism never was a mere orthopraxy and that it always involved orthodoxy, i.e., dogma. The issue then is not, does Judaism have dogma. Clearly it does. Rather, the issue is, beyond the question of what beliefs are and are not dogmatic, what does it mean to say that any doctrine is a Jewish dogma.

My account of the status of dogma within traditional rabbinic Judaism will focus on Maimonides' list of thirteen foundations of Judaism. Maimonides' way of expressing the beliefs is not the only legitimate expression, and his understanding of what is a foundation belief is not universally accepted by spokespeople for rabbinic Judaism. However, Maimonides' attempt to formulate essential Jewish beliefs occupies a place of importance in any statement of Jewish belief, at least in the sense that no formulation of what Jews believe can be adequate that does not take seriously Maimonides' judgments.

In his commentary on the tenth chapter of the tractate Sanhedrin in the Mishnah, Maimonides gives a list of thirteen foundation principles of the Torah, the first of which is the existence of the creator. Concerning all thirteen, Maimonides says that anyone who even doubts them "leaves the community (of Israel), denies the fundamental (KOFER BE-'IKKAR), is an EPIKOROS, and is one who 'cuts among the plantings.' "[4]

A close study of what Maimonides meant by calling them "foundations" and what other alternatives Jewish tradition offers on the status of belief in Judaism will yield the following conclusions.[5] It is reasonable to draw parameters of forms of Jewish belief, even though it always lies outside of those parameters to designate any single expression as *the* Jewish belief. Jewish dogmas function to specify what Jews ought to deny rather than what they ought to affirm. While this state allows considerable room for thought on any subject, that freedom is not unlimited. Not everything that is possible may be true, and not everything that may be true is true.

In the light of conceptual challenges to Judaism internally from the Karaite movement and externally from Islam, rabbis began to formulate precise statements on Jewish belief in the tenth century C.E. Initial attempts by Bachya, Judah Halevi and Abraham Ibn Daud in the eleventh and twelfth centuries culminated in Maimonides' first comprehensive formulation of his thirteen foundations. This effort is hardly noticed until the fifteenth century, when Maimonides' formulation was subject to sharp and careful examination in the light of the major Christian persecution of that century. With the decline of persecution in the next two centuries, interest in formulating Jewish dogmas declined. However, in modern times it again became important for Jewish thinkers to examine with care precisely what it is that Jews ought to believe and what it means to say that they ought to believe it.

Maimonides' list of principles of the Jewish faith were intended to define salvation from a Jewish perspective. They are distinctive beliefs in that those who deny them will have no share in salvation. However, in this critical respect Maimonides' position is unique. Abravanel would deny that this list of beliefs is any more central to Judaism than any other beliefs found in the Torah. However, what is more important, most authorities would argue that salvation has more to do with the intention of the believers than what they in fact believe. What is critical for salvation is that Jews accept the authority of God's word as revealed in the Torah, whether or not they are successful in interpreting that word. In his *Light of the Lord* (OR HA-SHEM) Chasdai Crescas presents the clearest picture of the place of dogma in Judaism. The salvation of Jews rests on their obedience or disobedience to the commandments of the Torah. However, God can only command what is voluntary, and belief is involuntary in the sense that no one can will to believe what they, for any reason, disbelieve. Judaism teaches true beliefs. The creation of the universe is such a belief. However, it can no more command this belief than it can command someone to have a healthy heartbeat. True belief can be acquired only through good education. It can neither be forced nor demanded. Consequently, while Judaism would claim that it is true that God created the universe, it cannot command anyone to believe it. One *ought* to believe this doctrine because it is better to believe what is true than what is false. However, "ought" in this sense does not entail an imperative. To make the same point in slightly different words, it is a dictate of morality that one ought to believe what is true, but this moral judgment cannot be expressed as an imperative. Not to do so is itself a dictate of morality.

It is true that few modern Jewish philosophers have emphasized creation. The two most notable exceptions to this generalization are Nachman Krochmal and Franz Rosenzweig. However, without question the author(s) of the Hebrew scriptures affirmed that the universe was created and considered this doctrine to be a principle belief of their world-view. The doctrine is not "merely" poetic. It purports to describe reality, and that claimed reality is critical in reading every other story in scripture. Similarly, no topic, including God's unity, is more fundamental to Judaism than the question of the creation of the universe. Only one pre-modern Jewish thinker, viz., Abraham Ibn Daud, did not discuss it, but none denied it. Everyone who ranked Jewish beliefs by any criteria, made creation fundamental. The first of Moses Maimonides' thirteen root principles of Judaism speaks of the existence of "the creator", and he later claimed that his fourth principle, that God is eternal, entails that God created the universe. Simon Ben Zemach Duran made creation the single most important principle, from which everything else follows. Bibago made it one of his only two basic principles. Finally both Chasdai Crescas and his student Joseph Albo relegated it to a lesser rank than Duran and Bibago, but, nonetheless, it remained for them as well a basic principle of Judaism.

Moses Mendelssohn turned Crescas' judgment that true belief, unlike right behavior, cannot be commanded, into an argument that Judaism is superior to Christianity because in Judaism there are no revealed dogmas. However, the proper conclusion from Crescas' separation of dogmas from commandments is not that Judaism has no dogmas, but that dogmas do not function in rabbinic Judaism in exactly the same way that they usually are thought to function in Christianity. In the case of Judaism dogmas function to set parameters for speculation. They specify what Jews ought to doubt rather than what they ought to believe.

METHODOLOGY

In one important sense this book is modeled on Hermann Cohen's approach to doing Jewish philosophy. Like Cohen my claims are intended to be rational, and I turn to Jewish sources to extract Jewish doctrine. However, my claims are in no sense absolute. While I would affirm that my conclusions are *a* Jewish position, and may be a more informed Jewish position than most, they are not the only possible informed Jewish position on creation. Similarly, while I would affirm that my conclusions are *a* reasonable view, and may be a more informed reasonable view than most, they are not *the* only possible reasonable belief about the origin and general nature of the universe. Furthermore, unlike Cohen, I will not examine the entire corpus of Jewish intellectual history. Rather, my Jewish sources are limited to the relevant texts in the Hebrew scriptures (interpreted in the light of contemporary biblical criticism as well as traditional rabbinic commentaries) and in the classics of (medieval and modern) Jewish philosophy.

All Jewish thought begins by studying biblical texts. I read them in the light of intellectual history because the meaning of any text at any particular time and place in history is informed by the meaning of that or any similar text at any other particular time and place, past or future, within or without the tradition of the initial text itself. Hence, breadth of vision in and of itself is a research virtue. However, it is also a vice, since breadth tends to be inversely proportional to depth analysis.

The virtue of depth in itself demands that any global approach to truth must be resisted. Resistance as a virtue, however, does not require avoidance; it only counsels caution. This book attempts to find an ideal middle path between the opposing virtues of breadth and depth while avoiding their mutually entailed vices. No single scholar can scrutinize all relevant texts to any major philosophical or scientific question. Human finitude demands that scholars impose reasonable standards for selection. The most obvious criterion for scholars choosing texts is to give preference to texts that have molded the scholar's own personal history, since these writings, whether or not they are consciously confronted, will affect the scholar's judgment. Scientific objectivity is not achieved by ignoring personal history; rather, it requires scholars to make their own subjective inheritance (i.e., their own personal, cultural tradition), itself a direct object of reflection. For this reason this book focuses on the writings of European and Jewish philosophers, while it ignores, for example, equally gifted Indian or Chinese Buddhist philosophers who thought about creation.

Even after this first level of selection, the list of relevant texts remains too broad. It is not possible for a single scholar to focus analysis with desired depth on everything of value in Western civilization or in rabbinic Judaism. A second standard of selection is necessary. In this case, because the focus of attention is a scientific/philosophical question, the criterion is comparative philosophical rigor. Texts were selected from the list of Judaica because of their historical importance and because they were judged to be in technical terms the best texts. They are the writings representative of their ages that are the clearest, most complete and best reasoned.

TEXT SELECTIONS

In conclusion, I am studying creation first because it is a fundamental dogma of Judaism, and second because it matters to me to determine, to the best of my ability, how the universe began. In all probability what I will conclude will be wrong in as yet unimagined ways, but I believe that it will be most reasonable, i.e., more likely to be true, than any recognizable alternative. In making up my mind I cannot consult everything. I will look at everything I can, but I must make reasonable choices about my texts as well as about my evaluations of the advice of my selections. Specifically in terms of the question of creation, my textual selections were made as follows:

1. Genesis 1. The Hebrew scriptures are chosen simply because the scriptures are the starting point of all Jewish thought. Genesis 1 is selected for emphasis precisely because it is the most detailed account of creation in all of the Bible.

2. Plato's *Timaeus*. This classic Greek text in philosophical cosmology and cosmogony is chosen because it was a major source of the conceptual schema used by medieval Jewish philosophers to interpret the text of Genesis.

3. The accounts of creation in the medieval Jewish philosophy of Abraham Ibn Daud and Gersonides. Attention is also paid to why Saadia could take creation for granted and Maimonides found it to be a problematic doctrine.

4. Philosophical interpretations of the text of Genesis 1 by classical rabbinic commentators, notably Rashi, Ibn Ezra, Nachmanides and Sforno. Consideration is also given to relevant early midrash, notably to the collection entitled *Genesis Rabbah*.

5. Creation in Franz Rosenzweig's *Der Stern der Erlösung*. Rosenzweig is selected because he more than any other major figure in modern Jewish philosophy singles out creation, along with revelation and redemption, as a fundamental concept of Jewish thought. Attention is also paid to Nachman Krochmal and why others – notably Baruch Spinoza and Martin Buber – do not discuss creation.

6. Contemporary discussions of the Big Bang Theory and other alternatives to a steady state model for the origin of the universe in contemporary Western astrophysics. It is in connection with these works that I want to answer the question, in the light of the advice of Jewish tradition, how was the universe created.

The above list appears in chronological order. However, this is not the order in which these texts will be examined in this book. We will begin with Rosenzweig's *Star of Redemption* as the best contemporary example of creation as a doctrine of Jewish philosophy. We will then explore the Jewish roots of his view in medieval Jewish philosophy and in classical rabbinic philosophical commentaries. This will lead us to examine the sources for their views, viz., Genesis and Plato's *Timaeus*. Finally, in attempting to judge the truth of what Jewish philosophy has to say about creation, we will look at contemporary physics.

Each text is considered in its own right, which means that I attempt in every case to open my mind to the selected written words with minimal prejudice from my inherited traditional and scholarly interpretations. In other words, each text was read initially under, to borrow a phrase from John Rawls, a "veil of ignorance." Only after this first exercise was the veil lifted gradually by determining what the isolated text means within the broader context of its own historical environment. It is recognized that what the author(s) of a text meant need not be the same as what the text in isolation means, and what the text means in isolation may differ from what it means as part of a larger whole. Hence, for example, the range of possible interpretations of Genesis chapter 1 need not coincide with that range within the context of the Book of Genesis and/or the Hebrew scriptures as a whole. Attention will be given to the meaning of the texts in both isolation and in context, but, as usually happens in judgments of case law, minimal attention will be paid to the original intent of the authors. Again, the interest is in the text and not the author. This is just another way of saying that philosophy, even liberal Jewish philosophy, is not history.

PART I:

The modern dogma of creation

Introduction

This book is a study of the doctrine of creation as it arose out of the texts of rabbinic Judaism with particular emphasis on those writings that are explicitly philosophical. I will begin with what in my judgment is the best contemporary statement of the dogma, that of Franz Rosenzweig in *The Star of Redemption*, and then work my way back through his sources in classical rabbinic commentaries on the first chapter of Genesis in the Hebrew scriptures. Of all of the material to be discussed in this book, Rosenzweig's text is the most difficult. Beyond the inherent complexity of the concepts themselves, his work is especially difficult for two reasons:

First, his language and style make his text simply difficult to understand in its own right. Rosenzweig wrote for an audience educated in Germany at the beginning of the twentieth century that was schooled in the philosophic tradition of people like Schelling and Hegel. Furthermore, he believed that good philosophy ought to be literate in the sense that its words should speak simultaneously at many levels. Furthermore, the sentences that express good philosophy should be rich in the sense that they contain allusions to the best that civilization has produced in every form of literature. All of this today functions as a barrier to readers (particularly those raised in an English language tradition) comprehending what Rosenzweig wants to communicate. Part of the problem is that most of us grow up in a tradition of philosophy modeled on empirical science, where non-literate and shallow language (generally appealed to under the slogan of "clarity and precision") are more valued. An even more important part of the problem is that Rosenzweig's allusions are to the literature of two different civilizations, neither of which formed part of our secular education. The one is that of rabbinic Judaism, shared only by the few among us who have been trained either in yeshivot or in rabbinic seminaries. The other is that of eighteenth- and nineteenth-century Germany, which is shared today by almost no one, including Germans.

Second, today most people believe that there are two "Jewish" views on creation. One is the belief of most orthodox Jews who claim that Judaism, like Christianity, affirms that the universe was created by God out of absolutely nothing at the first moment in time in a way that lies beyond any possible human understanding. Rosenzweig's concept of creation has little relationship to what they affirm. For him creation expresses an atemporal relationship between an element called "God" and an element called "the world" which is a constituent part of a schema through which everything about the temporal world is to be comprehended. The other is the view of most non-orthodox Jews who believe that Judaism, being an orthopraxis (i.e., it has definite requirements for what Jews ought to do but takes no firm stands on what they ought to believe), leaves such questions to philosophers and scientists. These Jews see nothing Jewish about the question and therefore wonder why a Jewish theologian would discuss it at all. (In fact most modern Jewish thinkers have not.) To the extent that they are interested in the question of the origin of the universe as students of science (philosophers also no longer are interested in the question), Rosenzweig's language, style and method appear to be totally foreign. The contemporary interest in the origin of the universe is limited to a second level subbranch (viz., cosmology) of a first level subbranch (viz., astrophysics) of a branch (viz., theoretical physics) of a single science (physics). It uses terms like "radiation" and "quarks" in connection with states like temperature and density. In contrast Rosenzweig uses terms like "God," "man"[6] and "world" in connection with relations like "creation," "revelation" and "redemption." The language disparity is so radical that it is difficult for any of the few readers who are familiar with contemporary physics, modern German philosophy and classical rabbinic thought to recognize how much both the modern cosmologists and Rosenzweig share in common in their discussion of creation.

It is too soon now to discuss that commonality. Here the concern is only to set up the problem. However, I will give you one hint. Beyond the simple human interest in trying to understand "How did it all begin and where did we, and everything else, come from?" is a concern with the implications of the modern mathematical formalism of dynamics, viz., of calculus, for understanding creation.

This judgment about the shared concerns of Rosenzweig's philosophy and contemporary scientific cosmology in itself points to one more problem for readers understanding Rosenzweig's doctrine of creation. His analysis presupposes a philosophical interpretation of the working language of modern science (notably, calculus) that Rosenzweig learned from his teacher, Hermann Cohen. Now, contemporary education has become so deeply compartmentalized that the kind of readers who know modern math tend to have no interest in creation in the language and style in which Rosenzweig discusses it, viz., as a humane interest. On the other hand, the kind of readers who share Rosenzweig's humane interests tend to be mathematically illiterate. Worse, they tend to be mathophobic, i.e., the mere use of logical/mathematical symbols, as opposed to the kinds of words you find in standard dictionaries, sends them into a shock (probably because it triggers traumatic memories of compulsory math courses in grammar school and high school) that causes them to close their minds completely. However, Rosenzweig's language throughout *The Star* is rich in its use of mathematics, both shorthand symbols for vectors and allusions to geometry, all introduced in the name of Jewish theology. From this perspective no other contemporary philosopher is more difficult to read than Rosenzweig (including A. N. Whitehead and Charles Peirce, whose problems-in-comprehension share a similar root).

Yet, to understand Rosenzweig's concept of creation is

inescapable. He must be confronted for at least two reasons: First, with the sole exception of Alfred North Whitehead, no other contemporary philosopher of religion has dealt as seriously as has Rosenzweig with the implications of modern science and mathematics for a viable schemata for understanding the universe (*Weltanschauung*) and dealing with life (*Lebenschauum*). Second, no other contemporary Jewish thinker has placed as much emphasis as has Rosenzweig on the doctrine of creation in presenting his theology. Rosenzweig's basic approach to theology consists in constructing a geometric picture of reality. Its shape is presented as a six-pointed star, viz., a MAGEN DAVID, constructed from two intersecting triangles. One triangle is defined by points called "elements" and the other by points called "courses." In fact the courses are vectors whose origins are the elements. Hence, Rosenzweig's universe is to be understood in terms of vector analysis (or, what Whitehead and his disciples called "process"), where the elements define the vector operations as asymptotes, i.e., endless movements towards a limiting point. These points, that are infinitely remote both as ends and as origins, define three motions. One of the three originates from God and ends at the world. It is this operation that is called "creation." Its explanation in relation to the other two courses (from God to man [revelation] and from man to the world [redemption]) constitutes Rosenzweig's philosophy of creation. The explication of that philosophy is the subject matter for the first section of the first chapter of this first part of the book. (The second section will deal with the way that philosophy emerges out of Rosenzweig's interpretation of the biblical text of Genesis.) But we are getting ahead of ourselves. The only point at this stage is to show the centrality of Rosenzweig's concept of creation to his philosophy of Judaism. Again, creation is for him one of the three main conceptions upon which his entire philosophy and Jewish theology is constructed – the other two being revelation and redemption. In giving this much emphasis to creation, Rosenzweig is unique among modern Jewish thinkers. Hence, an adequate doctrine of creation, especially one built from a Jewish perspective, must deal with Rosenzweig.

CHAPTER I

Creation in Franz Rosenzweig's Star of Redemption

The fact that creation plays a major role in Rosenzweig's philosophy is itself an indication of the Jewishness of his thought. No philosophical topic, including God's unity, was more fundamental to traditional Judaism than the question of the creation of the universe. Only one pre-modern Jewish thinker, viz., Abraham Ibn Daud (1110–1180), did not discuss it, but none denied it. Furthermore, everyone who ranked Jewish beliefs by any criteria made creation fundamental.[7] The first of Moses Maimonides' (1135–1204) thirteen root principles of Judaism speaks of the existence of "the creator" and he later claimed that his fourth principle, that God is eternal, entails that God created the universe. Simon Ben Zemach Duran (1361–1444) made creation the single most important principle, from which everything else follows. Bibago made it one of his only two basic principles. Finally both Chasdai Crescas (d. 1412) and his student Joseph Albo (lived between 1380 and 1440) relegated it to a lesser rank than Duran and Bibago, but, nonetheless, it remained for them as well a basic principle of Judaism.

Nevertheless, Rosenzweig is atypical as a modern Jewish thinker in placing such strong emphasis on creation. In fact, given the centrality of the concept in classical Jewish philosophy, that Rosenzweig is exceptional in this respect is surprising. The topics that most concerned Baruch Spinoza (1632–1677) were the nature and existence of God, ethics, divine providence, and the status of the Torah as divine revelation. Furthermore, his specific list of the dogmas of a true religion include that (1) God exists, (2) God is one, (3) God is

29

omnipotent, (4) God governs the universe absolutely, (5) to obey God is to love your neighbor, (6) obedience is the path to salvation, and (7) there is both repentance and forgiveness.[8] Notably missing from both his list and his attention is the dogma of creation. Similarly, the major topics in Martin Buber's (1878–1965) major work in religious philosophy, *Ich und Du* (*I and Thou*), are "basic words," "history," and "God." The topics discussed include language, anthropology, theology and history, with particular attention to the religious issues of ethics and divine providence. However, the book says nothing at all about creation. In terms of the basic categories of Franz Rosenzweig's religious philosophy, while Spinoza and Buber were concerned with God, man, the world, revelation and redemption, they rarely if ever mentioned creation.

A similar point can be made about most other modern Jewish theologians. While they affirm creation, they de-emphasize its traditional status. For example, according to Samson Raphael Hirsch (1808–1889) the fundamental ideas included in the laws of the Torah deal with God, the world, mankind's mission, and Israel's mission. While the idea of the world includes its creation, this aspect is not central. Rather, Hirsch's explication of the interrelationship between the ideas of God and the world concentrates more on divine providence than creation. Similarly, Moses Mendelssohn (1729–1786) believed that natural religion, his counterpart to Spinoza's true religion, primarily teaches that there exists one God, God governs the world, and the human soul has life after its body dies. Once again, Mendelssohn, like Hirsch, does affirm creation, but he relegates it to a secondary level of importance by comparison with divine providence and other religious beliefs.

Mendelssohn argued that all rational human beings, aided solely by reason, without the input of revelation and religious tradition, can discover that the one God creates the world by an act of his will that is always aimed at the highest good. In other words, creation is a belief that all intelligent people will share independent of the teachings of any divine revelation. However, Solomon Ludwig Steinheim (1789–1866), in *Die Offenbarung nach dem Lehrbegriffe* (*Revelation according to the Doc-*

trine of the Synagogue), took exception to this judgment. He argued that creation is one of the topics by which natural and revealed religions are distinguished. The natural religions of paganism teach that God is subject to the necessity of his own nature and is limited in creation by the nature of the matter with which he must work. In contrast, the revealed religions of the Hebrew scriptures teach that God is a creator who acts freely and creates out of nothing. Hence, the doctrine of creation has an importance in Steinheim's thought that is missing in the writings of Mendelssohn. As we shall see below, Rosenzweig stood between Mendelssohn and Steinheim on this issue. In agreement with Mendelssohn, Rosenzweig treated creation as a topic of philosophy rather than revealed theology. However, in agreement with Steinheim, he argued that creation lies outside of the domain of what Mendelssohn understood philosophy to be. Rosenzweig claimed that the old positive philosophy of idealism cannot understand creation; needed in its stead is a new negative philosophy that would extend rational thought beyond its old limitations to the very limit of revealed theology.

The two modern Jewish thinkers who most emphasized creation are Nachman Krochmal (1785–1840) and Franz Rosenzweig (1886–1929). Both were students of Jewish philosophy and mysticism as well as of the Christian thought of Georg Wilhelm Friedrich Hegel (1770–1831) and Friedrich Wilhelm Joseph von Schelling (1775–1854). Of the two Jewish philosophers, Rosenzweig developed the more detailed account of creation. Hence, this chapter on modern Jewish philosophy deals primarily with Rosenzweig's writings. However, Krochmal's theory deserves some attention, both because of the ways in which his absorption of Jewish tradition and German Objective Idealism parallels Rosenzweig's beliefs and because of their differences. Rosenzweig's major work is *Der Stern der Erlösung* (*The Star of Redemption*)[9]; Krochmal's is his *MOREH NEVUKHIE HA-ZEMAN* (*The Guide of the Perplexed for the Time*).

Krochmal interpreted the dogma of creation out of nothing to be a transition from the Absolute Reality of Objective

Idealism to the generated reality of finite things. He understood this transition to be an infinite process through which God fulfills himself. Furthermore, drawing from the Kabbalah, Krochmal identified the nothing of creation with God himself. In other words, God, who was nothing, created the world from himself in order to become God, who will become something. As we shall see below, Rosenzweig shared Krochmal's judgment that creation describes an infinite, atemporal process, but Rosenzweig rejected the claims that the nothing of creation is God, and, what is more important, that the process expresses a movement from God to the world. Rather, the process is just the opposite, viz., a movement from an infinite number of unique created nothings that fulfill themselves by giving content to God's essence. Whereas Krochmal identified creation with emanation, Rosenzweig saw them as diametrically opposed descriptions of how the creator and the creatures fulfill each other.

Rosenzweig discusses creation in the first chapter of the second part of his *magnum opus* in Jewish philosophy. The discussion takes two distinct forms. First, it is discussed as a philosophical concept. Second, it is discussed as a commentary on the first chapter of Genesis in the Hebrew scriptures. The two discussions are interdependent both in content and in method. We will consider first the philosophical discussion, then the biblical commentary, and finally why Rosenzweig had to divide his discussion in this way.

ROSENZWEIG'S PHILOSOPHY OF CREATION[10]

In a single sentence, for Rosenzweig the term "creation" names the relationship between the elements "God" and "world." This relationship is to be pictured geometrically as a vector (i.e., a linear movement with a direction) whose origin is at a point and whose end is another point. The terms "God" and "world," which Rosenzweig calls "elements," are the points. The end point is an asymptote, i.e., it is the point to which the vector constantly moves but never reaches. In other words, creation is an end-directed process whose duration is endless because its goal is infinitely remote. In this respect it is appropriate to think of all three of his elements (God, world and man) as points. They are in themselves nothing. They only become something in relationship. The something they are is what makes relations intelligible. God is the origin for two relations – revelation (directed towards man) and creation (directed towards the world). But in himself God has no end. Conversely the world, which is never an origin, is the end for two relations – creation (originating from God) and redemption (originating from man). Reality is to be understood in terms of the intersection or configuration (*Gestalt*) of these three continuous relations. They are continuous because, as we have said, they can never cease, since their ends are infinitely remote. But they are also continuous because they cannot begin either, for their origins are equally asymptotes, i.e., infinitely distant.

Rosenzweig's account of his dynamic (or, process) universe begins with a discussion of the elements. The mode of expression in this case is philosophical-scientific-mathematical. For him these last three terms are interchangeable. They express the same form of thinking, viz., linear thought that uses logic to form mathematical-in-form equations about objects. As we shall see, such language is not adequate to comprehend the dynamic relations themselves.[11]

Rosenzweig's theory of creation is an exercise in what he called the "new thinking." It begins at what he believed to be the outer limits of what can be grasped through algebraic thinking. He claimed that the new thinking points to a new kind of knowledge in religious belief whose own source, beyond the limited perspective of human science, is divine revelation. Hence, before we can begin to say what Rosenzweig explicitly says about creation, we must first say something about the theory of knowledge that underlies his claims and the sources for that theory in modern German philosophy.

Rosenzweig justified the need for a new method of cognition as follows: Philosophy begins with "the fear of death." By "philosophy" he means the tradition of Idealism that begins with Parmenides, continues through Plato and the Neoplatonists, and culminates in the philosophy of Hegel. By "fear of death" he means what Aristotle called "generation and corruption," viz., the sensible, always changing material world.[12] Philosophy's response to this fear was to restrict the application of the term "something" (*Etwas*) solely for ideal, conceptual, unchanging entities. By so doing it denied that what continuously changes[13] is something. Hence, it judged the individuals of the sensible world to be nothing (*Nichts*). But philosophy's somethings are not everything that is. Continuously changing individuals also exist. Hegel in particular called the domain of the objects of philosophy "Being" (*Sein*), and equated it with the "All." However, philosophy's "all," that exclusively encompasses universal, unchanging entities, is not everything that is real. The individual nothings also are real. However, knowledge of them lies beyond the comprehension of this kind of philosophy. Hence, a new way of thinking is needed to encompass all of reality – what is in flux as well as what is permanent, the particular as well as the general. In Rosenzweig's language, since knowledge is what falls within the domain of philosophy, cognition of the individual transcends knowledge.[14] While the subject matter of knowledge (*Wissen*) within the domain of philosophy and science (*Wissenschaft*) is all that is something, i.e., is being, the rest of reality, viz., all that is nothing, is the subject matter of belief (*Glaube*) within the domain of revelation (*Offenbarung*).[15]

The term "creation" (*Schöpfung*) expresses the relationship between God and the world. As such it rests upon two disciplines of the new thinking. The first, metalogic, is the movement of thought beyond mathematical logic in thinking about the world. The second, metaphysics, is the movement of thought beyond mathematical logic in thinking about God. In addition, there is a third discipline, metaethics, on which metalogic is based. Metaethics is the movement of thought beyond mathematical ethics in thinking about man.

Metaethics projects a human individual who stands outside of all that is encompassed in Hegel's philosophy and as such falsifies the claimed success of that philosophy to comprehend absolutely everything. Hegel presupposed that the All must constitute a unified totality, because otherwise thought (*Denken*) itself could not be unified. In direct opposition to Hegel, metalogic, building upon the insight of metaethics that the world is not unified, rejects the claim that thought is a unity.[16]

The apparent unity of form, matter and reality are all different. The world is a single material unit (*Eins*), and the form of thinking about it is a single process or function that can be called a unity (*Einheit*), but to be a unit and to exhibit unity are not the same thing. The material world is not a unity, and thought is not a unit. The very fact that thought is a process of defining something in terms of a genus and a specific difference itself shows that thought is not a unit. To state a genus is to unify some being, but the specific difference separates the very same being within itself. It says that what is one with other things also differs from them. In Rosenzweig's language, it says that the object of thought is "such and not otherwise," i.e., that what is one in being is diverse in thought, so that being and thought necessarily cannot be unified, i.e., they are not the same thing. This intrinsic separation between thought and being in itself requires a separation between logic and metalogic. Whereas logic deals with the world in general, metalogic deals with the concrete world of individuals. Whereas logic unifies thought about reality, metalogic deals with reality. Finally whereas logic expresses necessity and universal laws, metalogic expresses contingency and concrete content. It is this concrete contingency that is the content of metalogic's contribution to the account of creation.[17]

As knowledge of man as an individual lies beyond ethics and cognition of the individuality of the world transcends logic, so God as an individual lies beyond the grasp of scientific and philosophical physics. As metaethics encompasses but does not deny the ethical dimension of man, as metalogic encompasses but does not deny the logical dimension of the world, so metaethics encompasses but does not deny the physical dimension of God. However, as metaethics raises man above ethics, and as metalogic raises the world above logic, so metaphysics raises God above the physics of what is (*Sein*).[18]

Rosenzweig's metaphysics builds on the thought of Schelling and Nietzsche. From Schelling it accepts the judgment that while God's nature is his being, there is more to God than his nature. From Nietzsche it posits that God is free only to the extent that he is independent of his nature, and man in defiance of God resents the fact that God has in the world a freedom that man both lacks and wants. In other words, whereas God has freedom, man wills to be free, and each in this respect transcends his nature.[19]

Rosenzweig acknowledges that the model for his development of the new thinking beyond its historical sources is Hermann Cohen's use of infinitesimal calculus to do Jewish philosophy. From Cohen's influence Rosenzweig came to understand reality as something which is nothing that becomes something that it is not, i.e., as a movement from *Nichts* to *Etwas*. The result is that Rosenzweig's conception of the universe is pluralistic, in marked distinction to Hegel's monism.[20] Following Cohen's precedent, the new thinking begins not with negation in general, but with very specific negations. Each of the three metasciences begins by negating the positive objects of traditional philosophical sciences. Theology studies God, cosmology studies the world, and psychology studies man. These subjects are Rosenzweig's elements. They are inherited from the history of philosophy as something (*Etwas*), but they are only something indefinite or undefined. In each case metaphilosophy knows that the something that philosophy claimed to know is not really known as it is in full reality. This negation is the *Nichts* upon which metaphilosophy concentrates its thought. The three elements are studied as irrational objects rather than as the subjects of rational sciences. Each moves from its own specific *Nichts* of a positive element to an *Etwas* that is in effect an infinite judgment, viz., a negation of a negation. While the resulting *Etwas* is something, it remains something indefinite or undefined. In other words, in metathinking that remains independent of religious belief prior to the influence of revelation, we can know each element, but we can only know it as what in modern logic would be called a variable.[21]

Again, Rosenzweig explicitly says that his model for this kind of thinking is mathematical,[22] but the math itself is new. It is not the simple geometric model of Platonic thought based on a correspondingly simple notion of numbers and harmonics, limited exclusively to the domain of positive rational numbers.[23] Rosenzweig's universe, unlike that of Plato's *Timaeus*, is not constructed out of rationally knowable right triangles. Rather, it has more in common with an Aristotelian tradition, abstracted from its Platonic influences, where pure matter would take precedence over pure form. In other words, as the old philosophy of Aristotle's tradition began with pure form, the new philosophy begins with pure matter.[24] The something that follows from Rosenzweig's negation of nothing is not something positive. Its form is not Aristotle's substance/being (OUSIA). Rather, it is still something negative. It is the negation of what is negative; it is what Cohen understood Kant to mean by an infinite judgment. In all three cases – the movement from the *Nichts* to the *Etwas* of God, of man, and of the world, what emerges are the elements of reality, which are three relative nothings. This movement defines creation.

To summarize, Rosenzweig's conception of creation as an atemporal dynamic relation grows out of a particular understanding of the relationship between philosophy and theology that he developed out of his intellectual sources in modern German philosophy. On this view philosophy and theology are radically different disciplines. Philosophy deals with eternal unchanging objects which, as such, are best understood on a simple mathematical model of positive natural numbers and Euclidian geometric shapes. Theology deals with everything else. Its subject matter (like most of reality) is constantly changing, subject to time, and never something completely definite or positive. The link between the two realms are the elements in Rosenzweig's "new thinking." It is this kind of "meta-" thought that exhibits the inadequacy of the objects of the old thinking and points to the need for the revealed understanding of theology. As such the philosophical concept of creation focuses on the relation's two elements, God and world, as elements, that in the end points to what can only be grasped through a revealed text. However, this last claim is premature. It is the conclusion reached through the analysis of the metaphilosophical concept of creation that links the philosophy of creation to its theology. The metaphilosophy itself in this case consists of Rosenzweig's analysis of the elements God and world in themselves and, their mutual need for some form of connection.

God, or A = A

Traditional metaphysics – what Rosenzweig calls "physics" – was a positive theology in the sense that it began with the positive assertion that God is something (*Etwas*) and concluded with Maimonides' judgment that God is nothing (*Nichts*) of which we human beings have knowledge.[25] In contrast, Rosenzweig's metaphysics is negative theology in that it begins with the nothing of traditional metaphysics and, through its negation, arrives at something positive, viz., God's reality (*Wirklichkeit*). In other words, the new philosophy concludes with the judgment that nothing is itself really something. However, the nothing of the new philosophy is not a single general something like Hegel's Being (*Sein*). Rather, there are three quite distinct nothings – viz., the affirmed nothing of God, of man, and of the world.[26]

In Aristotelian science the form of the model sentence is "S is P," where "S" refers to a substance, "P" refers to a universal, and ". . . is . . ." predicates the universal of the substance. The model sentence in the new philosophy is radically different. Its form is "$y = x$," where the left-hand term "y" is a negative supposition (*Setzung*) that expresses both the grammatical and the semantic subject, while the right-hand term "x" is an affirmative determination (*Bestimmung*) that expresses the grammatical predicate and the semantic content, and ". . . = . . ." expresses the relationship between the two terms. In God's case both "y" and "x" are "A." By itself "A" is the unknown content of God, which entails an infinite negative judgment of all that is not God. However, as we shall see, this symbol (*Zeichen*) has a significantly different meaning within its equation, depending on whether it is the left-hand or the right-hand term.[27]

The starting point of metaphysics is the specific nothing of the knowledge of God. It begins by affirming that this nothing is itself something. The something is not a simple affirmation (*Bejahung*, viz., the act of saying *Ja*); rather it is a negation (*Vereinung*, viz., the act of saying *Nein*) of the nothing. In other words, what is affirmed is a "not-nothing," i.e., a *Nichtnichts*. By saying that "the nothing of the knowledge of God is not merely nothing" the left-hand nothing is set free in the sense that this nothing that is God is not circumscribed by anything. God by definition is absolutely free, because he is not limited by anything whatsoever. In this case the left-hand term[28] is what in the act of judgment is an affirmed negated-essence. It says that what God is is what is not anything finite. Furthermore, the right-hand term[29] is what in the same action is a negation of an infinite number of things, i.e., a negation of everything finite. The equation itself expresses that God is free from every limitation. Rosenzweig identifies this freedom as God's "eternality." God is eternal in the sense that everything that comes to be at all times – past, present and future – defines God by not being God. At every moment God remains one, free, infinite, eternal and nothing whatsoever.[30]

The first sentence of metaphysics expresses divine freedom. It functions as the left-hand term of the second sentence. The second sentence expresses the "vitality of God" (*Lebendigkeit des Gottes*). It says that divine freedom is divine force. The "force" is God's essence and absolute actuality. The "is" of this second equation expresses an asymptotic movement ("=") of divine freedom ("y") towards the specific idea of divine essence ("x"). Here freedom is a force with a vector direction while essence simply is, without direction. Freedom is potentiality, while essence is actuality. In this movement the original infinite freedom of the first equation becomes increasingly constrained and transformed into divine power and caprice, while the divine essence is transformed into divine fate and obligation. The freedom becomes constrained because the more God becomes what he is, viz., his essence, the less free he is to be anything whatsoever, i.e., to be capricious. This constraint is divine fate and obligation. In other words, divine freedom entails divine power and caprice, divine actuality or essence entails divine fate and obligation, and the second equation expresses a movement away from God's past absolute power and caprice towards an infinitely remote absolute fate and obligation.[31]

The conclusion of the negative theology of metaphysics is the logical expression "A = A." "A" in isolation is divine power. As the left-hand term of the equation, it expresses the source-word (*Urwort*) "*Nein*" of divine freedom. It says that God as the subject is free to be anything whatsoever because he is not anything that ever was, is or will be. Furthermore, as the right-hand term of the equation, "A" expresses the source-word "Ja" of divine essence. It says that God as the object is an eternally unchanging ideal that the endless collection of created individuals approximate. Thus, the concluding formula of metaphysics expresses what God is. He is a self-contained movement from what he is as subject to what he is as object. Because the expression is self-contained, the God affirmed is only an element. As such it says nothing about reality, for the left-hand term refers to an origin in nothing whose concluding right-hand term is an ideal something. It begins in an infinitely remote past and concludes with an infinitely distant future, but totally skips over the real present. As such it points to the need of something outside of itself to which it must relate to be real. That something is the world of created individuals.

At this stage no distinction need be made between individual things and individual human beings. The difference is important, as has already been noted by the fact that the new thinking has three – not one or two – negative origins, viz., God, world and man. However, once again, at this stage the difference is not relevant. Rosenzweig's source-words for metascience are "yes" (*Ja*) "no" (*Nein*) and "and" (*Und*). God and world are in a certain sense opposites. God is pure self-directed action from himself as nothing to himself as something. In contrast, as we shall see, the movement of the generated objects of the world has no direction of its own. In this sense whereas God only acts, the world only is acted upon, i.e., it is purely passive. Because God only acts there must be something other than himself to receive his action. Similarly, as the world only is acted upon there must be something other than itself to act upon it. In this way each is a "no" to the other's "yes." What bridges the gap between their radical separation is man, who is the passive recipient of God's action in revelation, and who acts upon the world to redeem it. As such man is the "and" to the "yes" and "no" of God and the world. However, this is only one perspective on the relationship of these three elements. In another sense it is the world that bridges the gap between God and man.

The formulas for each best exhibit this understanding of their interrelationship. As "A = A" expresses God, "B = B" expresses man. As the equation for God contains no terms but God, so the only terms in the equation for man is man. As such there can be no relationship between the two. They are as separate as "yes" and "no." It is only through the third element, "world," that they are related. Its expression is "B = A," i.e., the world moves from what as an isolated term is man towards what is God in himself, viz., absolute power. It is this movement that is creation. However, all of this is premature at this stage of Rosenzweig's account of creation. As was the case in the traditional philosophy of Spinoza, Rosenzweig's new philosophy began with thought about God, and that reasoning in itself leads to his second starting point in the nothing of the knowledge of the world.

World, or $B = A$

It would seem that there is no doubt about the world comparable to traditional philosophy's doubt about God, since almost every philosopher has in some sense considered the world to be a self-evident given. However, it is a fact that philosophers have always made either the ego[32] or God[33] the point of departure for their thought. By doing so, philosophy has implicitly rejected the claim that it is self-evident either that the world exists or that we know what it is. The starting point for negative cosmology – what Rosenzweig calls "metalogic" – is what is left over when either God or man is made the starting point of thought. That self-evident residue, which Rosenzweig calls an "infinitesimal residuum," is Kant's "Ding-an-sich." It is this thing-in-itself that is the negative starting point for Rosenzweig's analysis of the element, world.[34]

In contrast to Descartes, negative cosmology does not begin with a general doubt of everything. Rather it begins with three specific doubts, viz., doubt of God, of man, and of the world. As such the doubt is only hypothetical, and, as such, it already entails an affirmation. Again, as in the case of metaphysics (whose subject is God), both metalogic (whose subject is the world) and metaethics (whose subject is man) move beyond the concluding negations of traditional philosophy to affirm an infinite judgment. In this way the new thinking moves beyond science to belief. Descartes had assumed a radical doubt in the hope that belief could be raised to the level of knowledge, but that hope was vain. However, while science cannot make belief knowledge, neither can it justify disbelief. Rather it can justify it by explaining why we must have belief.[35]

Unlike metaphysics, where the affirmation of the original infinite judgment led to an affirmation of God's infinite essence, the initial affirmative infinite judgment does not lead to an affirmation of the essence of the element, world. Rather it leads to logos. Logos is world-thought (*das Denken*). This thought is an abstract hypothetical whose validity is demonstrated through its universal, and therefore necessary, application to absolutely everything in the world.[36] However, since logos in its origin is hypothetical, it itself is not in the world, and thought about this thought also is not in the world. Consequently, since logos is the counterpart in metalogic to metaphysic's essence (*Wesen*), the essence of the world is something that itself is not the world and is not in the world.[37] As such the symbol that refers to the world-logos in isolation from the equation that expresses the world is the "A" that expressed

God in the position of grammatical predicate in metaphysics. This logos has the logical status of the universal operator [the "(x)"] in an affirmative conditional proposition in the symbolic language of the *Principia Mathematica* of Russell and Whitehead. It is nothing in itself. This "A" asserts universal applicability *qua* applicability, independent of any word content. When it is placed in the predicate position in the equation of metalogic as " = A," it expresses the potentiality of the operator. It is the power of the logos to be applied to absolutely everything in the world.[38]

While logos is distinct from both Hegel's "world-spirit" and Schelling's "world-soul," it is what metalogic affirms as the spirit or order or nature or intelligibility of the world. However, beyond both logos and the world itself is the nothing (*Nichts*) of the world. It is this nothing that is the independent source of the infinite creativity that itself lacks any order or intelligibility. The world in itself endlessly and blindly generates each of its individual parts that constitute the world's fullness (*die Fülle*). This constantly renewed creation arises out of the nothing of the world. This nothing endlessly generates an infinite number of original, concrete particulars, each of which is uniquely itself and devoid of any ordering or inherent intelligibility. These particulars are in themselves nothing, and their nothingness is concrete rather than general. Every generated particular is a "not-otherwise," which means that it is created singularly nothing and consequently it is uniquely meaningless.[39] Their symbol (*Zichen*) is "B." B is the contrary of order and intelligibility. It is "not-otherwiseness," i.e., it is a complete, absolute particularity that is total distinctiveness.[40]

B is the act of negation (*das Nein*) of the nothing of the world. It, in conjunction (*und*) with the affirmation (*das Ja*) of God's nature, yields the particular something (*das einzelnen Etwas*) of the world. In opposition to Empiricists, metalogic assumes no given (*das Gegebene*) in the sense that there is no unchanging permanent thing. Rather, the distinctive (*das Besondere*) which particularizes each particular something is at every moment unique and unintelligible in itself.[41] In opposition to idealist philosophy, from Parmenides through Hegel, the thought

(*Denken*) that is the logos is not everything. As we have already seen, logos is the universal (*das Allgemeine*) which functions to make intelligible the meaningless, created, distinctive individual, and as such it accounts for diversity by affirming a category (*Gattung*) that the individual comes to exemplify.[42] However, this universal is not the All.

In Rosenzweig's case the process produced by the interaction of the distinctive created particular (*das Besondere*) and the universal (*das Allgemeine*) produces the individual (*die Individuum*) who is a particular subsumed under, or rendered intelligible by, a category (*Gattung*). The expression of this movement is "B = A." "A" in isolation is the universal. It is something purely passive whose need for application in some unexplained sense exerts a force of attraction (*eine anziehende Kraft*) on the distinctive particular. In direct contrast is the "B." It is the distinctive particular who, in its aimlessness, is drawn towards the A as an exemplification. In the process that "B = A" expresses, the distinctive particular becomes conscious that it is being attracted in the direction of the universal.[43]

At this stage the particular no longer is merely distinctive. It now becomes an individual (*Individuum*) in transition from being aimless to being dominated by the universal. In the end the individual becomes dominated; it itself becomes its category. However, now the category (*Gattung*) is no longer identical with the universal. It is transformed into a species, viz., an individual universal. As the particular has been transformed, so has the universal. In conjunction they are a distinctive universality (*ein individualisiertes Allgemeines, eine besondere Allgemeinheit*).[44]

Again, "B = A" does not express an Hegelian "all." Even when the particular's intelligibility is complete within its categorical definition, the "B" preserves a separation from the "A." While ". . . = . . ." expresses a movement from one term towards another, Rosenzweig's equals sign never asserts identity. B preserves its distinctiveness as a representative of a particular plurality (*Mehrkeit*) through two determinations (*Bestimmungen*). First, it is the mark of its species (*das Gattungsmerkmal*), and, second, it continues to possess its own peculiar properties (*Eigentümschaft*). In other words, the movement of the utterly distinctive particular towards the universal transforms the former into an individual and the latter into a category. The individual becomes intelligible as an instance of a species, and the category becomes identified by its exemplary individual, but both continue to be distinct.[45]

This relationship between the individual and the category is related to the world as essence and freedom are related in God. In the case of man it expresses the relationship between people and their communities, and in the case of objects it expresses how they are related to their concepts.[46] In general, the central equation of metalogic, "B = A," expresses the process of life. "B" stands in isolation for the distinct particular (*das Besondere*). It is the active content of the world that includes every singular thing, living or dead, animate or inanimate, material or immaterial. "A" stands in isolation for the universal (*das Allgemeine*). It is the passive form of the world that includes the order of the world as well as every conceptual category. The relation between them, expressed by ". . . = . . .," is a "nonreversible direction of penetration" in which B fills A.[47] As such it posits a cosmology that is directly opposed to Hegel's, whose formula is "A = B," where the active universal penetrates every individual.[48]

The difference between these two cosmologies is the difference between emanation and creation. "A = B" is the equation of emanation. It is the only viable way that philosophy, i.e., Idealism, can view the relationship between God and the world. In contrast, "B = A" is the equation of creation through which religious belief speaks of God's connection with the world from the vantage point of revelation. Whereas the metalogical world of belief is composed of things that are filling it, so that the world is always becoming fulfilled, the world of Idealism already is filled, because it is the whole (*das Ganze*) that fills its individual member. From the perspective of creation, each part (*Teile*) follows its own unique way. In contrast, in the world of Idealism each individual never really is individual, because it is fulfilled by the whole in a single, determined, universally applicable way.[49]

To summarize, creation is to be conceived philosophically as a relation between two elements, God and world. This relationship is atemporal and eternal. It is not something that takes place in time. Rather, it is a dimension in the geometric sense of the term, in which temporal events are located. This dimension extends from an infinitely remote origin in God to an infinitely distant end in the world. With respect to creation itself these elements are points. However, when each element is examined in its own right, they too are to be understood as asymptotic vectors. Each elemental vector is expressed in Rosenzweig's metaphilosophy as an equation whose form is algebraic. The equation for God is "A = A." It is God's eternal motion from himself as an empty subject to himself as a full object. Every equation in metaphilosophy expresses an asymptotic vector from an origin towards an end. In God's case, the origin is freedom and the end is essence. God is that entity whose essence is not to be any entity. In isolation God is nothing at all. In origin nothing defines him, even negatively. God only becomes what he is (viz., not being anything that is) through the world. As the world generates its somethings, God becomes what is not them. Hence, God's essence (viz., what God is in himself as an element) depends on what the world becomes. In contrast the equation for the world is "B = A." The origin of the world's motion is pure concrete, characterless difference. From the space of the world there arises an endless stream of things so concrete and so unique that in their origin they are absolutely nothing. As such they are simply what is "other than" anything else (*Besondere*). As each emerges it moves towards becoming something. As each nothing becomes a something, it becomes an individual. The world is constituted by an endless stream of these nothings becoming something. The something towards which they move is universal (*Allgemeine*). It is each particular individual as it becomes universal that defines the world. But this process also defines God. Hence, the endless flow of an infinite number of particular nothings towards general somethings is the ground for the relationship between God and the world. The world endlessly becomes itself, but because this process is endless the world never is the end towards which it moves. Similarly, God endlessly becomes himself by not being what the world has become, but because the world never becomes itself, neither does God become himself.

In one sense, this analysis of the elements God ($A = A$) and world ($B = A$) are sufficient to explain Rosenzweig's concept of creation as the process that connects God to the world. However, a word must also be said about the third element, man. The algebraic symbolism itself points to the need. Man as an element is also a vector whose equation is $B = B$. The world is $B = A$. The "A" in the equation of the world is in itself the vector for God, viz., $A = A$. Similarly, the "B" in the equation of the world is in itself a vector whose equation is "$B = B$," and this is the equation for man. In this sense, as creation is to be understood as a movement from God to the world, the world itself is to be understood as something whose end is God but whose origin is man. Hence, the account of creation requires an account of man.

Man, or $B = B$

We need not concern ourselves with Rosenzweig's thought about the element man in the same detail that we looked at what he had to say about God and world, because again, creation is a relation that holds directly only between the elements God and world. Still, man also is a creature, and to that extent Rosenzweig's analysis of man also is part of his doctrine of creation. As Maimonides' negative theology introduced doubt about God into philosophical physics in the Middle Ages, and Descartes' meditations introduced doubt about the world into philosophical logic in the seventeenth century, so Kant's transcendental Unity of Apperception introduced doubt about man into modern ethics. As initial doubt about God in metaphysics led to an affirming act (*Ja*) of the creation of God's nature, and initial doubt about the world in metalogic led to a negating act (*Nein*) of the generation of the distinctive particular, and these two directions in thought were conjoined (*Und*) through the structure (*Gestalt*) of the individual (*individuum*), so now we move in metaethics from an initial doubt about man to a new expression in algebraic symbols that entails the unspoken source-words – "yes" (*Ja*), "no" (*Nein*) and "and" (*Und*).⁵⁰

On the right-hand side of the affirmation we derive man's true being (*Sein*), which is the essence (*Wesen*) of his infinite double negation (*Nichtnichts*). As the essence of God is to be immortal, and the essence of the world is to be necessary, so the essence of man is to be transitory (*Vergänglichsein*). As the being of God is unconditional, and the being of the world is universal, so the being of man is to be distinctive (*Besondere*). Finally, as knowledge is below God and in or about the world, so knowledge is above man.⁵¹

From Kant's analysis of the ego as the Transcendental Unity of Apperception, man's essence is given as a contentless but still affirmed precondition for all knowledge. In every act of knowing, man knows himself as something there (*da*). However, at the same time, he knows that what he is is distinct (*Besondere*) from the very act through which this essence is revealed. He knows himself as that unique nothing that is the focal point of the world.⁵² As such man is so particular (*einzelnis*) and distinctive (*Besondere*) that he can recognize no other particular but himself. This radical distinctiveness (*Besonderheit*) is his essence.⁵³

This transitoriness is the source-affirmation (*Urja*) of man. It is his singularity (*Eigensein*). The source-word that expresses it is "thus" (*So*), whose equivalent in biblical Hebrew is "KHEN." It constitutes the character through which man is individuated in every act. Its symbol (*Zeichen*) is "B."⁵⁴

The B of man's limited existence – his positively affirmed singularity and distinctiveness – is related as the direct contrary of the A of God's infinite being. At the same time, while the term "B" of metalogic stands in some kind of relation to A, the B of metaethics is so unlike the A of "= A," i.e., of the right-hand term of the fundamental equations of metaphysics as well as metalogic, that man's "B" can no more be related to A than God's A in metaphysics can be related to B.⁵⁵

It is the symbol "B" that exhibits the relation between man

and the world, for it is what their respective expressions share in common. In the case of the world, as the affirmed left-hand term, B is the negative (*Nein*) subject that is the world's distinctive singularity. However, in the case of man, who also initially is a created object, the term also is the right-hand symbol which, as such, expresses the affirmative (*Ja*) predicate that defines man's permanent character.[56]

The affirmation that is achieved through the double negation (*Nichtnichts*) of man is his finite freedom to will. Man differs from other creatures in that he, like God, can will, and he differs from God in that his will, unlike God's, has limited power.[57] Hence, the left-hand expression of the equation for man is "B =," which expresses human free will. It is a purely intentional, directional act that in itself is nothing because it lacks content. It is similar in form but opposite with respect to God to the "A =" of metaphysics. Whereas God's freedom is his overt action, man's freedom is only will. In other words, whereas in God's case what he wills is what he does, in man's case the two are separate. Hence, in God's case the freedom to will is also a freedom to act that entails infinite power, whereas in man's case there is no such entailment. While man is free to will, he is not free to act. While man has some power, that power is limited radically by the world in which he lives.[58]

The recognition that his will is limited makes his free will defiant. Man's defiant will is his "thus" (*So*) that is the counterpart of the "Thus" (*Also*) that expresses divine power. Both are expressed, depending on context, in biblical Hebrew by the term "KHEN." It is this defiant will that takes on determination (*Bestimmung*) as character. These determinations are the affirmative content of his singularity (*Eigenheit*). Its symbol also is "B." The defiant will is conjoined (*Und*) to the character as the self. It is the product of free will taking on content by which that will is transformed into a living person (*lebendige Mensch*). Its expression is the equation "B = B."[59]

"Self," unlike "personality," is not a relational term. As its equation exhibits, the self is a self-contained uniqueness of the particular. It is not, like personality, a qualified affirmation of distinctiveness in relation to other human beings. In other words, man is both within and without the world. Insofar as he stands outside, he, like the creator, is defined by his will as a self, and insofar as he stands within, he, like any other creature, is defined by what he does in relation to every other member of his species.[60]

"B = B" as the expression of man as a self is logically comparable to the equation "B = A" of the world. In both cases the right-hand term defines and develops, but at the same time limits, the possibilities, i.e., the freedom of the left-hand term. In both cases the left-hand term is something distinctive (*Besondere*). The difference is that in the case of the world what the particular becomes is a universal (*Allgemeine*), while in the case of man the particular becomes character. The difference is significant. Unlike "B = A," "B = B" expresses the self as non-relational. The self becomes itself independent of any relation to another self or a universal. In contrast, man's personality is something in the world. It is defined by its relationship to its species and its individual members.[61] In this respect, the equation for man as a self is more like the equation for God. Both "B = B," viz., free will having become character, and "A = A," viz., divine freedom having become divine essence, express self-contained elements. The sole difference is that the former has limitations that do not apply to the latter.[62]

The isolation of God and of man as self from each other and from the world in itself emphasizes that the elements that are the subject matter of Rosenzweig's metasciences are only hypothetical constructs. For Rosenzweig just as much as for Buber reality itself is relational. Rosenzweig considers the move beyond silent, mathematical thinking about elements that reveal their structure[63] but not their reality to be a mystery (*Geheimnis*). Here all that he means by "mystery" is that to solve this problem requires a way of thinking that transcends the method of negative thinking so far employed.[64]

Beyond the merely hypothetical if (*Wenn*) that modifies every mathematical expression in metascience about the independently self-contained elements, Rosenzweig moves to a slightly higher level of epistemic claim where what can be said in spoken language is "perhaps" (*Vielleicht*) an expression of reality. At this level the elements are no longer viewed in isolation. Here thinking deals with possible forms of relation. For example, it is possible to view any one element as the generator of the other elements. It could be, as Plato believed, that God is the creator of the world and the revealer of man. Similarly, it could be, as Aristotle believed, that the world is the source of the knowledge of God and man. Or, finally, it could be, as the Sophists believed, that man is the measure, i.e., judge, of both the world and God.[65] However, these are only possibilities. No philosophy, including metaphilosophy, can provide certainties as long as thinking remains confined, as it must, to the factuality (*Tatsächlichkeit*) of the isolated, hypothetical elements of reality. It must move beyond them to discern the structure[66] of their relationship and motion.[67]

The elements emerged and developed from a nothing of knowledge (*Wissen*) to a something known prior to any visible reality. In metaphysics God's deed-of-power (*Machtat*) was transformed into his compulsion-of-destiny (*Schicksalmuss*). In metalogic, the world's birth (*Geburt*) was transformed into its category (*Gattung*). Finally, in metaethics, man's defiance-of-will (*Willenstrotz*) was transformed into his peculiarity (*Eigenart*). However, once again, these conclusions are only hypotheticals.[68] Reality is grasped only when its elements are set in a current of motion (*Bewegung*). This movement must originate within, but still be external to, the elements themselves. In other words, if truth statements express states of affairs that are analyzable in terms of two-term relations, such as, in the language of the *Principia Mathematica*, "Ra, b," R can be neither the same sort of thing as, nor an additional thing to, a and b.[69] To move beyond thinking of static elements to speech about their path or course (*Bahn*) in reality is precisely to think about creation (*die Schöpfung*) and revelation (*die Offenbarung*), the latter dealing with how the elements are known and the former with their birth.[70]

Rosenzweig believed that his new thinking was needed as much to replace the theology of his day, what he calls "historical theology" (*historische Theologie*), as it was needed to replace the old Hegelian tradition of philosophy. Whereas traditional theology grounded belief in the miracle of revelation, historical theology, having lost faith in the miraculous, attempted to replace that foundation with modern science. Historical theology is rooted in a belief in progress that entails that future readings of the scriptures will always be superior to past interpretations.[71] However, the past will not go away. Historical theology had to interpret the past in such a way that there could be nothing about it that would question the authority of present experience and its moral faith that the future will be better than the past. In other words, this kind of theology substituted faith in the scriptures, viz., the written record of divine revelation, for faith in modern science. The foundation of this belief is the scientific theory of evolution. However, Kant showed belief (*Glaube*) to be independent of scientific knowledge (*Wissen*). Hence, in principle science cannot provide the needed grounds for modern religious belief.[72]

As science is supreme to criticize its own scientific claims, so science equally was authorized by historical theology to criticize the claims of religion. Hence, for example, Schweitzer could use modern scientific methods to place in doubt the existence of a historical Jesus. As such, science as much in practice as in principle could not justify religious faith. A better philosophy than faith in the science of history was needed, a philosophy that separated philosophy from theology without making theology subservient to philosophy. Rosenzweig argued that the critical mistake of historical theology was that it overemphasized revelation in the present and, in so doing, neglected creation in the past. Hence, what was needed as a corrective is a new theology that would reconstruct theology in such a way that revelation will be built into its prior conception of creation, so that ultimately both revelation and redemption will themselves be part of creation.[73]

A theology that roots belief in present experience (*das Gegenwartserlebnis*) cannot build its foundation on knowledge (*Wissen*), since necessarily knowledge is grounded in the past (*die Vergangenheit*). Rosenzweig's alternative is to base belief on a concept of creation (*Begriff der Schöpfung*) which is both in the past and is the foundation principle (*der Grundbegriff*) of knowledge.[74] It is the basis of a new rationalism that can avoid the errors of medieval Scholasticism that either subjugated philosophy under theology or absorbed theology into philosophy.[75]

The new theology (*die neue Theologie*)[76] needs the new philosophy[77] to bridge creation and revelation so that theology itself can connect revelation and redemption. Note that the link between them is revelation. Philosophy cannot deal with redemption and theology cannot deal with creation. Where they overlap is in revelation. Philosophic thinking moves from the data of the past to the horizon of the present; theological faith projects a morally ideal future grounded in the present. Creation is the doctrine of this past, redemption is the concept of this future, and, again, revelation links both. From this perspective revelation is the most fundamental doctrine of Rosenzweig's *The Star*.[78] However, from a different, equally appropriate perspective, creation is more fundamental. As we have already noted above, theology will justify belief in such a way that, beginning with the philosophic concept of creation, its interrelated beliefs in both redemption and revelation are themselves thought of as part of creation.

The philosophy in which thought (*Denken*) is thought about creation provides the foundation for the content claims of theological belief without itself dealing with those claims. The content of theology consists of lived events rather than statements about static states of affairs. As such theology places more emphasis on the event of creation than on statements about it. Consequently for theology creation is not a conceptual element (*begriffliche Element*). Rather, it is a non-temporal, continuous, immanent reality (*vorhandene Wirklichkeit*) that informs all of theology's content.[79]

The content that both philosophy and theology know is the same thing. However, they know it in different ways. Philosophy knows it as creation while theology knows it as revelation. This created content of philosophy functions for theology as the precondition (*die Vorbedingung*) for the revealed content of theology in the same way that Christian theology understands the content of the so-called "Old Testament" (*sic!*) to function for the New Testament from the perspective of the Apostolic Writings. While the two are sharply separated from each other, philosophy provides the inner authority for theology by anticipating the miracle of revelation (*das Offenbarungswunder*). It does so, to paraphrase Rosenzweig, by recognizing within creation that the creature (*das Geschöpf*) is the subject that bears the visible seal of revelation on its face. The three source-words (*Urworte*) of philosophy – "yes" (*Ja*), "no" (*Nein*) and "and" (*Und*) – are the pre-linguistic words, i.e., the logical structures, that become language (*Sprache*) in the logic of conceptual thought. The inaudible elemental words (*die Elementarworte*) of philosophic metalogic – A, B, A =, = A, B =, = B – prophesy, i.e., are the foundation of, the reality language (*die wirkliche Sprache*) of the grammar of audible or spoken language. This spoken language of grammar is the organon of revelation (*das Organon der Offenbarung*) that links creation in cognition (*Wissen*) with redemption in volition (*Wollen*).[80] The word spoken by man becomes a sign for the word spoken by God. God's word is in content the same for both philosophy and revelation. However, from the perspective of philosophy this word is simultaneously revelation and the creation of the world, while, from the perspective of theology, it is simultaneously revelation and human redemption.[81]

Rosenzweig's discussion of creation as a philosophical concept operates simultaneously, perhaps inescapably, in terms of two disciplines – epistemology and ontology. In terms of ontology, Rosenzweig's account of creation contains a number of important claims, all of which are controversial. First, reality is to be understood dynamically in terms of interconnecting complex processes. Ultimately there are no things. Or, more accurately "things" are not names for static objects. Rather they express relations which, when examined in themselves, also are relations. Second, this reality is more negative than positive. The positive that is the subject matter of traditional philosophy and science consists of ideal ends that direct the flow of human and natural events but are not themselves part of it.[82] For the most part, insofar as we can speak of reality consisting of things (rather than processes or dynamic relations), things are relative nothings that arise from absolutely nothing who (consciously or willfully in the case of humans, but in every other case without will or intention) struggle, but fail, to become really something. Third, creation does not name something that an entity called "God" did at a particular moment in time. It was not the case that at first there was God, then God made something, and the something he made was the world. Rather, there always was and always will be both God and the world, and neither is a thing. Rather, the term "creation" expresses a form of dynamic relation between two otherwise distinct (but not independent) processes. Fourth, the product of creation is nothing definite. Rather it is itself a movement from absolute formlessness to increased form, but never realizing so much form that we can say of either God or the world, "this is what it is."

Rosenzweig's judgments on the epistemology of thinking about creation are no less controversial than his judgments about ontology. First, Rosenzweig makes a radical separation between two forms of thinking, one associated with philosophy and the other with theology. Philosophy tries to achieve knowledge, and in so doing it directs its attention through empirical science to positive things whose character is general and necessary. In contrast theology tries to achieve belief, and in so doing it directs its attention through revelation to negative things whose character is concrete and contingent. Second, the claims of philosophy and science are never more than hypothetical. Philosophy can only look at possibilities and decide among them as best it can. But it cannot on its own legislate which possibilities are actual. Actuality is known only through lived experience. (In this sense Rosenzweig would have considered himself to be "a radical empiricist.") Third, the revelation upon which theology stands is concrete lived experience, and as such, it, unlike philosophy, can make claims about reality.

What links the seemingly irreconcilable realms of philosophy and theology is creation. Creation is the subject of both, and, as such, points to the need for their interrelationship. For Rosenzweig there can be no adequate belief without empirical science, and there can be no adequate knowledge without revelation. This last claim shows that our discussion of Rosenzweig's theory of creation is not yet complete. So far we have considered it solely from the perspective of philosophy and its philosophic sources in Maimonides' doubt about God, Descartes' doubt about the world, and Kant's doubt about human consciousness. But creation also is informed by revelation. In this case the record of the revelation is the account of the origin of the universe in the first chapter of Genesis. In other words, creation is both philosophical and theological. As philosophy it is revealed logical analysis, i.e., mathematical thinking informed by a revealed text. Conversely, as theology it is a philosophical commentary, i.e., a textual midrash informed by contemporary philosophy and science. So far we have examined Rosenzweig's logical analysis. Now we will turn to his textual commentary.

ROSENZWEIG'S THEOLOGY OF CREATION

Rosenzweig understood theology to be a way of commenting on revealed texts. In the case of creation, the critical text is Genesis 1. The form of commentary is a detailed discussion of the grammar of key terms in the biblical narrative from which he draws major conceptual consequences. First I will simply summarize, mostly in his own language, what Rosenzweig says. Then I will draw out the implications of his exposition for his theology of creation in relationship to the philosophy discussed in the previous section.

The first term considered is the root-word "Good" (*Gut*, "TOV").[83] It expresses the divine affirmation of the form in which creatures exist (*göttlichen Bejahung des Kreatürlichen Daseins*). It is the conclusion-expression (*Schlussworte*) for each of the six days of creation. At each of these points God affirms that the thing (*Ding*) in question, produced by his labor (*gewirktes*, whose equivalent in biblical Hebrew is

"MELAKHAH"[84]), exists as something pre-existent (*Dasein als Schon-Dasein*). In other words, the work of creation establishes a non-temporal relationship between God as creator and the things created that together constitute other things (viz., the seven units of creation), that in turn together constitute the world as created. In other words, God's word of creation makes what already exists as hypothetical elements exist in reality as creatures. In other words, what exists hypothetically as elements, known as such through the unspoken language of science, becomes objective (in two senses of the term – as something actual and as an object [i.e., a substance]) through God's spoken word. Furthermore, the term is uttered as a single-term proposition (*Satz*) in a third person, past tense, narrative form. As such it expresses the double objectiveness (*Gegenständlichkeit*) already mentioned (viz., as actual and as substantial). It refers simultaneously both to a divine subject and a divine predicate. As such it affirms a form of unity between God and the world. The world is the work through which the power of God – in its absolute freedom and caprice, without determination – is ever becoming something essentially fulfilled from his inner necessity as a finished, structured, created universe.

God's first "good work" is the work of the first day. He divides it into two moments. The first moment (*Augenblick*) of creation is stated in Gen. 1:1–2, and the second in Gen. 1:3–4.

The First Day, First Moment (Gen. 1:1–2)

(1) BERESHIT BARA ELOHIM ET HA-SHAMAYIM VE-ET HA-ARETS.

"BERESHIT" introduces the first moment of creation. It only means "at the beginning" (*Am Anfang*) of creation. It says that what follows is the first moment, but the sense of moment here has nothing to do with time.

That "BARA ELOHIM," God created (*Gott schuf*), the earth only means that the earth "was" (ward). "BARA" is a creative word (*Schöpfungswort*) in the past tense, where this verb's tense indicates the atemporal creation as a whole in which the general substantiality of all things (*Gegenständlichkeit der Dinge*) is affirmed. The moment is set prior to the temporal coming-to-be of each particular thing (*jedes einzelne Ding*) in each subsequent present moment of creation. With respect to each of these particulars, "to be created" means to receive the specific determination (*Bestimmung*) by which the particular gains a definition (*Bestimmtheit*).

The use of the definite article (HA-) with both heaven and earth in conjunction (viz., "ET HA-SHAMAYIM VE-ET HA-ARETS") expresses the substantiality of the entire world considered as a whole. Furthermore, since the definite article is not used with the individual constituents of the world, the biblical text does not affirm their substantiality as objects.[85] In fact God's creation is unique (*Einzige*) in that the creation of the whole (*Ganze*) at the first moment precedes the multiplicity (*Vielheit*) of the world of things (*Dingwelt*) that are the parts of this whole.

(2) VE-HA-ARETS HAYETAH TOHU VA-VOHU VE-CHOSHEKH 'AL PENAI TEHOM VE-RUACH ELOHIM MERACHEFET 'AL PENAI HA-MAYIM.

"HAYETAH," was (*war*), is an event-word (*Geschehenswort*). That the verb is in the past tense affirms that its objects, "waste" ("TOHU" [*wuste*]) and "empty" ("BOHU" [leer]) both have being (*Sein*) and pre-exist (*Schonvergangenseins*) the act of creation. That they are the first particulars to be mentioned indicates that they are at the lowest level of things that exist as things. "CHOSHEKH," darkness (*Finsternis*), expresses this level of order, and not the presupposed chaos of emanationist philosophy. It is the state of the earth prior to the creation of light.

That "RUACH ELOHIM MERACHEFET," the spirit of God hovers (*Geist Gottes brütend*), over the waters expresses the beginning of the emergence of particular things (*Einzeldinge*) out of the mass of attributes as attributes (*Eigenschaften*), through divine action, at the lowest limits (*unteren Grenze*) of what can count as a thing and as an act. "MERACHEFET," hovers (*brütend*), is the lowest level of action. When stated in the adjectival form of a participle, as it is here, it expresses the lowest level of an action-word (*Tatwort*). As such it is "the dullest of all activities" (*dumpfste aller Tätigkeiten*). Similarly, "RUACH ELOHIM," God's spirit (*Gottes Geist*), is the lowest level of thing. It is God depersonalized as spirit, and, as such, is not really God. The depersonalization is made stronger by the fact that the "RUACH," spirit (*Geist*), is feminine. In other words, the biblical text here pictures the parent of the world as a mother, in the passive role, rather than as a father, in the active role appropriate to the subject of the narrative sentence.

What is created at this first a-temporal moment of creation is the world, but not what is within the world. In other words, what is here created is the space that the objects are intended to occupy. Consequently, space, which is not itself anything positive, is itself a real object whose existence is prior to the existence of the things that occupy space. The space itself is God's spirit.

For all Rosenzweig's protests against Idealism, the universe pictured here at the first moment closely resembles the initial universe portrayed in Plato's *Timaeus*, for Rosenzweig's divine spirit shares most of the characteristics of Plato's necessity (ANAGKE) and Rosenzweig's original universe most resembles Plato's receptacle (UPODOCHEN). As in Plato's account of the origin of the universe, Rosenzweig's interpretation of the first event in Genesis begins not with particulars but with the universe as a whole, a unity that precedes multiplicity. The reference to unity and multiplicity in itself alludes back to Plato. The account in the *Timaeus* proceeds from a set of ordered pairs – not the least of which are being (OUSIA) and becoming (GENESIS), limit (PERAS) and the unlimited multitude (HAPEIRON PLETHOS), mind (NOUS) and necessity (ANAGKE), as well as (from the Pythagoreans) male and female, active and passive, and good and evil – all of which have emerged in a similar role in Rosenzweig's commentary. In this sense Rosenzweig's interpretation of the Jewish dogma of creation appears to be more deeply imbedded in Platonism than in the commentaries of any of his rabbinic predecessors.[86]

God imposes order in the form of attributes onto existing, essentially disordered stuff. This ordering takes place through a series of events. The first event is the attribution of darkness (i.e., being pre-existent as empty and waste) to the world by God's depersonalized feminine aspect, which itself is not really God. Again, Rosenzweig's description of the beginning in Genesis could equally be a commentary on the role of the necessity of the receptacle in relation to the ordered design of the deity in Plato's *Timaeus*.

The First Day, Second Moment (Gen. 1:3–4)

(3) VA-YOMER ELOHIM YEHI OR VA-YEHI OR.

According to Rosenzweig Genesis 1:3–4 describes the work of the second moment (*Augenblick*) at the beginning of creation (*Anfang der Schöpfung*) on the first day. "VA-YOMER ELOHIM," God says (*Gott sprach*), is the first real action-word (*Tatwort*). The Hebrew text indicates that it is a higher form of action than "hovers" (MERACHEFET, *brütend*) by the fact that the tense of "says" is present whereas "hovers" is stated in the past tense.

"YEHI," let there be (*es werde*), is an imperative that as such states the first commandment. However, its form is impersonal and therefore passive. As such the act is separated from God. The word flows out of God's essence (*Wesen*) as a thing, as an it (*Es*), because at this stage God is only a spirit (RUACH) and not as yet a self (*Selbst*).

(4) VA-YAR ELOHIM ET HA-OR KI TOV VA-YAVDEL ELOHIM BAYN HA-OR U-VAYN HA-CHOSHEKH.

What is commanded to be is OR, "light" (*Licht*). It is what the whole becomes in place of darkness. Hence, the term expresses, as do the terms "darkness" and "good," an attribute (*Eigenschaft*). It is not a thing (*Ding*). Like "good," but not like "darkness," the term "light" is an utterly positive affirming valuation (*schlechtin bejahende Bewertung*). In contrast, "darkness" expresses an utterly negative valuation. However,

Creation in The Star 61

whereas "good" is applicable to volition (*Wollen*), "light" applies to cognition (*Erkenntnis*). Hence, the second moment of the work of the first day separates ("YAVDEL", *scheidet*), i.e., distinguishes, the disordered confusion of the attributes. The universe no longer is waste and empty. Now the individual (*einzelnen*) attributes become visible.[87] Their emergence completes (*vollendet*) the first unit of creation, i.e., the first day, by God uttering the spoken word (*tönender Laut*, . . . *Wort*) "TOV."

The Sixth Day (*Gen. 1:26, 27, 30–31*)

Rosenzweig's complete analysis of the creation relationship between God and the world of objects is contained in his commentary on the work of the first day. He turns his attention next to the work of the sixth day, viz., the creation of man. His commentary focuses on four verses – Gen. 1:26, 27, 30 and 31.

(26) VA-YOMER ELOHIM NA'ASEH ADAM BETSALMENU KIDMUTANU.

Man, "HA-ADAM" (*der Mensch*), is the final action of creation (*letzten Schöpfertat*). As such he is the limit between creation and revelation. The "NA'ASEH," let us make (*Lasset uns . . . machen*), indicates that man at first is part of creation and is not yet part of revelation. The "us" (*Uns*) constitutes the first step beyond objectivity (*Objecktivität*) to personality (*persönlichkeit*). It transforms the highest expression of the creative act into the lowest manifestation of the speech of revelation. As such it functions with respect to revelation in the same way that "MERACHEFET," hovers (*brütend*), functions for creation, i.e., it expresses the lowest limit of the specific relational act. The term HA-ADAM expresses the generic concept of a single man, but it also functions as a proper name. As such it is the first proper name, i.e., it is a term that expresses the existence of a person at its lowest limits.

(27) VA-YIVRA ELOHIM ET HA-ADAM BE-TSALMO BE-TSELEM ELOHIM BARA OTO.

The additional description (viz., that man is made "BE-TSELEM ELOHIM," in the image of God [*im Ebenbilde Gottes*]) emphasizes this point. Man is the only object created who is a person. Having personality (*Persönlichkeit*) is what the scriptures mean when they say that man was created in God's image. It means that man uniquely is neither mediated (*vermittelten*) through the universal of the category (*die Allgemeinheit der Gattung*) nor something in need (*bedürftigen*) of multiplicity (*Vielheit*). Man does not yet speak, which means that he does not as yet possess a soul (*Seele*). However, he already possesses the potential to achieve soul and speech. As such he is a wonder (*Wunder*) or sign (*Zeichen*) at the limit of prophecy (*Weissagung*).

(30) UL-KHOL CHAYAT HA-ARETS UL-KHOL 'OF HA-SHAMAYIM.
VA-YEHI KHEN
(31) VA-YAR ELOHIM ET KOL ASHER 'ASAH VE-HINEH
TOV MEOD.

When man's creation is complete, God does not merely say that the work of the sixth day is "good." Rather, he says that it is "TOV MEOD,"[88] very good indeed (*gut gar sehr*). The adjective "*gut*" is a comparative. The addition of "*sehr*" says that this earthly (*Irdischen*) creature is more than of the world. He is mortal and perishable, but he also is more. He has within him the potential to transcend death (*Tod*),[89] i.e., to move through the limit between creation and revelation to the next highest limit between revelation and redemption.

The necessity built into the concept of creation itself, that goes beyond creation to a second revelation, is contained in the sentence, "VA-YEHI KHEN,"[90] Thus let it be (*So ist es*). We have seen that God, in the acts (*Taten*) of creation, is no more than the origin (*Ursprung*), as creative power (*Schöpfermacht*), of an infinite number of individual acts of creation.[91] As such the seemingly revealed God of Genesis[92] remains hidden. To become revealed there must be a second revelation in which each created thing (*Ding*) becomes a testimonial (*Zeugnis*) to a revelation that occurs at every moment of creation. "VA-YEHI KHEN" (Thus let it be) expresses this need. It asserts the transformation of the mythical God's decree in creation of destiny (*Verhängniss*) into a compulsion (*Muss*) of the love (*Liebe*) of his eternal creative essence (*Wesen*) by a negation (*Verneinung*) of the negating act (*Nichts*) of the concealed God into a simple affirming act (*Ja*) of the lived-in moment of life (*unmittelbar gegenwärtig er-lebten Augen-blicks des Lebens*) in the created world.[93]

Hence, while creation is a three-dimensional concept, whose dimensions are the elements God, world and man, it is not complete in itself. As thought about each of the independent elements points to the need for a higher order of thought about relations, i.e., courses, that link them, so the courses themselves are not complete. Creation is the first concept of theology, but in itself it is not an independent notion. Its very conception reveals the need for a higher order of thinking where it is ordered with the other two courses (viz., revelation and redemption) into a unifying structure (*Gestalt*). Rosenzweig finds this higher order of thought in the language of participatory communal worship.

To summarize, for Rosenzweig the philosophically interesting biblical texts that deal with creation are the descriptions of the first and the sixth days. More precisely, the textual description of the first day contains everything that is involved in a philosophy of creation as such. But creation is not theologically self-sufficient. It points beyond itself to revelation and, through revelation, to redemption. In the first day we learn about the dynamic relationship between the elements God and world, while on the sixth day both are brought into a relationship to the remaining element, man. It is the creation of man that exhibits the insufficiency of creation and, by implication, the inadequacy of any purely philosophic/scientific conception of reality.

The creation of the first day is portrayed as an evolutionary process in which verbs, subjects and objects are seen to ascend in order of perfection. The first day is divided into two moments. The level of the first moment is so low that, even within the context of discussing creation as something atemporal (and therefore eternal), the event here described precedes creation. Verbs progress from being past tense, to being present tense, to being imperatives. Similarly their subjects/objects progress from being mere negative value attributes to being passive objects to being actors, first in the singular and then in the plural, with positive value attributes. The primordial waste and empty, whose value is the negative attribute darkness, does nothing at all. Then the passive mother form of deity as God's spirit passively hovers over a unified world that is nothing at all but space. While waste and empty are properly attributes and not objects at all (which means that initially there are no objects), space (the first object) is a non-object object. More properly it is that in which objects reside, so that, devoid of occupants, it is nothing. This empty space is that over which God's spirit hovers. It is not until the second moment of the first day that there is anything at all that acts and does something. Then God first speaks (in the present tense) and commands there to be something positive, viz., light. However, even at the end of the first day there is no positive object. Light is the counterpart to darkness. Both are attributes of objects whose creation precedes the objects they characterize. Darkness judges waste and empty to have negative value. In contrast, light judges the yet-to-be-generated individuals of the world to have, or to come to have, positive value. However, what light enlightens is not yet created. What God creates is a division of spatial regions and object prototypes which themselves generate the individuals who occupy them.

Everything that occurs on days two through five is built into God's action on the first day. What emerge on these intervening days are the models for the objects of the world. In other words, days one through five describe a single event, viz., God's differentiation of the unified space that is the primordial world into regions that host prototypes for an infinite number of particular nothings that strive through an inner necessity to become something. To the extent that they succeed they become objects. Hence, if creation ended on the first day, our universe would be the kind of world that science and philosophy would be sufficient to describe, viz., the world of physical objects. However, the first day is only the origin of creation. Its end is the work of the sixth day, viz., the creation of man.

On Rosenzweig's analysis the most critical feature of the grammar of the language that describes the sixth day is the movement from the singular to the plural number. On days one through five, the singular deity acts and orders the singular object of his action to act. However, the language of the creation of man speaks of "we" acting on something that is like "us," not "he" doing something to someone that is like "him." The change in number marks a change in the quality of what is being created. It symbolizes a movement beyond the generation of objects to the formation of persons. Rosenzweig's objects are movements from nothing towards an asymptote (i.e., towards an ideal that is constantly approximated but never reached). In this case the asymptote is a substance which, as a substance, is static and alone. It is a "thing-in-itself" and "for-it-self" (to use the langue of existentialism), i.e., it is something (ideally) self-sufficient (to use the language of Spinoza). Persons are more. They are not merely something; rather, they are a something that speaks to other somethings. Speech is relational or, in Rosenzweig's terms, "revelational." In other words, the substance that crowns God's creation is a kind of object whose asymptote is not to be self-sufficient, but to be relational. This kind of entity comes to be not merely by being spoken of, but by being spoken to. Hence, man the object, the creature, becomes man the person when God speaks to him. In this sense, man is the partner of God in the creation of man.

Just what it is that God says awaits a second, more advanced revelation, viz., the revelation at Sinai, for the creation of man as an act of revelation stands in relation to higher forms of revelation as hovering at the beginning of the creation narrative stands to other acts of creation. It is revelation at such a low level that it is barely recognizable as such. However, the conclusion of the creation narrative alludes to the higher expressions of revelation that will unfold in scriptures' subsequent narratives. The allusions are the expressions "very good" and "Thus let it be." These terms raise the level of the described created world from being something that merely is or is becoming (which is the universe as science and philosophy know it) to being something that is moral in the sense that its principle of change is a divine moral imperative. The primary object of the imperative is man, for he alone of God's created objects has the capacity to act as God does, viz., by a (moral) will. The concept through which this interaction of wills, divine and human, is to be understood is revelation. This revelation is expressed grammatically in imperatives. In general, what the imperative itself is is a generalized command to redeem the divinely created world.

At this point we can answer why Rosenzweig concludes that the description of reality must include theology. The answer is, because human beings exist. Were it not for man, there would be no need for theology. However, with man an additional element enters into the conception of reality that transcends any philosophic conception of creation. Man is more than a creature. He is also the object of a revelation that is a divine imperative to redeem the world. Hence, man's creation highlights that even in the act of creation itself God is a revealer. The creation of God's object in revelation crowns his production of the world, and that revelation transcends creation. In other words, creation is revelation, but it is the lowest level of the act. In other words, both creation and revelation admit of degrees. The activity of the divine spirit is the lowest form of creation, while God's creation of man is its highest. Furthermore, what is the highest expression of creation is the lowest expression of revelation. Similarly, as the highest kind of creation is revelation, so the highest form of revelation is an action that transcends revelation, viz., redemption. What God and man share in common is that they both act upon the world. As God its creator reveals himself to man, so man his recipient-of-revelation redeems the world. God creates the world, man redeems it, and revelation links the two.

As controversial as is Rosenzweig's philosophy of creation as philosophy, so controversial is his theology as biblical commentary. We have already mentioned that it seems to have much more in common with Plato's myth of the origin of the universe than it does with either the biblical account of creation or with the way that the rabbis interpreted the dogma. Not least among his debatable claims are the following two: First, the implicit ontology of the Hebrew scriptures is found in its use of the definite article. Everything either is a substance or an action. Substances are static objects, actions express dynamic, living persons. Second, the tenses of verbs in biblical Hebrew have no association with time. Rather they express different degrees to which an act can be called personal, where the more something is personal the more it is living and dynamic and the less it is dead and static. In this sense, being a substance or being substantial is the contrary of being personal. In ascending order from impersonal to personal, verbs can be past tense, present tense or imperative. Similarly, whatever the tense of the verb, passive forms express less that is personal than active forms.

These and other problems that arise from both Rosenzweig's philosophy and theology of creation will be discussed in the next chapter. The specific issues themselves fall into two categories. First, since this study claims to be a discussion of the doctrine of creation out of the sources of Judaism, in what sense can Rosenzweig's interpretation be called a "Jewish" position. Second, independent of how the position is classified in terms of institutional religious predicates, is it believable? "Believable" is used in this sentence in two ways. One, is it a plausible interpretation of the biblical text? Two, independent of its value as a commentary on a single piece of literature, is it a plausible way to think about the origin and nature of our universe?

CHAPTER 2

A critique of Rosenzweig's doctrine: is it Jewish and is it believable?

The discussion of the concept of creation out of the sources of Judaism began with modern Jewish philosophy. In this instance only two notable Jewish thinkers address the question with any degree of depth, Nachman Krochmal and Franz Rosenzweig. Except for the fact that the latter seems to have been influenced by the former, for our purposes there is no connection worth mentioning between them. In terms of development and depth of analysis, both with respect to philosophy and theology, there is no comparison. Clearly Rosenzweig's statement is the more important of the two. Hence, the descriptive part of the first section of this study (viz., creation in modern Jewish philosophy) focused almost exclusively on what Rosenzweig wrote. That discussion broke down into two parts – Rosenzweig's philosophy and Rosenzweig's theology.

The two parts of Rosenzweig's discussion of creation differ radically in their mode of presentation. The first is a philosophic analysis built upon his understanding of the history of philosophy. The second is a linguistic commentary on the Hebrew scriptures. However, the conception of creation that emerges through the two disciplines is a single view that in itself shows why Rosenzweig discussed the question in the way that he did. Creation is a philosophic/scientific concept that points to the insufficiency of both philosophy and theology to understand the origin of the universe. Its conception begins as a philosophical question but its solution lies beyond the scope of purely logical empirical thinking. Creation must be grasped as something that is revealed through the word of God as recorded in scripture. Hence, while creation begins as a philosophic doctrine, it ends as a theological dogma.

With respect to philosophy, creation expresses one of three dimensions of relations that determine the dynamic character of reality. Each of these dimensions is itself something dynamic. They are best understood in the same way that their occupants are, viz., as vectors whose direction is from an origin to an end, where the vectors' limits at both extremes are themselves vectors that function with respect to the dimensions they define as asymptotes. In the case of creation, the origin is God and the end is the world.

With respect to theology, creation is the first dimension of reality in the sense that it is the conceptual foundation of the other two dimensions. God's creation of the world, in its movement towards its end, overlaps God's revelation to man, whose own end overlaps man's redemption of the world. A God who is barely God (divine spirit) barely acts (hovers) over a unified universe (heaven and earth) that is really nothing at all (viz., an empty waste that is darkness); this negativity becomes minimally positive (light) as a passive object (space) that, through differentiation, leads through the generation of a series of substantive objects to the creation of man who, as an entity capable of hearing and speaking, becomes more than a mere object, a mere creature. The capacity for speech changes divine spirit into God and transforms (at least some) creaturely objects (viz., humans) into persons. But God as commander and human as person transcend creation. It points to the relationship that the trans-objective God and human define, viz., the path or way of revelation.

At this point I hope I have succeeded in showing that Rosenzweig's doctrine of creation is both intelligible and coherent. However, at this stage a critical reader ought to ask, what makes this view Jewish and why would anyone give credence to Rosenzweig's story. In the remainder of this chapter I will explain what these two questions involve. The next three parts of this book will attempt to answer them.

ISSUES OF JEWISHNESS

Much of the basis for denying that Rosenzweig's concept of creation is Jewish has been mentioned in the discussion of his theology. Simply stated, Rosenzweig's theory seems to have more in common with Greek philosophy, notably Plato's *Timaeus*, than it does with either biblical faith or the beliefs of the classic rabbis. To be sure, Rosenzweig's creation is not identical with Plato's account of the origin of the universe. The most notable difference is that, while both use mathematical tools to construct their models, the mathematics are radically different. Plato's universe begins with positive numbers that express figures in plate geometry. Conversely, Rosenzweig's universe is understood in terms of vector analysis and calculus. The former yields a picture of a static world of eternal objects, while the latter affirms a dynamic universe of endless intentional processes. At the same time the similarities between the two pictures are apparent. Most notably, Rosenzweig's primordial universe in relation to its primordial deity seems more like Plato's receptacle under the influence of necessity than it does like the heaven and earth that God creates in the biblical text. How can a view that seems to be so Greek and so non-Hebrew be called a "Jewish" conception of creation?

In a word, the answer is, because Rosenzweig's view grows directly out of the analysis of creation in classical (medieval) rabbinic philosophy and in the rabbi's commentaries on the first chapter of Genesis. The demonstration of this thesis will be the subject matter of the second part of this book.

ISSUES OF INTERPRETATION

At this point it should be objected that even if Rosenzweig's account is Jewish in the sense that it agrees with, and was influenced by, rabbinic tradition, that does not mean that it is what the biblical text in fact says. This is no trivial claim. Rosenzweig's interest in interpreting the Hebrew scriptures is not merely the interest of a literary critic or a historian of ideas. For him the scriptures are revelation, and as such they are important for determining truth. Furthermore, as I argued in the introduction, the validity of Rosenzweig's reading of the text of Genesis is important for our concern with truth as well. As a Jew I take seriously what Jewish tradition has to say about everything. Part of that tradition is classical rabbinic literature. But the foundation of that literature is the Hebrew scriptures. Now, the entire tradition need not be coherent. If the way that the rabbis read scripture differs from what scripture says, that in itself is important, and, independent of the testimony of the rabbis, there is reason to believe that what Rosenzweig says the biblical text says is not what it says. Most notably, Rosenzweig's interpretation is questionable in two major respects. First, his interpretation of critical terms in the Genesis narrative does not seem to be what those words in fact mean in their biblical context. In general Rosenzweig seems to make them far more sophisticated and complex than they are in fact. Put simply, God's spirit hovering over water in a world that is empty, waste and dark means simply that. It does not tell us anything about an implicit ontology of a universe that is negative and passive. In other words, few readers of the Bible will recognize terms such as "God," "world" and "man" as Rosenzweig discusses them. For most people – be they Jewish or Christian, philosophers, theologians or laypersons – these words name objects, and not (as Rosenzweig would claim) dimensions. Similarly, "light" is light, "good" is good, and "man" is man. The text says nothing about the nature of attributes or the relation between creation and revelation. Put simply, what Genesis 1 describes is the God of the world who at a particular moment in time brings into existence the world out of absolutely nothing, then wills into existence the occupants of that world, including the first human being whose name is "Adam." That is all it says. Anything more found in the story is in the mind of the interpreter and not in the mind of the author(s) of the biblical text. In the words of nineteenth-century biblical criticism, it is an exercise in "eisegesis," not "exegesis." In other words, the attempt to turn a biblical story into a philosophical statement is in itself a distortion of the text. Hence, how can a view that is so philosophical be seriously considered a viable interpretation of the Genesis account of creation?

The third part of this book is intended to be an answer to this question. There I will look in some detail at both critical texts involved here (viz., Plato's *Timaeus* and the first chapter of Genesis). I will argue for the viability of using the schema presented in Plato's account of the origin of the universe to conceptualize the cosmology and ontology implicit in the Torah's narrative about creation. Furthermore, I will argue that both works of ancient literature involve a similar understanding of what is the epistemological status of judgments about the origin of the universe and what is the style of thinking appropriate to discussing this question, viz., the use of a cosmological narrative or (in Plato's language) "myth."

ISSUES OF TRUTH

The most difficult set of questions is left for last. Even if we grant that Rosenzweig's conception of creation is Jewish in the sense that it grows out of and is coherent with the traditional texts of Jewish philosophy, and even if we grant that it is a viable interpretation of the biblical account of creation, why should we believe it to be true? There are a number of features of Rosenzweig's view that stand in marked opposition to the way that most of us (who are the products of twentieth-century European/American culture, whether we be philosophers or scientists or laypersons) think about our origin.

First, Rosenzweig posits that at first the universe is uniform and then, through divine will, becomes diverse. In other words, basically what God does is takes a unified universe and makes it diverse. Consequently, difference rather than unity expresses God's will. In contrast, most contemporaries at least share this much in common with Hegelian philosophy – they agree that unity is better than diversity, so that if there is a God and God governs the world, then he values unity over diversity, so that the world ought to become more and not less, a single universe.

Second, Rosenzweig presents an ontology in which space is

prior to relations in space which are prior to the terms of the relations. In other words, space is prior to actions and actions are prior to objects. Most of us consider the exact opposite order to be simple common sense. First and foremost there are substantive, objective things. These things enter into relations with other things, and it is the things that define the relations. Furthermore, things in relationship are located in space, but space is merely a way of relating things; in itself it is not anything at all.

Third, it seems to be common sense to claim that things either are or are not, absolutely, but it is not intelligible to speak about things being more or less. In other words, affirmations in ontology admit of only two values, being and not-being, and these values are discrete. In contrast Rosenzweig affirms a universe in which being and not-being constitute a continuum that admits of infinite degrees.

Fourth, Rosenzweig's reality is more negative than positive; in contrast, it seems common sense to believe that what is is something, and what is not is not anything at all. In other words, whatever reality is, it is something positive.

Fifth, Rosenzweig's universe has moral value built into it. The consequence of such a view is that a science that is valueless (i.e., morally neutral), is not capable of giving a sufficient account of reality. In contrast, most people today believe that, at least in principle, modern empirical science can give an adequate account of reality, which entails that in reality the universe is a-moral.

The concluding part of this book will attempt to defend Rosenzweig's position in all five of these instances. In doing so, Rosenzweig's doctrine of creation will be scrutinized in the light of contemporary physics. However, this is not to say that I will agree with everything that Rosenzweig says. For example, I want to argue against his radical separation of philosophy and reason from theology and revealed faith. This critique will be presented in the conclusion of this book where I look explicitly at the method for doing philosophy of religion that is implicit in this study.[94]

PART 2
A Jewish view of creation

Introduction

The charge that Rosenzweig's philosophical theology is not Jewish is both general and specific. In general the objection is that philosophy is foreign to Judaism. It is true that at one time in Jewish history – from the time of Saadia in the tenth century through the time of Albo in the fifteenth, rabbinic Jews engaged in philosophical speculation about the doctrines of Jewish faith. However, that period constitutes an exception to the major trends in Jewish civilization, where the emphasis is always on praxis and (possibly) spirituality, but not on conception. Philosophic speculation is a foreign import that infected Judaism for one period (viz., when Islamic civilization dominated the ways that Jews think), but it was precisely that – an infection. Fortunately Judaism cured itself and subsequently returned to its more characteristic pragmatic concerns with law and ethics. The only other period of infection was the late nineteenth-, early twentieth-century Germanic tradition of Hermann Cohen and those whom he influenced (Baeck, Buber, Rosenzweig, etc.), but this period also is an aberration. It occurred because of the highly assimilated condition of German Jewry under the dominant influence of liberal Protestantism.

Underlying this general objection are a number of assumptions about Judaism, two of which are relevant to our enterprise. The first is that only what is distinctively the product of Jewish civilization should be called "Jewish." The second is that certain periods are more definitive of Judaism than others. The claim is that because the Hebrew scriptures, the early works in Jewish law (viz., the Mishnah and the two talmuds) and (possibly) the early commentaries on the Bible (viz., classical midrash) are uniquely Jewish products, they define what Judaism is. In contrast, Jewish philosophy (as well as Jewish mysticism) developed in a later period under foreign influence. Hence, the topics in fifth-century C.E. Judaism (i.e., the century of the completion of the talmuds) define Judaism, but the modes of Jewish thinking that developed later are (at best) of secondary importance or (at worst) deviant.

Briefly, I would object to both assumptions underlying the general criticism of Jewish philosophy as follows: First, a consequence of the historical fact that Jewish civilization has always been subject to the dominant influence of foreign, empiric cultures is that there is no discernible pristine element in anything Jewish. In the case of Jewish philosophy, the external influences are overt. But they are no less present in the early works of Jewish law as well as in the Hebrew scriptures. The latter exhibit the influence of all the cultural and political forces present in the ancient Mediterranean world, including Egypt and Mesopotamia. Similarly, the Mishnah was composed under the influence of Greco-Roman law and the talmuds exhibit input from the Sassanid empire. Furthermore, I can conceive of no sound reason why the fact that something is pristine makes it more valuable than something that is hybrid. In the end, what seems to be correct (i.e., true and/or good) ought to be accepted, irrespective of the sources of the claim and/or activity. More specifically, if philosophy is a good thing for a civilization, then that culture ought to be concerned with philosophy, whether or not it involves external influences.

Second, to draw a line at any specific period in Jewish history and claim what comes before it is authentic while what originates after it is not, seems to me to be arbitrary. The Jewish people have lived in many places throughout their temporal history. While the time is continuous (i.e., there has not since its origin been a time when there was no Jewish people), the location and character of Jewish life is not. Radical changes in external environment have broken the continuity of what Jews do as well as what they believe. In coming to terms with these changes, the Jewish people have been influenced by what they had been, but they have also made (for a variety of reasons) major breaks with their past.

The rise of rabbinic Judaism itself is the product of such a rupture in Jewish history, viz., the displacement of the dominance of Egyptian and Mesopotamian culture (which influenced the Bible) by Hellenistic culture (which influenced the Mishnah and the Midrash). The rise of Jewish philosophy in the Middle Ages is an expression of the Jewish response to another rupture, viz., the displacement of Hellenistic and Sassanid influence by Islam and Roman Catholicism. Now the earlier crises have no inherent ground for primacy over the later ones other than the fact that some occurred earlier and others later in time. However, both the earlier rabbinic law and the later rabbinic philosophy are products of the same kind of historical dynamics and both produced literature essential to the life of the people. The former is the source of rabbinic Judaism's legal-moral-political system while the latter is the foundation of its systematic speculation about ultimate questions of meaning, including the origin of the universe.

The second, more specific objection to the Jewishness of Rosenzweig's conception of creation is more serious, and therefore deserves a fuller response. Even if we grant the general legitimacy of the enterprise of Jewish philosophy, especially philosophy based on early rabbinic texts, Rosenzweig's philosophy does not seem to qualify. It is only loosely connected to the meaning of the biblical text that it utilizes to express categories recognizable from Greek philosophy, notably from Plato, but unfamiliar in the classical medieval texts that defined rabbinic theology. More precisely, the dominant schema upon which the medieval rabbinic commentators drew to interpret the words of scripture and to define true Jewish faith was the science and philosophy of Aristotle. While some philosophically inclined rabbis drew their intellectual insights from other traditions (for example, Saadia from Kalam and Ibn Gabirol from Neoplatonism), the dominant tradition of Jewish philosophy was Aristotelian, beginning with Abraham Ibn Daud and culminating in the interpretations of Gersonides and the criticism of Crescas and his student, Joseph Albo. Hence, Rosenzweig's apparent Platonism in itself marks his interpretation of creation as something foreign to Jewish thought. For Rosenzweig, space has priority over relations that in turn have priority over objects. In contrast, the Jewish Aristotelians gave ontological priority to objects over relations (treating the former as substances and the latter as attributes [or modifications] of substances) and they denied that space itself had anything more than conceptual reality. Furthermore, for the Jewish Aristotelians reality is composed of clearly defined particulars, and not by anything like Rosenzweig's indefinite processes that endlessly unfold in space. Similarly, for the Jewish Aristotelians, lived reality is expressed in terms of finite states of finite objects, whereas for Rosenzweig it is expressed in terms of infinite movements from points of origin to asymptotes. Finally, whereas the Jewish Aristotelians affirmed the moral superiority of unity over diversity, Rosenzweig affirms diversity over unity.

The response to this most serious objection is that what Rosenzweig says about creation in all of these respects is coherent with what even the Jewish Aristotelians taught when they discussed the dogma of creation. The explication of this claim is the subject matter of this second part of this book. I will argue in the first chapter that what we have identified in Rosenzweig's philosophy to be a Platonic conception of cosmology and cosmogony is in fact the theory of creation expressed by most important Aristotelian Jewish philosophers, from Abraham Ibn Daud to Levi Ben Gershon (Gersonides). Then I will make a similar-in-form argument about the sources of Rosenzweig's theology in the classical rabbinic commentaries on Genesis. Here we will look specifically at four representative kinds of rabbinic interpretations – the midrash of the so-called sages whose words are recorded in *Genesis Rabbah*, the early northern European linguistic commentary of Rashi, the Muslim Andalusian influenced philosophical commentary of Ibn Ezra, the Roman Catholic European influenced Kabbalistic commentary of Nachmanides, and the late southern European philosophical commentary of Obadiah Sforno. Here it will be seen that Rosenzweig's particular readings of the Genesis text, while not identical with these commentators' interpretations in every respect (which would be impossible, since they do not always agree among themselves), is authentic, i.e., it is coherent and consistent with their diverse ways of reading the Hebrew scriptures.[95]

CHAPTER 3

Classical Jewish philosophy

This chapter is an argument for the claim that Rosenzweig's interpretation of creation is Jewish on the grounds that it is consistent and coherent with the view of creation that grew out of the development of medieval Jewish philosophy. The substance of the argument is a survey of the field on this particular question. After a brief introduction to set the historical context of the topic, it begins with the first Jewish Aristotelian, Abraham Ibn Daud, and concludes with the most rigorous, detailed Jewish philosophic interpretation of creation in this tradition – that of Gersonides. Particular attention will be paid to those topics on which Rosenzweig seems to depart most radically from Jewish tradition – the ontological status of space and its occupants, the nature of reality as positive and negative, the temporal modification of the act of divine creation, the relationship between ontology and ethics, and the epistemic status of the doctrine of creation.

The classical rabbis believed that the Hebrew scriptures were a record of the word of God. As such, the correct meaning of what the Bible says is true. The problem was to determine in every instance which of the many possible interpretations of the text was the correct one. One test was truth itself. Since the correct meaning of the scriptures is always true, any interpretation that is not true could not be the correct interpretation. To the extent that the claims of revelation were philosophical, e.g., the doctrine of creation, the rabbis used what they knew of philosophy to interpret their holy texts. In practical terms, accepted philosophical traditions provided the schemata through which the data from biblical texts were made intelligible.

The dominant philosophic tradition that rabbinic scholars absorbed in Muslim civilization was atomist until the end of the tenth century C.E. By then the generally accepted religious interpretation of creation was that (a) the universe was created out of nothing, (b) solely through an act of divine will, (c) at a first moment in time, (d) before which there was nothing whatsoever, (e) including time. The commonly accepted scientific atomism of this period raised no conceptual problems about the veracity of this interpretation. Hence, the early Jewish philosophers could accept this position, with little reflection, as the literal and true meaning of recorded divine revelation. A good example is Saadia's[96] Jewish philosophy in *The Book of Beliefs and Opinions*.[97] Saadia's primary issues had to do with the nature of God and the authority of rabbinic tradition. In both cases, but most explicitly in the former, his arguments rest on the veracity of a divine willful creation out of nothing.

This situation of easy compatibility between religious and scientific accounts of creation changed radically as Aristotelian science[98] gained ascendancy over atomism in the western regions of Muslim civilization. The new scientific conception of the universe seemed prima facie to entail that there had always been a universe and that the only sense in which the world could be said to have been created is that (a) an eternal deity, (b) involved solely in a single act of self-contemplation, (c) without any concern for or interest in anything other than himself, (d) necessarily but without intention, (e) serves as an ultimate end towards which everything else in the universe (f) perpetually strives without success to attain. Such a conception of creation seemed to be in fundamental conflict with what was by this time the standard interpretation of the root principle of creation in Judaism. In consequence of this apparent conflict, the Jewish philosophers who dealt most extensively with creation were those who most clearly identified scientific truth with some version of Aristotelianism. Our concern in this chapter is to state their general interpretation of what Genesis affirms as the doctrine of creation. That statement involves two distinct components – what the world as a whole looks like (cosmology) and how it came into existence (cosmogony). First we will consider cosmology and then cosmogony. For reasons that will become clear below, the focus for the former subject will be Ibn Daud and for the latter it will be Gersonides.[99]

ABRAHAM IBN DAUD'S COSMOLOGY

From the tenth century on there were Jews in Muslim civilization who were Aristotelians. However, none of them attempted to reconcile their religious beliefs with their scientific commitments until Abraham Ibn Daud published his major work in Jewish philosophy, *The Exalted Faith*.[100]

With the exception of Gersonides' *Wars of the Lord*,[101] Ibn Daud's discussion of cosmology is more elaborate than that of any of the Jewish philosophers who follow him. Those who came later wrote for an audience already familiar with Aristotelian astronomy. Furthermore, these later rabbis had nothing to contribute to the scientific side of the field itself.[102] Hence they said little about it. In contrast, because of its newness, Ibn Daud provides us with a basic overview of the discipline. Consequently, Ibn Daud is a better source for us than any of the later Jewish Aristotelians for a clear statement of cosmology.[103]

The sections of Ibn Daud's *Exalted Faith* that deal with cosmology are Book 1 chapter 8, and Book 2 Basic Principle 4 chapter 3. His statement in these two chapters is not totally coherent. However, based on what he explicitly says, we can construct the ordering of the cosmos as shown in Table 1.

The first principle or cause of everything that exists is the Absolute One, who is the god of Mosaic revelation and rabbinic tradition. The conception of divine unity as absolute dictates the way by which he causes all diversity. Ibn Daud noted four possible means by which a multiplicity can arise from a single source, only one of which is logically possible for a source that is absolutely one. The diversity of the universe cannot arise from different powers in God, because his unity entails that anything that is true of him ultimately is identical with him.[104] Nor can the diversity arise by God asserting a single

Table 1. *Ibn Daud's ordering of the cosmos*

	Mover (intellect #)	Resulting Existent (sphere #)	Characteristics
Pre-Cosmos	*Unmoved*		
	(1) The Absolute One = God	The Essence of the Throne = The 2nd Intellect	
Cosmos	(2)	(1) The All-Encompassing Sphere = The First Sphere = The Right Sphere	It is divisible into a Northern and a Southern (Inclined) Sphere. It contains the Motion of the Same. It produces the elementary forms.
	(3)	(2) The Eccentric Sphere = The Ecliptic = The Zodiac	It contains the Motion of Diversity. It produces common matter.
	Moved		
	(4)	(3) Saturn	
	(5)	(4) Jupiter	
	(6)	(5) Mars	
	(7)	(6) Mercury	
	(8)	(7) Venus	
	(9)	(8) The Sun	
	(10)	(9) The Moon	The Active Intellect is its soul

influence on a corresponding diversity of materials, because not all diversity is material. There exist different spiritual entities, viz., souls, whose existence cannot be accounted for in this way. Nor could the diversity arise from differences in the natures of the instruments that he employed in creation. This alternative requires that God performed the same act on a number of different occasions, once with each different instrument or celestial intellect. However, absolute unity entails that God can have only one act, identical with himself, that neither begins nor changes nor ends. The only remaining alternative is some version of the Neoplatonic theory of overflow.[105] In this case the universe follows from a variety of effects causally dependent on a variety of intermediaries through a causal chain from the single ultimate source. The first source creates (or, more accurately, is the logical cause of) a single thing, which in turn creates in the same way one or more other things, which in turn create(s) other things through a finite number of steps that ultimately results in the existence and persistence of our present physical universe with all of its diversity.

Ibn Daud summarized this causal chain with respect to two distinct but interrelated causal principles. First, everything that exists has a formal cause; second, every physical entity in the sublunar world has a material cause.[106] Hence, Ibn Daud presented two distinct chains in his resume of the structuring of the universe.[107] The formal order produces the diverse spheres of the heavens, as well as the different kinds of entities within them. Presumably this chain also includes the sphere of the sublunar world. However, formal causality is not in itself a sufficient principle to account for the different material particulars that reside within the sublunar world. In this case there must also be a material chain.

With respect to the formal order of creation, God produces a single entity that can be called both an "intellect" (in the language of Aristotelian science) and an "angel" (in the language of the Torah). This first created entity is one in the sense that its existence is necessary, i.e., it cannot cease to exist. Ibn Daud called it an "unmoved mover," by which he meant that the proximate cause of its existence is its own nature and not an external force. However, it is not absolutely one, because its nature is causally dependent on God's self-action. As a necessary being, it contains only form and not matter. However, because its nature is itself caused, it is a complex being in a way that is analogous to entities composed from form and matter. With respect to its existence having neither temporal beginning nor end, it is something like a form, but with respect to its existence being causally dependent on God's action, it is something like matter.

This first created intellect is called the "second intellect," and probably it is what Ibn Daud once referred to as "the Essence of the Throne."[108] From it overflow three entities, viz., the unmoved mover, the soul and the body of the "Right Sphere," which probably is the "All-Encompassing Sphere," i.e., the heavens of our universe.[109]

From this second generated unmoved mover, identified as the "third intellect," overflows a "fourth intellect" or unmoved mover, as well as the soul and body of the next sphere, which Ibn Daud identified with the sphere of "Saturn." In this case, the produced soul of the sphere is called a "mover" but not an "unmoved mover." Presumably this soul of the second sphere is the mover of the third generated sphere,[110] which differs from the mover of its own sphere in that it is not "unmoved." Unmoved movers are entities that, with respect to their essence, exist necessarily. Presumably, from this level on intellects still exist necessarily, but their necessity is a consequence of their external cause rather than their own nature. In any case, while at first glance it would seem that for each celestial level there exist three entities (viz., a mover,[111] a soul and a body), in fact there are only two entities, since each mover of a sphere is the soul of the proximately preceding sphere that contains it. This is what Ibn Daud means by calling every soul a "mover" of its body, and every intellect a "mover of its mover."

The fifth through eighth intellects with their corresponding spheres, souls, and bodies are not mentioned specifically by name. However, from related astronomical sources of the period, there is no doubt that they are the intellects of the spheres of planets Jupiter, Mars, Mercury and Venus. The ninth intellect is the mover of the sphere of the sun. The lowest sphere to have a mover is that of the moon, whose soul is the Active Intellect. The Active Intellect in turn is the mover, i.e., the proximate cause of the existence, of every human soul.

Note that spheres are circular vectors that differ in angular velocity. Once in existence each vector is governed by each sphere's soul. Furthermore, as the first cosmic sphere contains multiple spheres within it (viz., a Northern Sphere and an Inclined, Southern Sphere that, in virtue of being eccentric, generates an epicycle), so in principle each contained sphere in turn can contain multiple spheres. Just as the subspheres of the First, All-Encompassing Sphere mentioned in 1:8 are not listed in the resume of 2:4:3, so it can be assumed that each of the lower eight spheres named in 2:4:3 also contain unnamed additional spheres. Consequently, Ibn Daud should be seen to have asserted that celestial objects are subject to at least, but not at most, the nine explicitly mentioned spheres. These spheres should be understood as spatial currents in which celestial objects move.

With respect to the material order of creation, Ibn Daud distinguished between formal and efficient causes. Concerning the efficient causes of the material order, the Right Sphere produces the distinct forms of the four elements which inform a common matter produced by the Zodiac of Constellations within the Ecliptic. The result of this conjunction is the elements from which are constructed the complex material objects in the physical universe. Which form is materialized by which portion of the common matter is determined by the inclination of the Zodiac. Each composite entity contains a different harmony of these elements, i.e., irrespective of size, the nature of a physical object is determined by the specific ratio between each constituent element. These ratios are determined by the specific forms of the complex entities which, in turn, are caused by the rate of speed and inclination of the unique rotations of each celestial entity. Their angular velocity is determined by their relative distances from the sun.

Minerals are subject only to efficient causation. This causal order is a necessary but not a sufficient reason for the existence of physical life forms. This class of more venerable entities also requires formal causes. Concerning these causes of the material order, Ibn Daud enumerated a chain of intellects. The first is God, the Absolute One. Each of the other movers primarily directs its force towards unity with its approximate formal cause. In other words, the primary, essential function of each intellect is to conceive of God by means of conceiving of itself. At the same time these primary acts, called "motions," in themselves necessarily produce a number of effects. One of them is the creation of the next lower intellect. At the lowest level of this chain, the Active Intellect's essential function of self/divine contemplation both disposes matter to receive specific forms and material entities to receive non-essential forms through which each member of a species is individuated. Insofar as each individual is a member of a species, it embodies certain forms that define it, and insofar as each individual within a species is unique, it also embodies other forms that happen to be, but need not be, true for the entity to be the kind of thing that it is. In one sense nothing is accidental, since there are necessary and sufficient reasons for every individual characteristic and every unique event. Yet, in another sense every individual is an accident, because nothing essential to the particular determines those causes.

Ibn Daud's ontology includes six distinct kinds of bodies, three kinds of incorporeal substances that are dependent on matter, and three kinds of substances that are independent of matter. The six kinds of bodies are elementary, mineral, vegetative, irrational animal, rational animal and heavenly bodies.[112] The three kinds of materially dependent incorporeal substances are the common matter, forms, and ordering principles of the elements. The three kinds of substances that exist independent of matter are the Active Intellect, and the souls and movers of the spheres. In Ibn Daud's cosmology humans are the one kind of complex entity that link all of these diverse kinds of constituents of the universe. They stand at the end of the formal order of celestial, spiritual causation, and at the summit of sublunar, material causation. They possess matter, form, elements, the kind of soul that has much in common with both plants and animals, and an intellect that is similar in nature to the angels. They are like plants in having growth and reproduction, and like animals in having motion and sensation. They are different from them in that they can perform these life functions in a more venerable way. Similarly, although they attain these ends in a less venerable way, like the angels they can approximate knowledge of God and become immortal.

Ibn Daud considered the above cosmology to be the most reasonable scientific picture of the universe. However, he did not consider it to be knowledge. Undoubtedly he would not deny that we could increase our information about the heavens, but he did not believe that this information could substantially improve our knowledge of cosmology. Rather, he maintained that in principle human science could attain nothing more certain than probable opinions in this area.

There are differences in the details of the picture of the universe as a whole between Ibn Daud's account and those of other Jewish philosophers, notably Maimonides and Gersonides. But there are no differences on anything that matters for our subsequent discussion of cosmogony in comparison to Rosenzweig's philosophy of creation. The most important features that all of these cosmologies share in common are the following: (1) The picture of the present universe as a whole is a product of an interaction between divine intention and natural necessity. (2) Our source for any details in this picture is the empirically grounded science of astronomy, and not divine revelation. However, this science is not capable of giving us information that we can with certainty call knowledge. At best these judgments are true opinions. (3) The picture is primarily one of regions of space that have determinative characteristics. This has the consequence that in its most general features space is something real that has causal priority over any physical objects that occupy it. These real spaces are called "spheres."

Within these most general features, cosmology is a shared

given by all of the Jewish philosophers in their discussion of cosmogony. As we shall see in the next part of this chapter, no similar claim of agreement can be made about the origin of the universe. Here the situation is more complex.

GERSONIDES' COSMOGONY

Gersonides' philosophy is a distinguished conclusion to some four hundred years of rabbinic interpretations of Jewish religious texts in the light of a scientific-philosophic tradition whose sources are the known writings of Plato and Aristotle.[113] As someone who appears at the end of a tradition, his work reflects and builds upon those who preceded him. Hence, before we look specifically at Gersonides' cosmogony, a few general words need to be said about his Jewish sources in order to place his cosmogony within its proper context. This tradition of Jewish philosophy begins with Saadia.

Saadia was the first rabbinic philosopher systematically to list and explain the fundamental beliefs of Judaism. His list included creation, God's unity, divine revelation and the ethical implications of the commandments. As noted above, the logical source for his discussion of all of these beliefs was the doctrine of creation. In this respect, for Saadia creation was logically the most fundamental Jewish belief. His failure to give a detailed account of it had nothing to do with its lack of importance. Rather, it speaks to the fact that in his world the doctrine was relatively non-controversial. In contrast, Ibn Daud does not list creation as a principle at all. According to him, the six basic principles of Judaism involve the nature of God,[114] the veracity of the Torah,[115] and divine providence.[116] Like everyone before him and after him, he believed that Judaism makes claims about cosmology. Since angels are identified with the souls of the celestial spheres, cosmology is subsumed under divine providence.[117] Yet, he had nothing to say about cosmogony.

There is no way to say with any degree of certainty why he omitted cosmogony, since he says nothing about it one way or another. Yet, at least one answer immediately suggests itself.

He believed that no false claim could be a principle in Judaism, and, based on Aristotelian science, it could not be true (contrary to what his rabbinic predecessors had maintained) that the universe had a beginning. In any case, whatever he believed about cosmogony, it is most likely that he considered it to be something problematic. Certainly his immediate successor, Moses Maimonides, would have agreed. For Maimonides, creation is a basic Jewish belief. Yet it is also in fundamental conflict with the scientific Aristotelian consensus of his day. In *The Guide*[118] he was able to affirm belief in creation only by placing creation beyond the limits of scientific knowledge, and, when he formulated his own list of thirteen basic principles in his *Commentary on the Mishnah*,[119] no specific statement of creation was included.[120]

The pressure of the seeming contradiction between the dictates of Mosaic revelation and Aristotelian science with respect to cosmogony led Ibn Daud to omit creation altogether and Maimonides to be uncomfortable with it. It awaited Gersonides to present a new, clear and coherent picture of the origin of the universe through which the prevailing science of his day and his faith in Judaism could be rendered compatible. Hence, we will look to Gersonides for cosmogony.

Gersonides was a committed rabbinic Jew who also believed in the Aristotelian model for scientific inquiry. His major work, *Wars*, deals only with issues where he believes that his predecessors, notably Maimonides and Ibn Rushd, were wrong. Each of the six treatises that make up this work deals with a different topic. The sixth and final section, on creation, is the longest and most complex.

The sixth treatise contains two parts. The first consists of philosophical arguments for creation.[121] The second consists of arguments from revelation, viz., interpretations of the Hebrew scriptures. Only the first eight chapters of part two deal with creation in Genesis. In fact they constitute a systematic statement of the material Gersonides developed in a linear fashion in his separate commentary on the Book of Genesis. The concluding six chapters deal with miracles, and therefore need not concern us in this chapter. Hence, our discussion of creation in classical rabbinic philosophy will focus on part one and chapters 1–8 of part two of the sixth book of Gersonides' *Wars*. First I will summarize how Gersonides interpreted the biblical text, and then I will situate his commentary within Gersonides' general philosophy of creation.[122]

That[123] God created the universe means that God informed an original, indefinite matter with the potential to receive from him the definitions by which the universe became differentiated into its present composition of multiple bodies in motion. It is a consequence of divine unity that in God's case no distinction can be made between the actor and the act. Hence, God's act of creation is indistinguishable from God. Therefore, while it is an act of will,[124] because God desires[125] its creation, it is also an act of necessity.[126] The identity of act and actor in God's case entails that if God were to act otherwise then God would not be God. Divine unity also entails that God can have one and only one invariable act that has neither beginning nor end. Hence, the act of creation cannot occur in time. In other words, while the universe has an origin, it has neither a beginning nor an end. Therefore, although the world is created, it cannot be destroyed. While everything within the universe changes within time, the universe as a whole is timeless.

Since God's one act of creation is invariable, creation itself is a unity. However, as what God knows in a single act is knowable to human beings through a conjunction of multiple acts, so creation is knowable to everyone other than God through multiple stages. These stages can be ordered in terms of priority with respect to logical causes and natures. Each unit of logical priority is called a "day." Anything, a, created on an earlier day than some later thing, b, is prior in any one of the following respects. Either a exists for the sake of b, and/or a is more venerable than b, and/or a is a necessary cause of b.[127] However, it does not mean that it came into existence at some time before b. Hence, it can be said that God created the universe in seven days. However, that does not mean that creation involved seven divine actions or that one act occurred before another.¹²⁸ Rather, it means that from a human perspective the creation of certain aspects of the universe is logically prior to the generation of other aspects, and this differentiation with respect to causal order has seven levels. Scripture uses the expressions "day and night" to show that the entities created within each unit are themselves hierarchically ordered, and "evening and morning" to affirm that they constitute a unity. That the causal ordering is also a moral hierarchy is indicated by calling each complete unit "good." Similarly, scripture states that the overall ordering of all the logically distinct events is the best possible order by calling the completed universe "very good." In other words, this is the best of all possible universes.

The product of the first day is the mental and physical building blocks of the universe, viz., the separate intellects and the material elements. God informed prime matter (BOHU) with the last form (TOHU) that disposed the indefinitely dimensional absolute body (MAYIM) to receive specific dimensions. So informed, regions of the absolute body were differentiated into the four elements – earth (ARETS), WATER (TEHOM), air (RUACH), and fire (RUACH ELOHIM). Then God generated the separate intellects of the cosmos and the souls of the sublunar world, all of which are referred to by scripture as "light" (OR). The biblical term for the remaining original matter unspecified by form is "dark" (CHOSHEKH).

At this stage of creation the elements exist in their natural places with a sphere of earth encompassed by a sphere of water, encompassed by a sphere of air, encompassed by a sphere of fire, encompassed endlessly by indefinite matter (see Fig. 1).

On the second day the cosmos itself (SHAMAYIM) is produced. Some (but not all) of this peripheral, undefined matter is informed into a fifth element (RAQI'A) that constitutes the body of the then differentiated celestial spheres.

The work of the second day of creation is the heavens and the earth, i.e., the space of the universe. Now begins the creation of their "hosts." On day three the dry land and plant species are created to house and feed the land. On day four the stars and other bodies that occupy the supralunar world

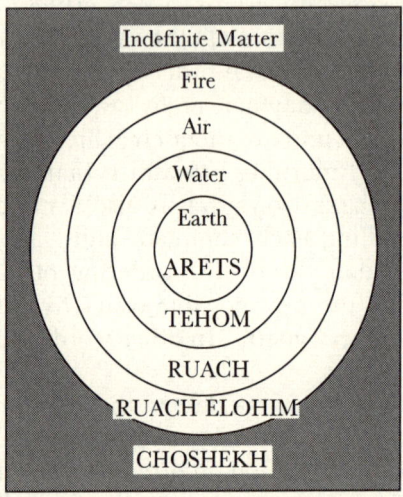

Fig. 1. Day One

(MEOROT) are created. The inhabitants of the water and air, the species of sea-life and flying-life, are created on day five, and the inhabitants of the dry land are created on day six. In general, the land animals perform the same natural functions as fish and birds, but they do them in a more excellent way. The standard of excellence in this case is natural endowment for survival. For example, while all three life forms reproduce by fertilizing eggs, the eggs of the land animals are carried within the bodies of the females. As such they are less subject to destruction than the eggs of fish and birds that are laid outside the bodies of their mothers. Of all of these creatures, the human species is the most venerable. It alone is a product both of the four elements and the souls that are the life forms of the sublunar world, but also of the separate intellects that govern the celestial spheres and bodies. In these respects (viz., the species' moral excellence and the way it participates in both the terrestrial and the celestial regions of the universe), scripture says that humanity is created in the image and likeness of God.

This concludes the summary of Gersonides' theological interpretation of Genesis on creation. Now let us examine its significance for his philosophy, notably its implications for the epistemic status of beliefs about creation.[129] In opposition to Maimonides, Gersonides believed that his interpretation of creation could be demonstrated. The dispute between them had nothing directly to do with religion; it was as pure a dispute about science as can be found in any text of medieval Jewish philosophy. Maimonides' reasons for believing that an answer to this question lies beyond the limits of human knowledge were purely philosophical. He explicitly said that if his judgment can be demonstrated to be wrong, and reason can determine how the universe originated, then Jewish law would dictate what in fact Gersonides does in the second part of the sixth treatise of *Wars*: he would determine the true meaning of what scripture teaches about creation by interpreting the text in the light of the appropriate scientific categories.

Gersonides and others suspected that Maimonides' motives for his judgments about the epistemic status of belief in creation were really religious.[130] However, these considerations are irrelevant, since Maimonides explicitly said that he set limits on human knowledge for scientific (and not religious) reasons. Hence, Gersonides' disagreement with him on this point is purely philosophical.[131]

Gersonides used the following general strategy to prove that the universe was generated and is not eternal.[132] Everything that is generated has three characteristics. They have a final cause; they possess accidental (i.e., non-essential) properties; and they essentially function for the sake of something other than themselves. Gersonides attempts to prove that both the heavens and the visible part of the planet earth[133] have these characteristics. Therefore, they must be generated, which in turn entails that the present universe in general was generated.

Gersonides' evidence for the claim that the heavens possess accidental properties are the very characteristics of the heavenly objects that Maimonides used to show that necessity is insufficient to account for the existence and nature of the sublunar world. Heavenly objects rotate from east to west rather than from west to east; they have different specific angular velocities; different spheres contain different exact numbers of bodies; and the circuits of the spheres are epicycles and/or eccentric rather than simple circles. These are facts that can be known by observation but not by unaided reason, and as such they demonstrate the limits of what is knowable by reason. So far Gersonides does not differ from his predecessors. However, Maimonides concluded from this data that there can be no human knowledge of the universe in general and the supralunar spheres in particular. Gersonides contends that this conclusion does not follow. It is not the case that we do not know these matters because of the limitations of the knower; rather, the limitations are in the object known. Everything contingent is to a certain extent formal and to a certain extent material. What is true of that entity in terms of its form is necessarily true of it and as such is knowable by both God and humanity. Conversely, what is true of that entity in terms of its matter merely happens to be true of it, and as such it is knowable neither by humanity nor by God.

The dispute between Maimonides and Gersonides on creation parallels their dispute about divine attributes. No facts are at stake in either issue. Rather the argument is over what those facts mean. The data led Maimonides to conclude that human beings have no knowledge whatsoever in these areas. Gersonides argues that his conclusion was too strong. Certainly there is a difference between human and divine knowledge. What God knows as a cause of the object known in a single act is knowable to human beings only as an effect of the object known through what could be expressible only in an infinite conjunction of distinct propositions. Nevertheless, what humans know about God and creation is knowledge, and insofar as both are in principle unknowable there is nothing to know.

Critical to this dispute is the assumption that only what can be affirmed with certainty to be true is knowledge. It is not the case that we have no information about the heavens. Through observation we know how celestial objects move and what are their velocities. However, by definition, this is not knowledge. It is merely true opinion. Maimonides assumed that in principle some higher form of entity could know what we merely believe, and concluded from this that we can know nothing about the origin of the universe. Gersonides rejects this assumption. That these truths merely happen to be true in itself shows that the heavenly objects themselves merely happen to have the particular properties that they have, which in itself demonstrates that they, like the bodies that occupy the sublunar world, are generated, contingent entities.

This central difference between Maimonides and Gersonides points to other, less central changes that from our perspective are as important as the dispute over epistemology. First, Gersonides has assigned a subtle but significant value to empirical observation for both scientific and religious thought that Maimonides did not. After all it is through observation that we can determine what in truth scripture teaches about creation. Second, the supralunar world is far less different from the sublunar world in Gersonides' picture than it is in Maimonides'.

Critical to Maimonides' argument against the Aristotelian view is the claim that no valid inference can be drawn from what is true of the sublunar world to the nature of the supralunar world. Critical to this logical move is that these two domains of the universe are radically different, and the crucial premise in this argument is that the physical world consists of contingent, material objects while the occupants of the cosmos are necessary, spiritual entities. Gersonides has rejected this distinction. The hosts of both domains are contingent and material.[134] Consequently, Maimonides' denial that inferences can be made from one domain to the other is illegitimate. In other words, what we know about creation is based on astronomy, and astronomy is fundamentally no different a human science than physics. Hence, reason can draw from its sources in sense experience to judge what is true of the heavens just as Maimonides and the entire tradition of medieval Jewish philosophy would grant the legitimacy of this move for all of the earthbound physical sciences.

This completes the summary of the doctrine of creation in medieval Jewish philosophy.[135] It is sufficient for us to make some (at least preliminary) judgments about the Jewishness of Rosenzweig's philosophy. There are certainly differences in the details of the two accounts of creation, but in general they have more to do with the linguistic and stylistic differences between medieval Mediterranean philosophy and modern European philosophy. In substance as well as in methodology, the similarities are striking. Most notably, the doctrine of creation is a product of insight gained from the testimony of a biblical text that is accepted as divine revelation, expanded upon by empirical-scientific observation that is considered to be reliable evidence, that is synthesized through philosophical categories of thinking. To be sure, there are differences in judgment about the relative reliability of each ingredient in this mix. That difference is important, and we will return to it at the end of this book, but it does not affect our consideration of the roots of Rosenzweig's doctrine of creation in classical Jewish philosophy. Rosenzweig stands with Maimonides in this instance, in opposition to Gersonides, in claiming that an adequate understanding of this dogma transcends the domain of rational thought based on observation. In its fulness the character of this belief is determined by theology, i.e., what is reported in scripture as God's word. Still, beyond this epistemological difference, the coherence between the modern and the medieval views of creation in Jewish philosophy is striking. Most notably, creation is seen as a single act set outside the bounds of time in which the universe is understood as a movement from God in the direction of a moral ideal, for this is what Gersonides means when he says that each conceptual stage (or day) of creation points to an increase in the moral perfection of the entity created.

None of these remarks are intended to cover over the apparent disparities between Rosenzweig's interpretation and the accounts of Ibn Daud and Gersonides. One important difference is that the classical sources emphasize that creation is about entities located in space, whereas on Rosenzweig's account creation is primarily about the differentiation of space itself. However, when we look more carefully in the next chapter at what the peers of these philosophers had to say when they examined the biblical text in detail, we will see that this apparent difference is only apparent and not substantial, for the created objects are really determinant regions of space.

More important are the implications of the different understandings of what scripture means when at the end of the sixth day the whole of creation is called "very good" (TOV MEOD). For Rosenzweig it expresses a movement beyond the created world towards a transcendent redeemed world; for Gersonides it says that the created universe is (to paraphrase Leibniz) the best of all possible worlds.[136] This difference has important consequences for ethics. For Rosenzweig the world as it is is never good. The value of what is always remains that of a nothing that points us towards what ought to be, and the ought to be never is. In contrast, while Gersonides' world can always be better, given what it is at any point in time, it is nonetheless something of positive value. In other words, what is is no mere nothing; it is also a something that as such is good. In other words, while both philosophers integrate their science with ethics, Gersonides' world is more "friendly" than Rosenzweig's. For Gersonides that the world is good is itself part of its description; for Rosenzweig it is a moral imperative to constant action that entails unending dissatisfaction with whatever is. As we shall see, in this respect Rosenzweig is closer to Plato than to Aristotle, and Gersonides is closer to Aristotle than to Plato. However, both remain Jewish philosophers. The issue here is not between Jewish and Greek philosophy. Rather the issue is between two different visions of the correlation of physics and ethics. The spokespersons for each view classically are Maimonides and Gersonides. Gersonides' view leads directly to Spinoza, while Rosenzweig's anti-Spinozistic view has its roots in Maimonides.

However, the above generalization is too extreme. In fact the differences between Maimonides, Gersonides and Rosenzweig are not as sharp as our analysis at this stage of the discussion suggests. A more careful examination of how

Maimonides and Gersonides explain God's act of creating the world out of an apparent nothing will close this gap.

Maimonides proposed that there are four ways to answer the question: What role does God play in bringing the universe into existence? – The way of the Epicureans (by which he meant the ancient Greek atomists), the way of the Torah, and the respective philosophies of Aristotle and of Plato in the *Timaeus*.[137] According to the ancient atomists, God plays no role; the universe originates through chance. However, "through chance" does not mean what we would today call "by accident." Rather, the claim of these philosophers was that the universe was generated through a motion of "conjunction and separation"[138] whose governing principles are purely mechanical. The random motion of elementary particles produced collisions that in turn resulted in a vortex motion. Within the vortex, like particles combined and unlike particles separated. The denser compounds, meeting greater resistance to their random motion, concentrated at the center of the vortex. Similarly, less dense compounds, meeting less resistance, spread out along the periphery of the vortex. Qualitative differences in the present universe are explained by mechanical laws that govern the ratio of elements in a compound. Differences in the elements are reducible to their different densities, and differences in density are a consequence of the rotation of the original vortex.

According to the Torah, God brings the universe into existence by an act of will and volition; according to Aristotle, God acts from necessity. In Plato's case, the way that God generates the world is analogous to the way that a potter works with clay or a smith with iron. On all of these views there never has been a time at which there was no universe. At the same time, all of them agree that the universe did not merely happen; its existence was determined by God. However, did this determination occur through an act of will and/or by necessity and/or by some other means? Again, the Torah says by will, Aristotle says by necessity, and the two claims seem to be contraries.

The disagreement itself cannot be clear until these alternatives are clarified. Given that the universe has in some sense a beginning, what can be said about it prior to its origin? The conventional understanding of the Torah said that it was absolutely nothing, while both Plato and Aristotle were interpreted to deny this possibility. In Maimonides' words, the Torah says that before creation the universe was "absolutely non-existent,"[139] whereas for Plato and Aristotle what comes to be can only come to be "from some-thing."[140]

However, the line between creation from "something" and from "nothing" is not as sharp as it might at first appear to be. Greek literature gives us two ways to say "out of nothing." Something could be said to come into existence either (G1) "EK TOU ME ONTOS" or (G2) "EK TOU ME EINAI," literally, out of what is no entity or out of what is not. These two expressions may function as synonyms or they may be used to express alternatives between being something and nothing. (G1) could be used to characterize something that exists but is not something like a substance. Similarly, (G2) could describe something that does not exist, but, because it is a thing, is something. For example, in the first case, a relation could legitimately be said to be "EK TOU ME ONTOS," and, in the second case, Pegasus could be said to be "EK TOU ME EINAI."

Out of these two ambiguous Greek terms there developed six Arabic expressions, four Hebrew expressions and six Latin ways to say "out of nothing." The Arabic expressions are (A1) "MIN AL-MA'DUM" (literally, from the lacking), (A2) "MIN LAYS" (literally, from [an is] not), (A3) "LA MIN SHAY" (literally, not from a thing), (A4) "MIN LA SHAY" (literally, from no thing), (A5) "MIN LA WUJUD" (literally, from no existent), and (A6) "BA'D AL-'ADAM" (literally, after [or, according to] the lack). The Hebrew idioms are (H1) "ME-AYIN" (from [an is] not), (H2) "ME-LO DAVAR" (from no thing), (H3) "ME-LO METSIYUT" (from no existent), and (H4) "ACHAR HA-HE'DER" (after [or according to] the lack). Finally, the Latin translations are (L1) "ex non esse" (out of [an] is not), (L2) "ex nihilo" (out of nothing), (L3) "non ex aliquo" (not out of some[thing or other]), (L4) "ex non existente" (out of no existent), (L5) "post non esse" (after [or, according to] [an] is not), and (L6) "non fit ex aliquo" (is not made out of some[thing or other]). In other words, from the relatively simple distinction in Greek between two different forms of contraries to "out of something" emerge six distinct expressions in Arabic, six in Latin and four in Hebrew. Furthermore, the relations of historical influence from the distinctions in any of these languages into the others are sufficiently complex that prima facie linguistic expression alone is insufficient to provide any insight into what a medieval philosopher meant when he denied that the universe was created out of something.[141]

Gersonides' analysis of the meaning of the clause "out of nothing" in the dogma of creation is the most developed in all of classical Jewish philosophy. The three elements in Gersonides' account of the origin of the universe are (1) first matter,[142] (2) first or corporeal form,[143] and (3) absolute body[144] or last form.[145] The general view that Gersonides inherited from his Aristotelian predecessors (most notably Ibn Sina, Al-Ghazali and Ibn Rushd) was that the four elements are a composite of elementary forms and absolute body, and absolute body is itself a composite of primary form and matter. However, his predecessors differed about the nature of primary matter.

According to Ibn Sina it is simply a potential disposition to receive dimensionality, i.e., primary matter is the principle that accounts for the fact that something otherwise defined occupies three-dimensional space. In contrast, Al-Ghazali argues that it is something actual. It is not a mere potential disposition for dimensionality; it is in fact the property of being dimensional, i.e., of having cohesion and mass. Ibn Rushd adopts a more subtle intermediate position. Like Al-Ghazali he claims that primary matter is something, namely dimensionality. However, like Ibn Sina, it is not definite dimensionality, since it is something common to everything that has dimensions. It is that property without which something could not have any kind of dimensions whatsoever. In other words, first matter is an actual indefinite dimensionality that disposes something to receive definite dimensions.

Both Ibn Sina and Al-Ghazli thought that primary matter is something definite. They did not imagine that it made any sense to claim that something could remain indefinite and in any sense exist. They differed only on whether this definite thing existed potentially or actually. In seeking a middle position, Ibn Rushd in fact adopted a more radical ontological position than either of his predecessors. With Al-Ghazali in opposition to Ibn Sina he agreed that only something actual can properly be said to exist. With Ibn Sina he agreed that the shared dimensionality of everything in the material universe could only be a potentiality for material determination. He resolved this apparent contradiction by claiming that the first material element of the universe is an actual indefinite thing through which all subsequent composites have the potential to receive actual determinate dimensions. However, the problem remains how to make intelligible the claim that something, neither actually definite nor potentially indefinite, actually is indefinite.

Gersonides' contribution is his concept of an absolute body. It is the building block of the universe. It is Ibn Rushd's composite of prime matter and prime form. The form disposes the body for definite dimensions that it does not possess. The matter makes it something actual whose dimensions are indefinite. Gersonides' critical conceptual step beyond Ibn Rushd is to disassociate the essence of this body from any notion of form. In other words, the form of absolute body, properly speaking, is not really a form. In fact it is nothing at all. Like God it is something that can only be known PROS HEN equivocally. In other words, the composite universe of form and matter has two ultimate principles. The formal principle is God and the material principle is absolute body. However, this body is not a second deity worthy of worship. In fact, it is nothing at all. God is the sole principle of what is. At the same time it also is true that only God is completely. Ours is not the world of being. It is the world of becoming, which, as such, both is and is not. Absolute body is the principle of the non-being. Precisely because it is by nature indefinite[146] (i.e., it is not precisely anything at all), it is the nothing out of which God created the world of becoming. As such it is the indefinite stuff that God, through an act of will that is identical with his nature, informs and (by so doing) gives definition and purpose. For those who cannot imagine what such a stuff would be, Gersonides suggests that scripture itself provides an example. It is water. Water is a definite substance that, unlike other substances, in itself has no definite shape. Similarly, absolute body is a definite body that, like water and unlike other bodies, "does not preserve its shape."[147]

The ontological parity between Gersonides' absolute body and Rosenzweig's indefinite created objects is self-evident. In other words, Gersonides provides us with a conceptual source for understanding Rosenzweig's assertion that the universe is composed of an infinite number of formless nothings whose origin is a radically characterless particularity and whose end an infinitely removed ideal of becoming a fully determinate something, viz., a something that only God can be in his perfection. The origin is Gersonides' absolute body which, devoid of divine purpose (by accident in the ancient scientific sense of the term described above), is the source of all particularity in the universe. Furthermore, purpose or intention enters the world only as an end, and not an origin, in the ideal being of God. Hence, the earlier proposed difference between Gersonides and either Maimonides or Rosenzweig (on what it means to say that the world was created "out of nothing"), on more careful inspection of the texts, is not as great as it at first seemed to be. No Jewish philosopher was more committed to the use of Aristotelian science to understand the universe than was Gersonides. Yet, even he, on final analysis, adopts a schema that is more closely aligned to Plato's dynamic creation story in the *Timaeus* than it is to Aristotle's static cosmology of substances.

In conclusion, the classical rabbinic philosophers used Aristotelian and (more importantly) Platonic science to clarify what the fundamental religious duty to believe in creation means. On their interpretation(s), the revealed biblical text itself affirms that our positive universe originates (1) out of negative materials, (2) in a spatial nothing, (3) through a single, atemporal, divine act, (4) that is both meaningful[148] and (5) purposeful.[149] These affirmations function as boundaries on otherwise free intellectual inquiry, and set the framework for the discussion of creation in modern Jewish theology.

On a superficial reading of both classical and modern Jewish philosophy, the former best exemplified in the arguments of Gersonides and the latter best represented by the assertions of Rosenzweig, seem to be very different. The one uses Aristotelian terms like "form," "matter" and "body," in grammatically static sentences about substances, while the latter uses "meta-Hegelian" words like "the All," "the utterly distinctive" and "the individual" in dynamic sentences about processes. However, a deeper reading of both Jewish sources exhibits a close affinity between their respective interpretations of creation, as the preceding summary indicates. Furthermore, both ways of interpreting the dogma of the Genesis account of creation seem closest in origin to Plato's picture of the universe in the *Timaeus*. Furthermore, the two philosophical theologies of the origin and general nature of the universe are similar not only in their ontological conclusions, but in their epistemic method as well. For both representatives of Jewish philosophy make the following affirmations: (1) Philosophy and theology are distinct modes of analysis that yield coherent conclusions about religious dogma. (2) Philosophy in itself can make no truth claims. Rather, its function is to provide possible schemata by which the data given in science and religions are to be made intelligible. In Gersonides' language, the domain of philosophic knowledge is what can be known with certainty, but all of the particulars in all of the created world (events and subjects in both the supra- and the sub-lunar worlds) are contingent, which means, in Rosenzweig's language, that philosophy deals only with the possible, not even with the probable, and certainly not ever with the real. (3) The domain of science is the contingent; its source is sensible experience. (4) The domain of theology is revelation; its source is revealed scripture. (5) We are commanded to strive to know creation. As such pure study, in both science and religion, is as much a religious obligation as are the study of law and the pursuit of right action. Medieval Jewish philosophy expressed this religious dimension by calling creation a dogma, which translates in Rosenzweig's language to mean that creation leads to revelation that leads to redemption. However, this knowledge is itself an asymptote which we may at best approximate (again) from the evidence of empirical science and revealed theology made intelligible through the schemata of philosophy.

However, these conclusions are premature. The first point listed above (in reference to epistemology about the similarity of both classical and modern Jewish philosophy) was that thought about creation requires both philosophical speculation about scientific claims and theological analysis of the words of scripture. So far we have only examined the classical Jewish philosophy of creation. Now, in the next chapter, we will turn to its theology, i.e., its biblical commentary. We have already seen how Gersonides interprets the text of Genesis 1. Now we will compare and contrast his interpretation with other major representatives of classical rabbinic commentary – the sages of the midrash (specifically, of *Genesis Rabbah*), Rashi, Ibn Ezra, Nachmanides and Sforno. Here the question becomes: Is Rosenzweig's theology as Jewish as his philosophy? – which means the following: Granted that Rosenzweig's doctrine of creation is coherent with classical rabbinic philosophy and with Gersonides' biblical criticism, how Jewish is the way all of them read the biblical text? In other words, is the textual interpretation of the Jewish philosophers representative of (or [more accurately], coherent with) the way that the standard classical rabbinic commentaries on the Torah interpret creation? The answer lies in comparing and contrasting both sets of biblical commentaries.

CHAPTER 4

Classical rabbinic commentaries

At the beginning of this book we presented the picture of the universe that Franz Rosenzweig developed in *The Star* as the leading candidate for a text on which to base a Jewish philosophy of creation. The central question of this (the second) part of this book is whether or not that philosophy can be called "Jewish," which we interpreted to mean, does Gersonides' account of creation follow from classical texts of Judaism and is it coherent with them. We began to answer that question by examining what classical Jewish philosophy has to say in comparison with Rosenzweig's philosophy. For this comparison we referred to a number of works in philosophy in the Aristotelian tradition of Jewish thought, beginning with Abraham Ibn Daud's *Exalted Faith* and culminating in Gersonides' *Wars*. That survey and comparison has brought us to a tentative affirmative answer to our question.

The reason that the affirmation of the Jewishness of Rosenzweig's doctrine of creation is only tentative is because the tradition of Jewish philosophy, from Saadia through Albo, is only one major source for classical rabbinic thought. Another, equally important source, is the collection of classical rabbinic commentaries on the biblical text itself, particularly (for our purposes) on the opening narrative in the Torah. We looked first at the statements of the Jewish philosophers because these works, precisely because they are philosophy, provide the clearest and most detailed statements of what the rabbis believed about the creation of the universe. Hence, they are the clearest basis for drawing our comparison. However, logically, the theological commentaries have priority over the philosophical works. This judgment is a consequence of how both Rosenzweig and the classical philosophers agree about the epistemic status of this belief. Reason alone can determine nothing more about the general nature of the universe (cosmology) and its origin (cosmogony) that determine different schemata, any of which may (but only may) be true. To decide which is probably the case requires an external source of data. One source is that testimony of experience that is presented to philosophers by empirical science, and the other source is the testimony of the divine word through revealed scripture. Hence, from a purely theological perspective, the critical concern becomes the testimony of scripture as its words are interpreted with the schemata of philosophers and expanded by the data reports of scientists.

The consequence for our purposes of the logical judgment – that the witness of how the classical rabbis understood the words of scripture has epistemic priority to how the rabbis of this same period schematized their doctrine of creation – is that no final judgment about the Jewishness of Rosenzweig's account of creation can be made without comparing it with the classical commentaries on the Bible. More specifically, at this stage of the argument the question becomes the following: First, are the ways that Gersonides' philosophy of creation is coherent with the cosmology and cosmogony of the Jewish Aristotelians themselves coherent with the interpretations of the classical biblical commentators? Second, what positions do these commentators take on the issues on which Rosenzweig and the Jewish Aristotelians disagree?[150]

In general, the following agreements and disagreements between Rosenzweig and the classical Jewish philosophers emerged from our discussion in the last chapter. They share in common the epistemological judgment that knowledge of creation is not possible solely on the basis of reason, so that conclusions are at best probable but never certain, and that the sources for these claims are observational science and revealed scripture. Furthermore, the product of this interaction of philosophy, science and religion is a view that says that both God and space are fundamental to an adequate understanding of reality, and space itself, while real, is not anything that is. Furthermore, reality is best understood in terms of processes (rather than things) whose duration is infinite (rather than finite), where what is real is indefinite (rather than definite). Furthermore, the physical order of the universe is also spiritual order in which purely mechanical principles of causation are necessarily conjoined to teleological principles of moral intent. In other words, an adequate physics is not morally neutral. Furthermore, the moral order of the universe is associated with divine will, and somehow that will involves taking a pre-existent unified universe and making it diverse. Hence, in some sense diversity is more to be valued than unity from a divine perspective.

Where Gersonides and (subsequently) Rosenzweig part company with the creation philosophy of the earlier medieval Jewish philosophers (and Spinoza as well) is in emphasis. For all of them the universe is to be understood as an interplay between form and matter, necessity and contingency and/or volition, souls and/or intellects and spheres and/or bodies, statics and dynamics, spiritual and material causation, and (finally) divine intent and natural necessity. In the case of Gersonides' and Rosenzweig's accounts of the origin of the universe, the role of all the left-hand terms (viz., form, necessity, souls, intellects, statics, spiritual causation, and divine will) declines while the role of all of the right-hand terms (matter, contingency, spheres, bodies, dynamics, material causation and natural necessity) increases. In other words, insofar as our attention is on the origin of the universe in creation (as opposed to its end in redemption), the dominant role is that of a real nothing whose purposeless action initiates a direction towards an ideal, purposeful something. The question now becomes, where do the classical Jewish exegetes stand on these questions?

In this chapter we will summarize the conception of creation in Genesis 1:1 through 2:3 of the Hebrew scriptures as it was interpreted in classical rabbinic commentaries. We will not look at every possible commentary. However, I believe that the selection made here is representative of the diversity of interpretations in rabbinic tradition and, with one qualification, sets the parameters for determining their theology. The qualification is that we will not give an equal voice to all genres of rabbinic interpretation.

Every rabbinic commentator on the Hebrew scriptures sought to explain the biblical text in any or all of the following ways: He explained its simple and/or its hidden meaning.[151] The former dealt primarily with linguistic questions, viz., semantics and grammar. The latter was homiletic, philosophical and/or mystical.[152] All four kinds of interpretation are important to understanding how the rabbis understood scripture. Often these different approaches produce contrary explanations, and most commentators recognized the contradictions. However, for most[153] rabbis this diversity of meaning was not problematic. God expresses his truth in multiple ways in his written word. While one kind of hidden meaning may not seem to agree with another kind, the conflict is not real. The difference lies only in the mode of expression. Just as the statements "$1 + 11 = 12$" in base 10 and "$1 + 11 = 100$" in base 2 look as if a different answer is being given to the same question, when in fact the problems are different and the statements (when set in their appropriate context) are coherent, so a homiletic and a philosophical statement (for example) may seem from their language to be dealing with the same question and reaching different conclusions, when in fact each kind of statement is dealing with a different question, so that there need not be any conflict between them. This is not to say that the rabbis advocated any kind of multiple truth theory any more than modern mathematicians believe that clear mathematical problems have multiple, incoherent answers. Without exception these rabbis believed that the one God of the universe is the source of only one truth. However, this epistemological unity has diverse expressions. Consequently, within each kind of commentary there is a need to determine, in keeping with the logical rules of that language, coherence and consistency. Hence, two philosophical interpretations that violate the law of the excluded middle cannot both be true. However, to give a reason is not the same thing as to give a homily, and what the language of a text explicitly says[154] or what that explicit statement logically entails[155] need not be consistent with what the text alludes to or how the text is used in a homily. Allusions or hints[156] are subject to their own distinct kind of grammar.

This chapter is primarily interested in what the classical rabbinic commentators determined to be the linguistic and philosophical meaning of Genesis' account of creation. In the final part of this book we will scrutinize what both Judaism and contemporary Western science teach about creation. What the commentators say as linguists and as philosophers is comparable, but what they say as preachers and mystics is not. In the latter cases the languages simply are too different. In this instance a comparison would be the proverbial error of comparing apples and oranges. Hence, midrash and Kabbalah will largely be ignored. However, there is no intent to diminish the importance of this kind of rabbinic literature, and it will not be ignored entirely. All of the commentators to be discussed below were familiar with and used midrash, and at least one of them[157] emphasized Kabbalah. To the extent that these materials relate to the commentators' consideration of the linguistic and philosophical meaning of creation in Genesis, they will be dealt with in this chapter. This is particularly true with midrash, where the single most important collection of early rabbinic discussions of the meaning of the biblical text is *Genesis Rabbah* (BERESHIT RABBAH), in which the linguistic and philosophical levels of interpretation are formally inseparable from the homilies.

The body of this chapter will explore the interpretations commentators recorded in *Genesis Rabbah* and in the MIQRAOT GEDOLOT. The former text records statements attributed to the Tannaim and the Amoraim.[158] The rabbis mentioned in *Genesis Rabbah* all lived during the last five generations of Tannaim or the first five generations of Amoraim. Whether or not the statements attributed to these rabbis were in fact said by them is an important question of scholarly research in the field of midrash, but it need not concern us in this chapter. Whenever *Genesis Rabbah* was written, it is our earliest collection of rabbinic commentaries on the Bible, and no matter who in fact authored the interpretations presented in this collection, they are the oldest authorities we have for what scripture means when it describes God as "the creator."

The MIQRAOT GEDOLOT are the standard printed rabbinic bibles that present the Hebrew text of scriptures together with the Onkelos Aramaic translation and selected commentaries. While the commentators presented in this set of volumes have changed over the centuries, and "making it" into an edition is in no formal sense a process anything like canonization, still, these commentators have a special status in consequence of their inclusion that testifies to their legitimacy as major voices in determining what is the Jewish understanding of scripture. Whether they have this authority because they were included in these editions or they were included because they have the authority is an issue of scholarship that once again need not concern us here. Whatever the historical causes are, it is legitimate to look to these commentaries on Genesis as expressions of the rabbinic understanding of creation. The four presented are Solomon Bar Isaac,[159] Abraham Ben Meir Ibn Ezra,[160] Moses Ben Nachman Gerondi,[161] and Obadiah Ben Jacob Sforno.[162]

Each commentator presents a distinct kind of biblical interpretation that reflects the best of Jewish thought about creation at the time that he lived. Rashi was an eleventh-century French rabbi who survived the first crusade.[163] His commentary has little philosophic content, but is a critical source for understanding the traditional rabbinic simple and homiletic interpretations of the biblical text. Ibn Ezra was a twelfth-century Andalusian poet and philosopher, a contemporary of such philosophic giants as Judah Halevi, Joseph Ibn Zaddik and Abraham Ibn Daud. Although his commentary was written near the end of his life when he travelled in the southwestern regions of Christian Europe, it represents the best of Jewish philosophical thought in the Muslim world. Nachmanides was a twelfth-century European rabbi, steeped in Kabbalah, who played an important spiritual and political role in the so-called "Maimonidean controversy"[164] and represented the Jewish community in the disputation with Pablo Christiani in Aragon in 1263. Nachmanides reflects a sophisticated European rabbinic reading of scripture informed by the commentaries of Rashi, Ibn Ezra and Jewish mysticism. Finally, we encounter in the early sixteenth century physician-rabbi Sforno an excellent representative of how a committed and informed rabbinic Jew of the Italian renaissance interpreted creation in the light of the entire tradition of classical rabbinic commentaries.

This chapter will not examine every aspect of these commentaries. Rather, they will be discussed in detail only to the extent that they deal with creation, i.e., with a picture of how the universe came into being and what the initial universe looked like. First the sages and Rashi will be discussed, who are our primary source for the rabbi's simple and homiletic meaning of the text. Then we will turn our attention to the philosophical interpretations of Ibn Ezra, Nachmanides and Sforno. Finally, these two discussions will be used to draw a composite picture of what they say about creation. This general picture will be used to answer the above set of questions that constitute the primary concern in this chapter: Is what modern Jewish philosophy (exemplified by Rosenzweig) says about creation, in the light of medieval Jewish philosophy (best represented by Gersonides), in the light of the rabbis' understanding of the Hebrew scriptures, a Jewish view of creation?[165]

THE COMMENTARY OF THE SAGES AND RASHI — HOMILY AND LANGUAGE

Day One (Gen. 1:1–5)

BERESHIT BARA ELOHIM ET HA-SHAMAYIM VE-ET HA-ARETS. (1:1)

(In the beginning God created the sky and the earth.)[166]

Of all the words in the Genesis narrative of creation, the first verse has produced the most discussion, much of which has focused on the meaning of the first word, BERESHIT. In many ways it is the most difficult in the entire text. The problem is that it is in a construct form and there is no obvious noun present in an absolute form for it to modify. Rav Oshaya interprets[167] the word to mean that with what is first, viz., the Torah, (God created the earth and the sky). Rashi repeats this homily while noting that the simple meaning of the verse as a whole is an introductory clause to the next verse that says, "At the beginning of (this account of) the creation of the earth and the sky . . ."

According to *Genesis Rabbah*, the Torah is not the only thing whose existence preceded the Genesis account of creation. Besides the Torah there actually existed the Throne of Glory.[168] Furthermore, the thought[169] of the patriarchs, the nation of Israel, the Temple and the Messiah preceded creation.[170] Rav Menachem and Rav Joshua Ben Levi read Prov. 8:22 in such a way that six things are said to precede creation, viz., TOHU, BOHU, CHOSHEKH, MAYIM, RUACH and TEHOM. At the same time Rav Gamaliel provides proof texts to show that all six were created. For TOHU, BOHU and CHOSHEKH he cites Isa. 45:7, for MAYIM Ps. 148:4-5, for RUACH Amos 4:13, and for TEHOM Prov. 8:24.

Prima facie the text seems to report a conflict between Rav Menachem and Rav Joshua Ben Levi on one hand, who interpret scripture to claim that the universe was created out of these six entities, and Rav Gamaliel on the other hand, who interprets scripture to be committed to creation out of nothing. However, the same text of *Genesis Rabbah* also suggests how both claims can be reconciled. While these six entities, as the Genesis text suggests, already existed at the time of the creation of This-World, that does not mean that they are eternal and uncreated. On the contrary, when Genesis is read in the light of the rest of the Hebrew scriptures, it is apparent that they also were created, which means that the creation described in Genesis is not the first creation. In fact there was at least one creation prior to this one.

Rav Judah Ben Rav Simon draws an identical conclusion in his interpretation of "And there was evening" in verse five. Rav Abbahu bases this inference on "it was very good" in verse thirty-one at the end of the sixth day of creation. I.e., while the earlier created worlds were good, because God could only create what is good, no world before this one was good enough for God to preserve its existence. However, neither scripture nor reason provide any way to know anything about any creation before the generation of our universe.

We can infer *that* such a universe(s) existed, but not *what* it was (they were) like. Hence, while Bar Kappara and Rav Judah ben Pazzi, in opposition to Rav Jonah and Rav Levi, grant permission to describe the events of creation, there is a general consensus that no one can and/or should attempt to explain what existed prior to creation.

While it may be the case that there were multiple creations prior to this one, the sages agree that the creation of this earth and sky was a single divine event and not a series of distinct occurrences spread out over six or seven days. Shammai places heaven before earth, Hillel places earth before heaven, and Rav Simeon reflects the general consensus that they were created simultaneously.

VE-HA-ARETS HAYTAH TOHU VA-VOHU VE-CHOSHEKH 'AL-PENAI TEHOM VE-RUACH ELOHIM MERACHEFET 'AL PENAI HA-MAYIM. (1:2)

(And the earth was empty and void, and darkness was on the face of the deep, and the wind of God hovered over the face of the water.)

Rab, Bar Kappara and Rav Huna note that it is not objectionable to claim that this verse asserts that God created the universe out of pre-existing TOHU, BOHU and CHOSHEKH. The concern is whether or not such an interpretation would diminish the grandeur of God's act of creation, and they are confident that it would not. According to Rab, on this interpretation God's creation of the world out of TOHU, BOHU and CHOSHEKH is comparable to a king[171] building a palace[172] out of sewers, dunghills and garbage.[173] Again, the issue here is not whether this is the correct interpretation of what in fact occurred. They only assert this to show that the text does not raise any serious question of belief either about God or the root belief in creation. What is critical is the claim that God and God alone created the universe, and this interpretation of the verse would not entail any problem with this fundamental affirmation.

Rav Abbahu and Rav Judah Ben Rav Simon claim that "TOHU" is an expression of astonishment or amazement, and Rashi agrees that this is the simple meaning of the term. However, the sages and Rashi disagree about the cause of the wonder. According to Rashi, the amazement is directed at BOHU, which means emptiness or void. However, the former sages treat the term "BOHU" as a synonym of "TOHU," and claim that these two words function as active verbs whose subject is the earth. In other words, these rabbis read the first half of this verse as saying that the earth was amazed and astonished at the priority of sky to earth[174] even though they were created simultaneously.

According to Rashi the "TEHOM" is the water that is above the earth, and "RUACH ELOHIM" is the Throne of Glory standing in the air supported above the water by wind from God's mouth. Consequently, according to Rashi this verse says that "(At the beginning . . .) the earth was in the amazing condition of being empty, above the earth was water, and above the water in the dark was the Throne of Glory supported by the wind blown from God's mouth." In other words, from an empty universe God creates the earth, the water, the dark, and the Throne of Glory, from which he molds the earth and sky of This-World.

VA-YOMER ELOHIM YEHI OR VA-YEHI OR. VA-YAR ELOHIM ET HA-OR KI TOV VA-YAVDEL ELOHIM BAYN HA-OR U-VAYN HA-CHOSHEKH. (1:3-4)

(And God said, let there be light, and there was light. And God saw that the light is good, and God divided between the light and the darkness.)

Rav Judah Ben Rav Simon and Rav Berekiah interpret "And God said, Let there be . . . and . . . was . . ." to mean that everything God created came about solely by God's word without any work or effort on his part,[175] and Rav Judah and

Classical rabbinic commentaries

Rav Nehemiah claim that the creation of the primordial light preceded the creation of the world. Rav Samuel Ben Nachman, who is identified as a "master of HA-AGADAH," said that we learn from Ps. 104:2 that the light was created by God wrapping himself in a robe from which the light shone forth. However, it is noted that this explanation was stated in a whisper in answer to a specific question from Rav Jehotsadak, with the implication that such explanations should be given only in this way in this context.[176]

In general, the sages claim that this first light is not the sunlight of our world. Rather it is a light intended for the Messianic Age. Rashi picks up this interpretation, which he identifies as the text's homiletic meaning, viz., that God separated out of the waters the "upper waters" to be used by the righteous (HA-TSADIKIM) in the World-To-Come ('ATID LAVO). At the same time, Rashi notes, with Rav Jochanan and Rav Simeon Ben Lakish, that the simple meaning of the verse is that God separated the light and the dark in order to limit the day. In other words, the simple meaning of the text is that the light created is the light of our world, while the homiletic meaning is that this is an entirely different light that will only exist in the World-To-Come.

VA-YIQRA ELOHIM LA-OR YOM VE-LA-CHOSHEKH QARA LAYLAH VA-YEHI 'EREV VA-YEHI VOKER YOM ECHAD. (1:5)

(And God named the light day and he named the darkness night, and it was evening and it was morning on the first day.)

Rav Jochanan, basing his interpretation on Ps. 104:3, and Rashi, basing his interpretation on Rav Jochanan, assert that the angels were created on the second day. Rav Chanina uses Gen. 1:20 and Isa. 6:2 to claim that they were created on the fifth day. Rav Luliani Ben Tabri suggests that this dispute is of no importance. What is critical is to deny that they were created on the first day in order to preserve the critical affirmation that God acted alone in creating the universe.

Day Two (Gen. 1:6–8)

VA-YOMER ELOHIM YEHI RAQI'A BE-TOKH HA-MAYIM VIHI MAVDIL BAYN MAYIM LA-MAYIM. VA-YA'AS ELOHIM ET HA-RAQI'A VA-YAVDEL BAYN HA-MAYIM ASHER MI-TACHAT LA-RAQI'A U-VAYN HA-MAYIM ASHER ME'AL LA-RAQI'A VIHI KHEN. (1:6–7)

(And God said, let there be a firmament in the middle of the water and it will divide between water and water. And God made the firmament, and it divided between the water that is below the firmament and the water that is above the firmament, and it was so.)

In general the sages identify the RAQI'A as a roof composed of water, while Rav Judah Ben Rav Simon describes it as a lining. It was formed by solidifying the middle layer of the waters created on the first day. According to Rav Chanina, Rav Jochanan, Rav Judah Ben Rav Simon, and Rav Abba Ben Kahana, the solidification of the liquid was caused by heat from the upper layer of primordial fire, while Rashi simply says that it occurred in an amazing way through God's word. Rav Oshaya and Rav Phinechas add that an emptiness[177] surrounds the RAQI'A that separates it from the earth below and the upper waters. In general the sages say that the RAQI'A is as thick as the earth, but Rav Chanina says its thickness is that of a metal plate, Joshua Ben Rav Nehemiah says it is two fingers, and the son of Pazzi says that its volume exceeds that of the seas by thirty xestes.[178]

According to Rashi, "'ASAH" means that he set the RAQI'A in its proper space in the developing universe. Consequently, Rashi reads verses 6 and 7 together as follows: Solely by God's word, in an amazing way, what God made on the first day out of the pre-created waters became solidified, so that the liquid water was separated by the solid material composed of the same substance. Then God set the solid water in its proper place above the liquid destined for use in This-World while the upper liquid was hidden away for use in the World-To-Come.

VA-YIQRA ELOHIM LA-RAQI'A SHAMAYIM VIHI 'EREV VIHI VOKER YOM SHENI. (1:8)

(And God named the firmament sky, and it was evening and it was morning on the second day.)

According to Rab, Rav Abba Ben Kahana and Rashi, the term "SHAMAYIM" expresses a substance that is a composite of the elements water (MAYIM) and fire (ESH).[179] Rav Phinechas and Rav Levi say that it is made of water. Rab's judgment that it is solidified water is cited by Rav Isaac to show that there is no substantial difference between these two views. In other words, as in the case of the RAQI'A, the sky is water solidified by the heating of elementary fire. In other words, the sky is just another name for the RAQI'A. Rashi accepts these interpretations and notes that the word means that water (MAYIM) is there (SHAM).[180] Hence, on Rashi's reading, God took the portion of water solidified by its mixture with the element fire and set it in its place. Then he collected together the liquid under the solidified mixture of water and fire to form the dry land out of the seas. Then and only then is the work said to be good.

Rashi's reading is intended to explain why the work of the second day is not called "good" until the middle of the third day.[181] Other answers to this question are given by Rav Jose Ben Rav Chalafta and Rav Chanina. The former says it is because the second day is when Gehenna was created. The latter says it is because a separation, viz., between the upper and lower waters, occurs on the second day. However, Rav Nachman and Rashi both note that the simple meaning of the text is that the expression "KI TOV" designates that a particular work of creation was completed. Since the work begun on the second day is not completed until the third, the expression does not occur on the second day, but occurs twice on the third. In other words, God's first work (on the first day) was the creation of light, and his second work (on day two and the first half of day three) was the creation of the dry land and seas on the earth.[182]

Day Three (Gen. 1:9–13)

VA-YOMER ELOHIM YIQAVU HA-MAYIM MI-TACHAT HA-SHAMAYIM EL MAKOM ECHAD VE-TERA-EH HA-YABASHAH VA-YIHI KHEN. (1:9)

(And God said, let the water beneath the sky be collected into one place, and let the dry land appear, and it was so.)

According to Rav Acha, Rav Isaac, Rav Nathan and Rav Berekiah, the waters under the sky were gathered together, in order to separate the dry land and the seas, by God telling the land and sky to expand into the territory of the water, and then telling them to stop expanding. In other words, the planet earth was formed in the same way that the universe was created, viz., solely by the word of God, without any effort of God's part, in an amazing way.

VA-YIQRA ELOHIM LA-YABASHAH ERETS UL-MIQVEH HA-MAYIM QARA YAMIM VA-YAR ELOHIM KI TOV. VA-YOMER ELOHIM TADSHE HA-ARETS DESHE 'ESEV MAZRI'A ZERA' 'ETS PERI 'OSEH PERI LE-MINO ASHER ZAR'O BO 'AL HA-ARETS VA-YEHI KHEN. VA-TOTSE HA-ARETS DESHE 'ESEV MAZRI'A ZERA' LE-MINE-HU VE-'ETS 'OSEH PERI ASHER ZAR'O BO LE-MINE-HU VA-YAR ELOHIM KI TOV. VA-YIHI 'EREV VA-YIHI VOKER YOM SHELISHI. (1:10–13)

(And God named the dry land earth and the collection of the water he named seas, and God saw that it is good. And God said, let the earth sprout grass, herb yielding seed, fruit tree making fruit according to its kind in which is its seed, on the earth, and it was so. And the earth sprouted grass, herb yielding seed according to its kind, and tree making fruit in which is its seed according to its kind, and God saw that it is good. And it was evening and it was morning on the third day.)

According to Rashi, "DESHE" means a ground cover, "ESEV" are its roots, and the "'ETS PERI" are trees that themselves taste like fruit. In this last instance Rashi is repeating the interpretation of Rav Judah Ben Rav Shalom. Only as a consequence of God's curse on the first human did trees change and receive their current nature. In other words, the nature of the physical universe prior to the expulsion from the Garden of Eden was different from the present nature of the universe. In this connection Rav Nathan notes that creatures such as gnats, insects and fleas also came into existence because of the curse on the land. In this sense the beginning of Genesis speaks about the creation of a universe that preceded our own and not about our world.[183] The creation of this vegetation, at the end of the third day, is God's third work.

Day Four (Gen. 1:14–19)

VA-YOMER ELOHIM YEHI MEOROT BIRQI'A HA-SHAMAYIM LE-HAVDIL BAYN HA-YOM U-VAYN HA-LAYLAH VE-HAYU LE-OROT UL-MO'ADIM UL-YAMIM VE-SHANIM. (1:14)

(And God said, let there be lights in the firmament of the sky to divide between the day and the night, and they will be for signs, seasons, days and years.)

According to Rashi, everything created was made on the first day. However, while the MEOROT (celestial objects) were already created, they were not suspended in the sky until the fourth day. Furthermore, not everything created and located during the days of creation exists in This-World. Some of them were set aside for use by the pious in the World-To-Come. Among them are the OR and CHOSHEKH of the first day. In other words, both the OR and the CHOSHEKH are substances not found in our world. The MEOROT are our heavenly bodies.

Commenting on the statement that the heavenly bodies were created "for signs, seasons, days and years," the sages identify "seasons" with the Sabbath, "signs" with the three pilgrim festivals, "days" with the ritual associated with the beginning of the month, and "years" with ROSH HA-SHANAH. In other words, the primary purpose to the existence of the heavenly bodies is to enable Israel to calculate its major ritual observances. According to Rashi, what scripture means by saying that they were created to be "signs" ("LE-OTOT") is that they will function in This-World as evil omens.

VE-HAYU LIMOROT BIRQI'A HA-SHAMAYIM LE-HA-IR 'AL HA-ARETS VA-YEHI KHEN. VA-YA'AS ELOHIM ET SHNEY HA-MEOROT HA-GEDOLIM ET HA-MAOR HA-GADOL LE-MEMSHELET HA-YOM VE-ET

HA-MAOR HA-QATAN LE-MEMSHELET HA-LAYLAH VE-ET
HA-KOKHAVIM. (1:15–16)

(And let the lights in the firmament of the sky exist to enlighten the earth, and it was so. And God made the two large lights, the large light to rule the day and the large light to rule the night, and the stars.)

According to Rav Jochanan, Rav Chanina and Rav 'Azariah, only the sun gives light; the moon's value is its use in calculating seasons. Hence, at least Rav Jochanan distinguished between MEOROT that are stars, i.e., objects which emit light, and other heavenly objects that only reflect light. However, Rashi follows Rav Berekiah and Rav Simon in rejecting this distinction, as is apparent in Rashi's discussion of the relative size of the sun and the moon. According to Rashi, when they were created they had the same size. However, as a punishment for protesting this equality, the moon's dimensions were reduced. To compensate for this loss of light at night, the other stars were created. In other words, at first the only stars were the sun and the moon, and the others were subsequently created as a consequence of the moon's sin of jealousy.

In connection with Gen. 2:1[184] Rav Efes and Rav Hoshaya list the heavenly bodies and specify the length of each's cycle – 30 days for the moon, 1 year for the sun, 12 years for Jupiter, 30 for Saturn, and 480 for Mercury, Venus and Mars.[185]

VA-YITEN OTAM ELOHIM BIRQI'A HA-SHAMAYIM LE-HA-IR 'AL
HA-ARETS. VE-LIMSHOL BA-YOM U-VA-LAYLAH UL-HAVDIL BAYN
HA-OR U-VAYN HA-CHOSHEKH VA-YAR ELOHIM KI TOV. VA-YEHI
'EREV VA-YEHI VOKER YOM REVI'I. (1:17–19)

(And God set them in the firmament of the sky to enlighten the earth. And to enlighten the day and the night, and to divide between the light and the dark, and God saw that it is good. And it was evening and it was morning on the fourth day.)

In general the sages say that the sun is a globe set in a sheath of water to dim its heat so that the earth would not burn up.[186] Rav Jannai and Rav Simeon Ben Lakish add that Gehenna is nothing other than the end of This-World when the sun will be unsheathed to destroy the wicked. However, they stand in opposition to the other rabbis of *Genesis Rabbah* in making this claim. Rav Judah says that the orbits of the sun and the moon move behind and above the sky. The rabbis say that they are behind and below, and Rav Jonathan reconciles the two claims by saying that the orbits are behind and above in the summer and behind and below in the winter.

A number of sages[187] interpret the statement that God set the sun and the moon "in the RAQI'A of the sky" to mean that there exist a number of different heavens. In this context "(BI)RQI'A HA-SHAMAYIM" should be translated as "(in) a heaven in the sky." Rav Abbahu and Rav Pinechas place the sun and the moon in the second heaven above the earth, each of which is separated from the other by "a five hundred year journey," i.e., the height of each heaven is a distance of 500 w, where "1 w" is one walking year, which is the distance that an average man could walk across in one year.

Genesis Rabbah offers no more detail in its picture of a multi-heavened universe. However, set within the context of the whole body of AGADAH, the following elaboration is possible.[188]

The creation of the universe is divided into eight distinct sets of events on which the following objects were created: (1) The elements of creation are said to exist prior to creation itself. They include the great garment,[189] sky, earth, the TEHOM and fire.

(2) During the two thousand years prior to the first day of creation God generates the principal regions of the universe. Both the Torah and Hell are generated out of the elementary fire, while Paradise, the Throne of Glory and the heavenly Temple are created out of nothing. While the sages discuss distinct things, all of which are pictured in spatial terms, it also is clear that none of them are conceived as physical entities. Hence, the spatial descriptions are at best metaphors intended to express what are in reality non-spatial concepts. In any case, at the center of the universe is the Torah. To its left is Hell, to its right is Paradise, the celestial sanctuary is in front, and the divine Throne is above it.

(3) Day one is when the first differentiation of space into spheres takes place. The dark is formed out of the TOHU, and the waters are formed out of BOHU, which itself is formed out of the primordial TEHOM. The light and the wind of God are created on this day as well as 996,000 alternative universes. At the center of the universe is the stone of the Temple mount on which subsequently Abraham will bind Isaac and the Holy of Holies will be constructed. It is surrounded by the Temple mount itself, which in turn is surrounded by the sites of Jerusalem, the inhabited world of the descendants of Noah, and an uninhabited world which is composed of a mixture of wilderness regions and great seas. Beyond the circle of the uninhabited world are four distinct regions. To the east is Paradise with its seven heavens. To the south is a land ruled by a winged angel named Ben Netz that contains storehouses of fire and caves of smoke. To the west is the Great Ocean, beyond which are the Steppes. The Ocean is dotted with islands on each of which lives a distinct species of descendants of the first human. These Adamites are genetically different from the descendants of Noah.[190] The Steppes are a region without any form of vegetation, in which reside scorpions and serpents. Finally, severed from the rest of the sphere of the universe, there is the northern region. Its outermost territory is the place for Hell itself. Its innermost space, between the uninhabited world and Hell, is a hellish region in which will reside hellfire, hail, ice, dark, evil spirits and devils. Each subspace of the universe is thought of either as a circle or as a circle segment. The original distance from the center of the universe to the circumference of the Temple mount is one third of the distance from the center to the outer periphery of the inhabited world, which is 500 w, and the distance from the inner to the outer periphery of the uninhabited world is another 500 w.

(4) Day two is when the second differentiation of space takes place, viz., Hell is divided into seven divisions, each of which is further divided into seven unnamed subregions, and the waters are sectioned into their three parts. This day also marks the creation of the universe's first inhabitants, viz., the angels, who, together with the RAQI'A, are formed from the primordial

fire. Each angel is assigned to rule a specific differentiated region of space.

(5) Day three is when the third and final differentiation of space into cubes takes place,[191] and the universe's second inhabitant – vegetation – is created. God differentiates the elementary sky into seven heavens, while the elementary earth combines with the upper and lower waters to form dry land, mountains, hills and seas. In addition Eden, as distinct from Paradise, is created with 310 worlds.

(6) On day four the third inhabitants of the universe, the celestial objects, are created. This classification includes the sun, the moon, the stars, and the wheels to which these bodies are attached in order to make their regular orbits through the universe. The moon is reduced to one-sixtieth of its original size. The scattered pieces of its original body still are seen in space as debris. The sun has two faces, one of hail and the other of fire.

(7) The fourth group of inhabitants – sea- and air-life forms – is created on the fifth day. Both the fish and the birds are generated out of the dry land and the seas themselves.

(8) The sixth day is when the fifth and final group of inhabitants of the universe is generated, viz., land-life forms, which includes both mammals in general and the human species in particular.

Day Five (Gen. 1:20–23)

VA-YOMER ELOHIM YISHRETSU HA-MAYIM SHERETS NEFESH CHAYAH VE-'OF YE'OFEF 'AL HA-ARETS 'AL PENEY REQI'A HA-SHAMAYIM. VA-YIVRA ELOHIM ET HA-TANINIM HA-GEDOLIM VE-ET KOL NEFESH HA-CHAYAH HA-ROMESET ASHER SHARTSU HA-MAYIM LE-MINE-HEM VE-ET KOL 'OF KANAF LE-MINE-HU VA-YAR ELOHIM KI TOV. (1:20–21)

(And God said, let the water swarm a swarm of living creatures and a fowl that flies above the earth on the face of the firmament of the sky. And God created the large sea-serpents and every living creature that creeps which the water swarmed according to their kind, and every winged fowl according to its kind, and God saw that it is good.)

According to Rashi, "SHERETS" (swarm), includes all fish and all other kinds of living things whose bodies are not high above the ground, whether they are flying things such as flies, or creeping things such as ants, beetles and worms, or animals such as moles, mice and lizards. "REMES" (creeps) names the subcategory of SHERETS that live on the dry land.

Rav Phinechas and Rav Idi say that the "TENINIM" are the sea-serpent Leviathan and the great beast Behemoth, neither of whom have mates. Rav Mattenah and Rav Huna say that they are multi-colored creatures. Rashi says that the simple meaning is large sea fish and the homiletic meaning is Leviathan and his mate.[192] However, the female was slain in order to prevent this species from reproducing.[193] Since nothing in This-World could destroy it, the species would eventually slaughter every other living thing, and in so doing, destroy the world. It might be argued that if that was the case, God should not have created the female sea-monster to begin with. Rashi's answer is that its meat is preserved for food for the righteous in the World-To-Come.

VA-YEVAREKH OTAM ELOHIM LAMOR PERU URVU U-MIL-U ET HA-MAYIM BA-YAMIM VE-HA-'OF YIREV BA-ARETS. VA-YEHI 'EREV VA-YEHI VOKER YOM CHAMISHI. (1:22–23)

(And God blessed them saying, be fruitful and multiply and fill the water in the seas, and let the fowl multiply on the earth. And it was evening and it was morning on the fifth day.)

According to Rashi, the statement, "be fruitful and multiply" in this context is a blessing rather than a commandment. It is not so much that the multiple forms of life ought to reproduce, as God will see to it that they will be fruitful and multiply. This blessing is given to compensate for human beings receiving permission[194] to hunt them for food. Rashi reasons that from this perspective the land animals should have been separated out from the others for a special blessing of procreation. Because humans will eat land-life more than they will eat sea-life, land creatures are the most endangered species. However, they did not receive the additional blessing, i.e., the increased capacity to naturally reproduce above and beyond the power for survival granted to sea-life, because of the sin of the serpent.

Day Six (Gen. 1:24–31)

VA-YOMER ELOHIM TOTSE HA-ARETS NEFESH CHAYAH LE-MIN-AH BEHEMAH VA-REMES VE-CHAYTO ARETS LE-MIN-AH VA'YEHI KHEN. VA-YA'AS ELOHIM ET CHAYAT HA-ARETS LE-MIN-AH VE-ET HA-BEHEMAH LE-MIN-AH VE-ET KOL REMES HA-ADAMAH LE-MINE-HU VA-YAR ELOHIM KI TOV. (1:24–25)

(And God said, let the earth bring forth a living creature according to its kind, cattle and creeping thing and beast of its earth according to its kind, and it was so. And God made the beast of the earth according to its kind and the cattle according to its kind, and every creeping thing of the ground according to its kind, and God saw that it is good.)

Rav Hoshaya the Elder identifies the BEHEMAH with the serpent. Rav Chama Ben Rav Hoshaya claims that there is a discrepancy between verses 24 and 25. Whereas verse 24 enumerates four kinds of animal souls, viz., those of a NEFESH CHAYAH, BEHEMAH, REMES and CHAYTO-ARETS, verse 25 lists three kinds of land animals, viz., every one except the NEFESH CHAYAH. He and Rabbi then claim that the missing subcategory of creatures with souls[195] is that of demons. Of course the argument is specious. Within Rashi's own conventions of interpretation "NEFESH CHAYAH" expresses the general term for souls of the three subspecies of animals enumerated. However, as is always the case in midrash, it is not altogether clear how to evaluate homiletic arguments. In any case, what is significant for our purpose is that verse 24 is understood to refer to the formulation of the souls of the creatures whose bodies are generated in verse 25, and among the creations of the sixth day are demons.

According to Rashi, what it means to say that God makes[196] something is that he fixes[197] it in such a way that it receives a certain character or nature[198] and it has a definite height or stature.[199] Hence, none of the species of land-life were created[200] at this time. Like everything else, they were created on the first day, but their different natures and places in the universe are made definite on this sixth day.

VA-YOMER ELOHIM NA'ASEH ADAM BE-TSALME-NU KIDMUTE-NU VE-YIRDU VIDGAT HA-YAM UV-'OF HA-SHAMAYIM U-VA-BEHEMAH UV-KOL HA-ARETS UV-KOL HA-REMES HA-ROMES 'AL HA-ARETS. VA-YIVRA ELOHIM ET HA-ADAM BE-TSALMO BE-TSELEM ELOHIM BARA OTO ZACHAR U-NEQEVAH BARA OTAM. VA-YEVAREKH OTAM ELOHIM VA-YOMER LAHEM ELOHIM PERU URVU U-MIL-U ET HA-ARETS VE-KHIVSHU-HA URDU BIDGAT HA-YAM UV-'OF HA-SHAMAYIM UV-KHOL CHAYAH HA-ROMESET 'AL HA-ARETS. VA-YOMER ELOHIM HINEH NATATI LAKHEM ET KOL 'ESEV ZERA' ASHER 'AL PENEY KHOL HA-ARETS VE-ET KOL HA-'ETS ASHER BO FEREI 'ETS ZORE'A LAKHEM YIHYEH LE OKHLAH. UL-KHOL CHAYAT HA-ARETS UL-KHOL 'OF HA-SHAMAYIM UL-KHOL ROMES 'AL HA-ARETS ASHER BO NEFESH CHAYAH ET KOL YEREQ 'ESEV LE-OKHLAH VA-YEHI KHEN. (1:26–30)

(And God said, let us make man in our image in our likeness, and let them rule over fish of the sea and fowl of the sky and cattle and all the earth and every creeping thing that creeps on the earth. And God created the man in his image, in the image of God he created him, male and female he created them. And God blessed them, and God said to them, be fruitful and multiply and fill the earth and conquer it and rule over fish of the sea and fowl of the sky and every creature that creeps on the earth. And God said, behold I gave to you every herb yielding seed that is on the face of all of the earth, and every tree in which is the fruit of a seed yielding tree will be for food. And every creature of the earth and every fowl of the sky and every creeping thing on the earth in which is a living creature and every green herb is for food, and it was so.)

The first critical question about the first verse in this set concerns the referent of "us" in "Let us make man." Rav Hila says that it is no one; the first person plural in this case is merely a royal we. In other words, the passage simply is about a king (of kings) saying, "I will make man." Similarly, Rav Ammi and Rav Jassi say that God is speaking to himself. Rav Berekiah adds the qualification that it is God in his aspect of mercy.

Rav Joshua Ben Levi says it is earth and sky. Rav Levi and Rav Joshua of Siknin say that it is the souls of the righteous. Rav Simlai says it is the first man who (in verse two) is called "God's wind."

Of particular interest are the interpretations of Rav Samuel Ben Nachman and Rav Chama Ben Rav Chanina who say that "us" refers to the five prior days of creation. Rav Chama Ben Rav Chanina claims that each day is 1,000 years, and the Torah itself was created two days before the beginning of the six days of creation. Hence, when the sky and earth are created, the universe already is 9,000 years old. Clearly for this sage creation takes place in time, but the dating need not be taken literally. It may only be a way of saying that the universe already is "very old" when the biblical account begins. To this discussion Rav Leazar and Bar Sira cryptically add that people ought not to talk about what they know nothing.

According to Rav Chanina, Rav Huna, Rav Aibu, Rav Jonathan, Rav Samuel Ben Nachman and Rashi, the "us" with whom God makes the human are the angels, who are identified as God's "family." In this case Rashi characterizes his explanation as homiletic. He says that because of modesty, God sought permission from the angels to create man. Hence, "let us make man" means that, although it was not required, God decided first to seek the consent of the angels in giving the human species its specific nature. The text does not mean that the angels in any sense assisted in God's action. God sought this consent because in making humans he gave them certain power that otherwise would fall within the authority of the angels.

It is in this context that the sages construct a debate between the angels over whether or not God should create the human species. Love and righteousness are in favor while truth and peace are opposed. Note that angels, viz., a class of entities second in power and perfection only to God, who are created on the second day, are identified with abstract virtues.

According to Rashi, without any qualification humans are given power over sea-life. However, the human's dominion over land-life depends on whether or not it is worthy[201] of it. The term "KIDMUTE-NU" expresses what being worthy means in this case. It refers to the ability to have knowledge and understanding. In other words, humans resemble God and the angels in having the ability to gain rational knowledge. To the extent, but only to the extent, that this potential is realized does this species resemble the heavenly beings and have dominion over the animals. Consequently, Rashi reads verse 26 as follows: God said to his immediate subordinates in governing the world, "although I do not need it, I would like to have your consent in my plan to give something that I created the nature to have, without qualification, authority over sea-life, and, insofar as it realizes our ability to be rational, dominion over land-life as well."

VA-YAR ELOHIM ET KOL ASHER 'ASAH VE-HINEH TOV MEOD VA-YEHI 'EREV VA-YEHI VOKER YOM HA-SHISHI. (1:31)

(And God saw all that he made and behold it is very good, and it was evening and it was morning on day six.)

Rashi notices that the biblical text says "day six" rather than "the sixth day," in marked contrast to the designation of the first five days. He explains that the difference in language points to a major respect in which this day differs from the others. Since the sixth day completes the intended work of creation, it is the end or goal towards which the earlier works were dependent. As such, insofar as the term "day" expresses completed work, the sixth day is properly day one of the universe. Rashi specifies just what day it is. It is the sixth of Sivan, which also is the day on which the Torah was revealed at Sinai.

Day Seven (Gen. 2:1–3)

VA-YEKHULU HA-SHAMAYIM VE-HA-ARETS VE-KHOL TSEVA-AM. (2:1)

(And the sky and the earth and all of their host were finished.)

Several sages say that God created the universe from a number of different "balls." Rav Chama and Rav Joshua Ben Rav Chanina say six, and Rav Chanina says four. Rav Jochanan says two, and identifies them as fire and snow. In some sense the question seems to be about what are the elements out of which God created the universe. If this interpretation is correct, then Rav Chanina probably intends the four primary elements – fire, air, water and earth. Rav Jochanan's "snow," which other homiletic interpretations set beneath the Throne of Glory, may be the RAQI'A, since both are solid water.

VA-YEKHAL ELOHIM BA-YOM HA-SHEVI'I MELAKHTO ASHER 'ASAH VA-YISHBOT BA-YOM HA-SHEVI'I MI-KOL MELAKHTO ASHER 'ASAH. (2:2)

(And God finished on day seven his work which he made, and he rested on day seven from all of his work which he made.)

The work of creating and determining the nature of This-World was complete in every respect except one. Genibah and most of the other sages say that it lacked the Sabbath. Rav Judah Ben Rav Simon says tranquillity, ease, peace and quiet, which Rashi, in agreement with him, calls rest. Its addition on the seventh day completes the work of creation.

VA-YEVAREKH ELOHIM ET YOM HA-SHEVI'I VA-YEQADESH OTO KI VO SHAVAT ME-KOL MELAKHTO ASHER BARA ELOHIM LA-'ASOT. (2:3)

(And God blessed day seven and sanctified it, because on it he rested from all of his work which God created to make.)

In commenting on "all of his work" (KOL MELAKHTO), Rav Chama Ben Rav Chanina and Rav Levi presents the following summary of the works of creation on each of the seven days: (1) Sky, earth and light. (2) The RAQI'A, Gehenna and angels. (3) Trees, vegetation and the Garden of Eden. (4) Sun, moon, and constellations. (5) Birds, fish and Leviathan. (6) REMES, Adam and Eve. Rav Pinechas adds cattle, beasts and demons to the sixth day. (7) All the sages seem to agree that the biblical text does not mean that God actually performed work on the seventh day.[202] Noteworthy omissions from their list are TOHU, BOHU, CHOSHEKH, MAYIM, RUACH and TEHOM. In other words, these are not "works."

In summation, there existed a number of things in the universe before its creation. Different commentators suggest different ones, but the most common references are to TOHU, BOHU, CHOSHEKH and MAYIM as created in a prior world out of which God constructs this world. Creation itself is a single event which (at least according to Rashi) occurs on the same day that the Torah was given to Moses on Sinai, viz., 6th of Sivan. Consequently, the text's references to days are not literally days. Rather they express conceptual divisions in the single act of creation. Days 1–3 deal with the creation of the physical world, days 4–6 deal with its occupants, and the seventh day with rest.

According to the sages, the light of the first day is not the light of our world; rather it is light for the World-To-Come (= the Messianic Age). Also intended for the World-To-Come are the upper waters that were formed on the second day by the sky (or, RAQI'A) dividing the pre-existent waters. The sky is a result of the interaction of the elements water and fire to form a solidified liquid. Again, its purpose is to separate two regions of water, one for our world (the source of rain) and the other for the World-To-Come. This however is not the only separation that the RAQI'A is intended to make. It makes a second division on day 3, viz., the separation between dry-land and the seas on the surface of the earth. In all of these acts of partition, the sole actor or cause is God's word.

The creation on day 3 of vegetation completes God's production of the physical universe that his yet-to-be-described creatures occupy. Note that, as the universe was originally constituted, it differs from what it is now. For example, the trees themselves, and not just their fruit, were edible. Also, the sun and the moon were equal in size and there were no stars. All these changes are judged to be imperfections that result from sin. Furthermore, note that many heavens/skies were created and not just the one of our world, whose primary purpose is to enable Israel to fulfill the ritual commandments of the Torah, viz., to know when to observe the Sabbath, the pilgrim festivals, the new moons and the new year.

Finally, of the living things in the seas and in the air (on day 5) and on the surface of the earth (on day 6), the most important is the human being, who, like God, has the ability to know and to govern the world. Candidates with whom God is associated in creating Adam are himself, his aspect of mercy, the souls of the righteous, the RUACH ELOHIM, or the angels.

In general, the above (linguistically based) homiletic account of creation has the following distinctive characteristics in relationship to the major points we have emphasized in the modern and classical philosophical accounts of creation:

(1) Creation is a single event. The list of days designates a logical order and not a set of distinct acts. The expressions "it was so" and "it was good" have a different function in the text than subdividing the event. They state something about the object created, viz., that it is complete, but nothing about the act of creation itself. So far what we have found in these commentaries is coherent with Jewish philosophy.

However, (2) the event of creation described is temporal. Still, the time frame is not ours. In fact the world created in this foreign time also is not our world. Just as the time is a time outside of our time, so the space created also is outside of our space.

(3) The question whether the universe was created out of nothing or from pre-existent material tends to be resolved by affirming both views. The device through which these seemingly contrary claims are made coherent is the positing of at least one universe prior to this one in which entities ultimately created from nothing provide the material out of which God constructs our world. This pre-creation universe is populated by things named in the Genesis account,[203] as well as other entities not explicitly named.[204] In some cases it is possible to deduce textual reasons for the additions. For example, if we assume that the text of Genesis is consistent with the rest of scripture, which would necessarily follow if all of the Hebrew Bible records the word of God, then the interpretation of Gen. 1:1 in the light of Prov. 8:22 leads to the conclusion that the Torah existed prior to creation. In other cases it is not difficult

to deduce logical reasons for the additions. For example, to picture elements of creation in terms of cubes is intelligible in the light of Plato's creation theory in the *Timaeus*. Similarly, it would be natural for anyone committed to any form of Hellenistic cosmology to add fire and air to the list of elements explicitly identified in the biblical text.[205] Other additions[206] are not so easy to explain, and more often than not Rashi includes them on the authority of the midrash, which itself takes this addition to its ontology for granted.

(4) Among the things introduced by these commentators in the creation narrative of Genesis are the angels. As we have seen, the classical Jewish philosophers identify them with the separate intellects of the astronomy of their time. In the case of the sages and Rashi, these substances, who are second in power and perfection only to God, turn out to be reified virtues. The sages are careful to exclude them from the list of pre-creation entities in order to avoid any association of their views with polytheism. It is interesting to note that no similar fear prevented them from positing the prior existence of other things.

(5) Our world is only one of several worlds that were created by God. Minimally there was at least one before ours, viz., the Garden of Eden.[207] However, many of the sages suggest that there were many more. Furthermore, ours is not the only universe that exists now, for many of the sages suggest that there are multiple worlds that co-exist with ours. Furthermore, there will be at least one more world when ours comes to an end. In this connection, not everything said to have been created is created for our world. For example, OR and CHOSHEKH are not our light and dark; they are entities intended for the World-To-Come at the end of (our world's) days. In this sense our world exists for the sake of realizing a world that transcends it.

(6) A major concern in all of these early commentaries is to spell out, in considerable detail, the human and divine moral and spiritual character of their view of the universe. In the language of the accounts of creation in Jewish philosophy that

we have already examined, the predominant emphasis in the commentaries of Rashi and the sages is on the role of souls, angelic intellects and divine will in a spiritual picture of creation. There is some concern for the differentiation of space and the bodies that occupy it, but clearly the balance is weighted in favor of spiritual over material causation. Similarly, the universe is conceptualized more in terms of ethics than in terms of physics.[208] In this sense what the Jewish philosophers did in their interpretation of creation differs significantly in what we find in these earliest commentaries on the biblical text. In the view of Rashi and the sages, the universe exists primarily through an act of divine will and the world's intelligibility is primarily ethical. The philosophers in no way deny divine intention or the ethical component of understanding. On the contrary, both Jewish philosophers and commentators see the universe in terms of a teleology whose defining end lies beyond the created world.[209] In other words, what Rashi and the sages present as the relationship between This-World and the World-To-Come is clearly mirrored in the philosophers' understanding of the correlation between creation and redemption through the mediation of revelation. However, the role that the philosophers give to physical necessity and chance makes no appearance in the commentaries of Rashi and the sages.

Perhaps this difference has more to do with the kind of commentaries that Rashi and the sages wrote than with incoherences between the views of the biblical commentators and the Jewish philosophers. Necessity and physical determination play a minor role in what the early commentators say because their concern (beyond language analysis) is homiletic.[210] Hence, it is reasonable that their emphasis would be spiritual and ethical. Consequently, no final judgment can be made about the relationship between Jewish philosophy and theology until we look at the classical philosophical commentaries of the rabbis who succeeded the sages and Rashi and were contemporaries of our classical Jewish philosophers.

THE EXPLANATION OF IBN EZRA, NACHMANIDES AND SFORNO – LANGUAGE AND PHILOSOPHY

The linguistic, homiletic Bible commentaries of Rashi and the sages present a significantly different view of the general nature of the universe and its origin than what we found in the writing of the Jewish philosophers. The former commentators exhibit little of the philosophers' interest in ontology. Hence, the critical topics raised in the commentaries of the philosophers – space as a determining principle of physics, and reality judged to be more negative, infinitely dynamic and indefinite than positive, finitely static and determined – barely enter into the consciousness of the sages and Rashi. However, that does not mean that the two approaches to a Jewish understanding of creation are incoherent. It only says that the philosophers as philosophers were not concerned with homiletics, and the preachers as preachers were not concerned with philosophy,[211] and each explored the linguistic implications of the biblical text out of their own interests. Ultimately the sole value for considering Rashi and the sages of *Genesis Rabbah* with respect to the primary question of this part of the book – viz., is Rosenzweig's philosophy and theology of creation Jewish? – is that they inform the views of the later rabbinic commentaries. The real point of comparison between philosophy and theology only becomes clear when we turn to the writings of these latter commentators who, as commentators, were influenced by midrash, and who, as theologians, were influenced by philosophy.

With both Jewish philosophy and midrash in mind, we now turn to the theological commentaries of Ibn Ezra, Nachmanides and Sforno.

Day One (Gen. 1:1–5)

According to Ibn Ezra, verses one and two constitute a single sentence that says, when[212] the revered primary judge[213] together with the next level of judges,[214] all of whom are immaterial entities, decreed his desire[215] that limits be imposed on the pre-existing materials[216] in order to transform them into something new,[217] these materials became this physical universe. God's "desire" is the element, air.[218] TOHU and BOHU are pure capacities of the element, earth. The World-To-Come is an immaterial, eternal region in which reside incorporeal, unchanging angels. Like the angels and the World-To-Come, these capacities are not created, but neither are they real or of any value. Solely through his desire, God actualizes them into the yet undifferentiated material universe,[219] whose subsequent physical inhabitants will be subject to birth and death. This domain consists of three primary territories. There is a region composed from the element earth, that is covered by a region of water, that itself is covered by air. Subsequently these three spaces – earth, sea and sky – will be further divided into distinct areas. Light will be made explicit on the first day, sky on the second, vegetation on the third, stars on the fourth, and the souls of the living things on days five and six.

Basing himself almost exclusively on the grammar of the terms within the Hebrew scriptures, Ibn Ezra rejects the claim that "BARA" must mean bringing something into existence out of nothing. His analysis of the term "ELOHIM" reconciles the grammatical data that the form of the subject is plural and the form of the verb is singular. The verb is singular because the referent of the subject is God. The noun is plural because its sense makes reference to the angels. It is not a proper name. Rather it is a disguised description for the master of the masters who more directly govern the physical universe. On his view there are only two worlds – the World-To-Come,[220] which is an immaterial, unchanging region in which the angels reside, and This-World[221] of material change.

In contrast, according to Nachmanides, at the very first moment of[222] the universe,[223] the power that is the source of the power of everything[224] brings forth matter from absolutely nothing at all by an act of will without the imposition of any intermediary.[225] At this stage the universe was nothing at all except space predisposed to be made into something. As such it was a first body composed from prime matter[226] and form[227] into the elements fire,[228] air,[229] water[230] and earth.

According to Ibn Ezra, the physical universe becomes a unified, revolving sphere of earth, surrounded by water, surrounded by air. The sphere in turn becomes encompassed by an area of light that blends into an area of dark. As the sphere of the universe revolves through these regions, it passes from a period in light without dark to a period where both are indistinguishably present, to a period in dark without light, to a period where both are again present but now distinct. These periods are named in corresponding order, "day," "evening," "night" and "morning." During the first complete rotation of the physical universe,[231] God did two things. (1) God desired light into existence, and (2) he thought to name the pre-existent dark "night" and the created light "day."

In contrast, Nachmanides asserts that during this first twenty-four hour period of the universe, God creates out of nothing a first form and matter from which he produces the space of the physical universe. This space is immediately differentiated into upper and lower regions. The lower region then is differentiated into the four elements. In addition, God creates wisdom, the Throne of Glory, and the element light that fills the upper region. At this stage the universe is a large sphere at whose center is a sphere of earth, that is surrounded in consecutive order by rings of water, air, fire and light.[232]

On Sforno's reading, the text says that at the beginning of time,[233] in no time whatsoever, God, an eternal incorporeal necessary being who is the source of the existence of absolutely every other existent, made, out of absolutely nothing,[234] a perfectly spherical, finitely large space. At this stage of creation the universe consisted solely of this space. The sphere is called "heavens." At its core is "the earth" that, at this initial stage, is merely a point at the geometric center of the sphere. It is a compound of prime matter[235] and prime form.[236] Next God creates his celestial intellects[237] and differentiates the earth into a series of rings, each composed of one of three elements of the physical universe – earth, surrounded by water,[238] surrounded by air[239] – by the rotation of the entire sphere on its axis as a vortex. The motion of the particles at the circumference of the ring of air produces a friction that ignites them. Elementary fire consists of these inflamed air particles. The universe as a whole is a rotating sphere whose elements move at a velocity directly proportional to their distance from the earth center of the sphere.

An apparent difference between these philosophical commentators is that Nachmanides takes the term "YOM" literally to express the same period of time as an earth day,[237] while Ibn Ezra and Sforno relativize the word. As an earth day marks the time required for the sphere of the planet earth to complete one rotation on its axis, so the term day in the account of creation marks the time required for the sphere of the universe to complete one full rotation cycle. Prima facie, given that the radius of the universe is vastly greater than that of the earth and that scripture itself says nothing about this question, there is no reason to believe that these two time periods are the same. However, this difference may only be apparent. There is no reason to assume that the velocity of these two spheres is the same. Nachmanides may accept Ibn Ezra's definition of a day and still assert that the two time periods are identical. In fact this is what Plato reports in his *Timaeus*, viz., that the length of the rotation of the motion of the Same, which rules the natural motion of the universe as a whole, is twenty-four hours.[238]

In this respect Sforno's commentary is more radical than that of any of his predecessors. It is clear on his interpretation that the time referred to in the biblical creation story is not our own. His universe rotates through regions of light and dark, but the "light" is not our light, and the "dark" is not our dark. They have nothing to do with the presence or absence of the light of the sun and stars. Rather, the dark is the pure element, air, and the light is an element that existed when our world began that will return at the end of days, but is found nowhere in This-World. Consequently, while creation takes place in time, it is in no sense our time.

Nachmanides' commentary does take issue with Ibn Ezra in a number of critical respects. (1) Nachmanides' claim that "ELOHIM" expresses the power of the powers of everything can itself be read as an elaboration of the sense that Ibn Ezra assigns to the term. However, there is one significant difference. The "powers" of which Ibn Ezra speaks are the angels. For Nachmanides they are the first form and matter whose combination produces the four primary elements of earth, water, air and fire. In this respect Sforno's commentary is closer to Ibn Ezra's. For Sforno the term "ELOHIM" is a general term for the species of all immortal, immaterial intellects, which includes both angels and God.

(2) According to both Nachmanides and Sforno, first form and matter respectively are "TOHU" and "BOHU." The entire physical universe is composed of a single uniform stuff that has a single essential nature. TOHU is that stuff and BOHU is its nature. Its creation begins the universe. In contrast, for Ibn Ezra "TOHU" and "BOHU" are synonyms. However, all three agree that these terms name the pre-existent material out of which the material universe is actualized.

(3) All of the commentators read the biblical text as saying that God created elements from which the universe was formed. However, they differ in how the Bible expresses it. According to Ibn Ezra, the text implies the existence of fire and explicitly mentions the other three classic elements. ERETS is earth, MAYIM is water, and RUACH is air. Furthermore, he notes that the light is an additional element that is generated on the first day.

Nachmanides has a number of problems with this interpretation. First, he considers it to be an unnecessary confusion to have the same Hebrew term, "ERETS," function in different contexts to mean both an element and a region of physical space. In his commentary this term and the word "SHAMAYIM" combine to name the undifferentiated space that subsequently is distinguished into the lower and upper physical realms. He introduces the term "AFAR" for the element earth. Second, he objects to Ibn Ezra's inclusion of "OR" on the same level with the pre-existent elements. It is clear that light is something formed out of the created first matter and not itself one of the elements. Third, he notices Ibn Ezra's failure to account for the terms "TEHOM" and "CHOSHEKH" in the text. Nachmanides identifies the former with elementary water and the latter with elementary fire.

Sforno's commentary once again is more radical than both of his predecessors. In agreement with Ibn Ezra and in opposition to Nachmanides, "MAYIM" names the element water and "ERETS" names the element earth. In other words, given a choice between the two interpretations, Sforno chooses the more literal account. "TEHOM" is the name of the ring of water that, at the initiation of the universe, encircles the earth. However, since he identified "RUACH" with the angels, instead of associating "CHOSHEKH" with fire, he identifies it with air, the object to which Nachmanides referred the term "RUACH." Whereas Ibn Ezra and Nachmanides follow the accepted scientific Aristotelian tradition of their day in listing four primary elements as the building blocks of the terrestrial world, Sforno lists three. On his analysis fire is itself something generated from the elements, viz., from the natural motion of the air. Hence, in spite of their difference in terminology, neither Ibn Ezra nor Nachmanides in this case add anything to the accepted Aristotelian cosmology of their day, whereas Sforno makes a radical departure. Sforno's physics is closer to that of the Stoics, whose active principle of the universe was PNEUMA (a mixture of fire and water), than it is to the physics of the Aristotelians.[242]

(4) According to both Nachmanides and Sforno, "BERE-SHIT" clearly expresses the beginning of the time of the universe. It is not a mere grammatical formalism for initiating the account as it is for Ibn Ezra. This issue points to the main substantive difference between their interpretations of the first day of creation. Ibn Ezra's explanation that the verb "BARA" means to ordain limits on materials presupposes that those materials already exist, and, as we already have noted, he is aware that this interpretation is controversial. While Ibn Ezra believes that the "out of nothing" modification of "to create" is not essential, Nachmanides and Sforno see this characteristic as the essential feature of what creating is. In terms of the language of the biblical text, the issue expresses itself in how these commentators differentiate "BARA" from "'ASAH." They agree that the first verb applies to God's action on day one, and the second verb expresses what occurred during the next five

days in consequence of God's act. For Ibn Ezra, what God ordained at first as an indefinite potentiality subsequently was made into a definite actuality by the angels. For Nachmanides and Sforno, God first brings forth matter out of absolutely nothing at all solely through his desire or thought.[243] Hence, while Ibn Ezra tells us that for God to "say" something means that God wills the angels to make something potential actual, in the case of Nachmanides and Sforno it is God himself who brings forth what is potential to actuality. In other words, for Nachmanides and Sforno "to say" and "to create" are synonyms. On their account, with respect to the first day, God's one creative act produces (1) the first form and matter that constitute the undifferentiated sky and earth, as well as (2) the potentialities within space from which subsequently actual (different) objects arise. The earth at its inception contains the elements from which the different species of entities in the sublunar world are generated, and the sky contains the light from which the different heavenly objects emerge.

It is of interest to note that, whereas Nachmanides introduces some products of the first day solely on the authority of the midrash[244] and Ibn Ezra does not, in fact angels play a dominant role in Ibn Ezra's cosmology that is lacking in Nachmanides' counterpart. In this respect, in spite of Nachmanides' commitment to both rabbinic homily and mysticism, his account ultimately is more naturalistic than Ibn Ezra's. Beyond the origin of the universe on the first day, whereas the latter's scientific cosmology is dominated by the causal influence of non-material entities from the heavens, the former's cosmology sees all subsequent causal explanation strictly in terms of the inherent nature and stuff of the fundamental matter of the universe itself. In this respect the astronomy of the mystic Nachmanides is closer to modern astrophysics than the counterpart of the rationalist Ibn Ezra. For this reason it is not surprising that the commentary of the Renaissance rabbi Sforno more resembles Nachmanides than Ibn Ezra. First, Nachmanides, unlike Ibn Ezra, can differentiate between heavenly bodies that reflect rather than emit light.[245] Second, what is far more important, Ibn Ezra's stars are created by angels on the fourth day, while Nachmanides' celestial objects arise naturally on that day as solidified forms out of their uniform material of elementary light. Like Ibn Ezra's universe, Sforno's contains angels, which are identified with the celestial intellects. However, like Nachmanides' angels, they play little role in the origin of the physical universe.

Day Two (Gen. 1:6–8)

According to Ibn Ezra, on the second day, in a single unit of time, the sky is produced, the waters are separated from the earth, the sky is named heaven and earth is named earth. In other words, it is not the case that the heavens are created on the second day while the dry land and the seas are distinguished on the first. These seemingly distinct events are the product of a single act. This is what scripture means by saying "it was so" and "it was good," i.e., they express distinct, single units within creation. In this case, the air[246] is stretched out between the water and the earth, transforming them into three distinct regions of dry land, bodies of water, and sky. It is not that these regions did not already exist. As noted above, everything that will be already is created on the first day. Rather, on the second day these regions are confined within their proper limits. What happens is what had been indefinite now becomes definite.[247]

According to Nachmanides, on the second day two events occur simultaneously. The upper region of space becomes filled with a primary material, called "sky,"[248] from which the heavens[249] are formed. The occupying sky itself is a ring of congealed water stretched out from the exact center of the region of elementary water formed on the first day. The space occupied is the heavens. On the first day it was a mere geometric point that contains a specific potentiality that now, through the filling motion of the sky, is actualized as the distinct location for all heavenly objects.

Sforno's explanation closely parallels that of Nachmanides with the following differences. The sky is composed from a combination of elementary air and water called "mist." God stretches it out between the ring of water and its ignited circumference at the world's periphery. Then he compresses some of this air within the mist into a separate ring that he forces between the sphere of earth and water encircling it at the center of the physical universe.

Except with respect to the differences noted in their commentaries on the first day, the descriptions of Ibn Ezra, Nachmanides and Sforno substantially differ in only three respects. First, Nachmanides and Sforno reject Ibn Ezra's judgment that the differentiation of the earth into regions of dry land and seas takes place on the second day. Following the literal order of scripture, these events occur on the third day. The element water[250] is separated from the element earth,[251] the space occupied by the former is called "seas," and the dry land occupied by the latter is called "earth."

Second, the ontology of Nachmanides and Sforno includes five rather than four elements. Although all of them include the four classical elements under different names – fire,[252] air,[253] water[254] and earth[255] – Nachmanides and Sforno add a fifth celestial element[256] for the heavenly objects.

Third, all three disagree about the meaning of the terms "KHEN" and "TOV." Ibn Ezra and Nachmanides agree that "KHEN" signifies that what preceded is a single unit, but they disagree about "TOV." Ibn Ezra treats the term as a synonym for "KHEN." According to Nachmanides, "TOV" means that the created unit will persist forever. Similarly, while Ibn Ezra says that when God "sees" he sees with his mind, according to Nachmanides, the verb signifies that God makes his object permanent. In contrast, Sforno asserts that the term that expresses permanence is "KHEN." Furthermore, "to see" means "to desire," and "TOV" indicates a yet unrealized end that is expressible in terms of theoretical, moral knowledge.

Nachmanides and Sforno endow the elementary structure of the universe with a permanence that Ibn Ezra's description lacks. Their optimism is a consequence of the inherent goodness of the creator. However, Nachmanides did not make clear the sense in which God's product will endure. As Sforno indicates, its goodness does not reside in what the physical world *is*. Rather, it lies in what the universe through natural knowledge must and/or ought to and/or will *become*. It also is true that Sforno makes it clear that while species cannot become extinct, neither can they evolve. Still, his biblical cosmogony is a moral physics.[257] The absolute goodness of the universe lies in its operation as a whole from its beginning to its end, and not in any part of the universe at any separate period of time. On his view to reshape what was created in order to benefit human civilization is superior to merely preserving the given state of nature. That nature can be improved through human intellect is itself a law of his physics.

Days Three and Four (Gen. 1:9–19)

According to Ibn Ezra and Nachmanides, the vegetation arises naturally from a power[258] placed into the earth on the first day. In the same way water generates the things that swarm and fire generates the stars. According to Nachmanides, the heavenly objects are generated through the interaction of light and the RAQI'A. As such these objects are composites of the distinct celestial element[259] set in a region of solidified water.

There seems to be only one substantial new difference in their descriptions of the work of days three and four. Both Nachmanides and Sforno are aware that while some heavenly substances emit light, others receive the light and then reflect it. In contrast, Ibn Ezra does not seem to recognize any distinction between stars and other kinds of celestial objects that do not emit light. This seems to be the implication of his claim that the designations "great," "greater" and "lesser" of the heavenly bodies refer to the intensity of their light and not to their relative size.

Days Five and Six (Gen. 1:20–31)

Ibn Ezra says that the souls of the living things[260] are created on days five and six. Nachmanides and Sforno say that it is the things themselves.²⁶¹ The life created is formed either out of elementary water or earth. The former are created on the fifth day and the latter on the sixth. The products of the fifth day are SHERETS,²⁶² REMES,²⁶³ things that fly, and TENINIM.²⁶⁴ The products of the sixth day are BEHEMAH,²⁶⁵ CHAYTO ARETS,²⁶⁶ other kinds of REMES, and man.

All of the commentaries agree that the charge to be fruitful and multiply is a statement of what will happen and not a command, since every material life form, excluding man, has no choice in this matter. They copulate through instinct rather than through reason.

Ibn Ezra and Sforno interpret the "us" in "let us make man" to refer to the angels, and they follow Saadia in claiming that "in his image" means in the image of God, and that God and man are similar with respect to possessing and using theoretical wisdom to govern themselves as well as others. This capacity is taken by Ibn Ezra to be the referent of the term "glory"²⁶⁷ as it modifies both God and man. Hence, near the end of the sixth day God commands his angels to form one more land animal whose soul is unique in comparison with other physical creatures in that it resembles God with respect to being able to rule through the use of abstract reason in determining judgments. The biblical term that expresses this point of identity between God and man is "glory."

Ibn Ezra also mentions the homiletic interpretation of "He created them²⁶⁸ male and female" to mean that the first human being had two faces set on his (her, its) body back to back.²⁶⁹ However, this explanation baffles him.²⁷⁰

According to Sforno, God created the paradigmatic human being as an actual mortal animal²⁷¹ who possesses choice and the potential to become an immortal intellect.²⁷² It is this potential that uniquely defines the human species. Insofar as the potential is not achieved, the human is a mere animal; insofar as it is achieved the human belongs to the species of deities²⁷³ whose other members are God and the angels. Angels are unique in that they are perfect by nature and not by choice. Humans are unique in that they become perfect by choice but not by nature. Only God by nature chooses his perfection.

There is general agreement between Ibn Ezra and Sforno in interpreting this section on the creation of man. However, since Nachmanides interprets the entire event of creation to be God's single activity, he cannot refer the "us" to angels. On his view, God performs a single act on the first day that sets potentialities into the as yet undifferentiated space. From that point on, all differentiation and actualization emerges out of the space itself. That space was identified as heaven and earth. Hence, on Nachmanides' interpretation, the "us" is heaven and earth.[274] What first emerges are the souls of the different life forms. Then their bodies arise. Next, the human soul is distinguished from the others by receiving special powers,[275] and only then is the human body made definite for this special human soul.[276]

Day Seven (Gen. 2:1–3)

"To bless" means to note that a subclass of a species possesses a capacity for its benefit not shared by the other members of the species. Living things are unique among physical objects in possessing the ability to procreate; man is unique among living things in possessing the ability to be rational; and the seventh day is unique among days in that on it all forms of labor are prohibited.

Both Ibn Ezra and Nachmanides reject the inference that God finishing his work on the seventh day means that God labored on this day. Ibn Ezra emphatically says that to conclude an action is not itself an action.

According to Sforno no time passes between the first and the sixth days. In effect they are a single moment. The "rest" of the seventh day refers to the realization of God's end for the universe. However, with respect to God's role, creation ceases, because his end is accomplished. From this point on creation becomes a human activity. Humans are now to begin the process of civilizing nature towards the end of actualizing all of God's knowledge. As such the Sabbath at one and the same time is a remembrance of God's completion of his end at the beginning of This-World with the creation of physical nature, and an expectation of man's completion of his end with the actualization of spiritual nature at the beginning of the World-To-Come.

This completes our survey of what classical rabbinic Judaism had to say about the dogma of creation. Now we should be in a position to answer the question with which we began part two – Is the single most important conception of creation in modern Jewish philosophy, that of Franz Rosenzweig, coherent with classical rabbinic thought? However, the results of the survey themselves raise a new problem. A comparison of the two – modern and classical Jewish philosophy – presupposes that each on its own is coherent, and it is not obvious that the collective rabbinic view in itself is coherent. The problem arises because, as the above survey makes obvious, classical rabbinic Judaism permitted a very wide range of disagreement in the way that rabbis could (and did) interpret the biblical text. Again, when we take together all of the classical rabbinic views considered – philosophical and theological, early and late – it is difficult to discern any continuity of interpretation,[277] since every commentator has his own distinct interpretation of the most critical terms (for our philosophical purposes) in scripture.

My conclusion is that the classical explanation of the dogma of creation is coherent both within itself and with Rosenzweig's philosophical theology. Let us focus on a single detail of both to explain my judgment. The detail is what they said about the elemental universe.[278] Table 2 illustrates the apparent incoherence of the classical explanation.

Every classical commentator, like Rosenzweig, claims that the infinite multitude of particulars that occupy the space of this created world originates out of a particularity that in itself is not anything at all. Beyond this origin, Rosenzweig says that each particular moves in a direction of continuous change into being increasingly something, where the asymptotic end of this movement is identity with God's essence. As the medieval philosophers share Rosenzweig's view of the origin, they share his view of the asymptotic end of the process as well. However,

Table 2. *The elemental universe*

	Sages and Rashi	Ibn Ezra	Nachmanides	Sforno
Pre-elemental material	TOHU, BOHU deep, fire, wisdom, Throne of Glory	TOHU, BOHU	TOHU, BOHU	TOHU, BOHU, earth, sky
Elements				
fire	—	light	dark	—
air	—	wind	wind	dark
water	water	water	deep	water
earth	earth	earth	dust ('AFAR)	earth
celestial element	light	—	light	light

beyond this most general level there does not seem to be any agreement about the meaning of the terms in the biblical narrative which describe the origin, viz., TOHU and BOHU. According to *Genesis Rabbah*, "BOHU" is something undefined that was created before the seven days of creation. In contrast, Rashi says that it is not anything at all; the term merely expresses the state of being empty. Ibn Ezra says it is synonymous with "TOHU," which is the pre-existent matter from which God created the universe. At the same time everyone else distinguishes between the two. However, whereas for Gersonides "BOHU" is first matter and "TOHU" is a form, Nachmanides and Sforno say the opposite. For them, "BOHU" is first form and "TOHU" is first matter. Furthermore, as discussed above, Gersonides' interpretation of what this elementary form is differs radically from his predecessors. However, again, at a slightly more general level there is no significant difference between these commentators. All of them agree that there are prior materials out of which God willfully and purposefully creates the universe which are initially nothing and then become something identifiable as elements.

Every case of difference can be analyzed in this way. For example, whereas Rosenzweig does not spell out in detail any of the intermediary steps between origin and end, the classical commentators do. In general, all of them affirm that at the first stage beyond origin what emerges are the elements. However, there is almost no agreement among the commentators about which biblical terms name which elements. For elementary fire, the midrash and Rashi say that the Bible calls it "fire," but Ibn Ezra says it is "light," Nachmanides says it is "dark," and Gersonides says it is "God's wind." Concerning air, the midrash and Rashi say it is "air," Ibn Ezra, Nachmanides and Gersonides say it is "wind," and Sforno says it is "dark." Concerning water, everyone says it is "water," except Nachmanides and Gersonides who say it is the "deep." Everyone agrees that elementary earth is "earth," except for Nachmanides, who says it is "AFAR." The most agreement at this level of detail is about the fifth celestial substance. Almost everyone says that it is "light." The one exception is Ibn Ezra. He says that "light" is elementary light. However, he also is the only one not to list a fifth celestial element.

Ibn Ezra's difference in this case shows how similar all of these commentaries are once we step up to the first level of generality. From this perspective we can say that all of them agree that the universe was created out of a set of elements which are four in the case of the sublunar world and, excluding Ibn Ezra, a fifth in the case of the supralunar world.

This agreement should not be exaggerated. Do not think that if we exclude Ibn Ezra we will find a general consensus on all or most questions about creation. For every issue there are exceptions, and each commentator at some point departs from the consensus. For example, only Rashi does not list "TOHU" and "BOHU" as pre-elemental materials of creation.

In general, we can find precedents for each of Rosenzweig's specific interpretations in different individual classical commentaries, but this influence in no way detracts from the obvious originality of his views. Furthermore, that originality in no way counts against the Jewishness of his account of creation. What can be learned from our comparison of the different classical commentaries is that while the Torah sets limits on interpretation (and Rosenzweig's interpretation falls within them), it does not determine the belief. Rather, within its boundaries, total freedom is given to speculation. More accurately, free speculation within the boundary of the revealed law is itself a duty of the law.[279]

It is the definition of the boundaries of interpretation that defines the criteria by which any modern interpretation can be called "Jewish," and Rosenzweig's interpretation falls clearly within them. The boundaries determined so far are that (1) there exists something that is nothing out of which God, (2) through an act of will, (3) creates eternally and/or continually a universe.[280] Furthermore, (4) the created universe, in virtue of God's intention, has meaning and moral value.

Of all these points of commonality, the last is the most important for any Jewish interpretation and the most incoherent with any modern scientific account of the origin of the universe. First and foremost a biblically based Jewish conception of creation affirms that any adequate scientific and/or philosophical account of the universe must include a moral and/or teleological component, whereas modern science in principle admits only quantitative judgments and mechanical causes. Furthermore, the apparent conflict between a Jewish and a modern scientific understanding of creation cannot be that the former is only concerned with what Genesis says while the latter is concerned with what really is the case. What is at stake in all of these Jewish commentaries is not just an accurate reading of an ancient text. Rather, that text is revelatory of something fundamentally true about the universe itself.

It is commonly assumed in modern academic circles that whatever value Jewish philosophical commentaries have, it is not that they explain the actual meaning of the biblical text. If we treat scripture as literature with minimal attention to the Bible as a source of truth, undoubtedly this assumption is correct, if for no other reason these commentators attempted to make clear what is ambiguous in the text. However, if we were to read the scriptures as they (including Rosenzweig) did, viz., as literary expressions of truth, then the correctness of this assumption is less evident. While it is not possible to make what seem to be truth claims in scripture as precise as scientific statements, i.e., scripture taken on its own in almost every instance admits to more than one possible interpretation, the words of the biblical text do set parameters on what are legitimate and illegitimate clarifications. In particular, whatever else the biblical account of creation may mean, the following is clear: Almost none of the key words used in the creation story can be simply understood in a literal, common sense way. However, in order to interpret them the commentator must introduce some kind of conceptual schema that is not explicitly given in the scripture themselves. Creation is a fundamental belief in terms of which, when combined with an understanding of God, every other critical religious belief in biblical faith must be interpreted. While the universe originated through an act of God, that act did not occur at any particular moment in time, and not everything in the universe is a product of the creation. Some things are prior to creation and other things are posterior to it. Among the entities that are posterior are all concrete particulars. In other words, beyond the generation of space, the things of creation are species and not individuals. In other words, the creator, the act of creation, and the object of creation are all eternal, i.e., they are not subject to temporal modifications. However, they are subject to moral attributes. That creation is called "good" entails that an adequate scientific understanding of the universe must also be a moral understanding. In other words, an adequate schema for interpreting scriptures' story of creation must be one in which categories of the ought are essential to understanding what is. In other words, no physical understanding of the origin and nature of the universe can satisfy the Jewish dogma of creation that does not make it possible to understand the universe in terms of ethics. In this central respect Rosenzweig's conception of creation is quintessentially Jewish.

This ends our study of what Judaism as such has to say about creation. What has emerged is that the Jewish doctrine of creation is essentially a product of reading scripture with Platonic eyes. The final question that we will consider will be, is this reading believable? This question will be discussed directly in part four of this book. However, before we turn to it,

I want to consider a question that the above discussion clearly raises. The issue is partially about the history of ideas and partially about philosophy[281] – Is it in fact the case that the cosmology and cosmogony of Plato's *Timaeus* are coherent with the creation narrative of Genesis? In the next, penultimate part of this book, we will look first at Genesis and then turn to the *Timaeus*.

PART 3:
The foundations for the Jewish view of creation

Introduction

Our survey of classical Jewish philosophy and theology suggested that of all of the schematic alternatives available to the rabbis, the Platonic schema most easily fits the Genesis account of creation. That judgment left open the question whether or not the rabbis had a correct understanding of both texts that served as the foundation (at least with respect to the history of ideas) of the Jewish dogma of creation. Now we will look at these core texts themselves. Our concern in this third part of the book is to determine whether or not Plato's *Timaeus* and the Genesis account of creation together constitute a coherent picture of the origin and general nature of the universe. First we will look at Genesis and then we will turn to Plato's dialogue on cosmology and cosmogony.

CHAPTER 5

The account of creation in Genesis

This chapter will draw a picture of the origin and general nature of the universe as it is presented in Genesis 1:1–2:3. To a certain extent the picture was informed by statements about creation in other parts of the Hebrew scriptures,[282] but these texts were utilized only to fill in blanks or ambiguities in our primary text. In general, it is the Genesis account of cosmogony and cosmology that served as a primary datum for what revelation teaches about creation, and not these other texts.[283] We will read Genesis in much the same way that the sources discussed in the first two parts of this book read it, viz., as a unified statement that makes a philosophical/theological claim. However, I will try, to the best of my ability, to set aside any of the interpretations of the text that I have inherited from two thousand years of close readings and interpretations by my predecessors in Jewish tradition and Western civilization. In other words, the text will be examined for what it says in itself. The summary presented below is the result of this effort.[284]

The narrative of creation is divided into seven distinct units. Each is marked off as a "day." In fact this is how the term "day" should be understood. In no way can it mean what we normally take the term to mean – viz., either a period of twenty-four hours or a single cycle alteration between light and dark on the planet earth, or a temporal measure of one cycle of the earth's rotation on its axis. The most obvious sign that these interpretations are not correct is that the earth is not created until the third day and there are no stars (including the sun and the moon) until the end of the fourth day. In fact, nothing in the text suggests any temporal frame for any of the events described.

What would mislead the modern reader into thinking that Genesis presents a narrative set in time are the verbs. In both modern Hebrew and English all verbs have a temporal sense, viz., the action they state is in grammatical forms that modify the act as past, present or future. However, this is not the case with biblical verbs. It is not that the Bible doesn't express time. It does. However, it does so either by context or by using time specific adverbs. In our narrative there are none of these adverbs and there is nothing to suggest a temporal context.

The verbs in our text appear as infinitives, participles, or in the imperfect and perfect tenses. Furthermore, the imperfect and perfect uses of the verb sometimes occur with the "waw" of conjunction and sometimes they do not, and it is generally accepted by scholars of biblical Hebrew that the use of the waw consecutive[285] modifies the verb's meaning. However, the standard interpretation of how this consonant modifies the verb does not seem to me to be satisfactory. The standard view is that when an imperfect verb is used with the consonant then the verb functions as if it were perfect, and, conversely, when a perfect verb is used with the waw then the verb functions as if it were imperfect. However, it seems to me that if the author(s) wanted to use the perfect (imperfect) then he (she, they) would have done so. That the perfect (imperfect) is used with the consonant of conjunction suggests that the usage expresses something different than merely expressing the action in the imperfect (perfect).

My reading of the creation text suggests the following meaning of these forms: In general, what all of the verbs express is an end directed process. Now, the perfect tense expresses a stable act – i.e., a form of action that is constant and does not change, and conversely the imperfect tense expresses a dynamic action – i.e., a form of action that itself changes.[286] In both cases, what defines the character of the action is its end. In some cases the end is finite and in other cases it is infinite. I propose that without the consonant of conjunction the verb expresses a movement towards a finite end, and conversely,

Creation in Genesis

Table 3. *The logic of the biblical verb forms*

	With Consonant of Conjunction	Without Consonant of Conjunction
Perfect tense	Stable act with limit	Stable act without limit
Imperfect tense	Dynamic act with limit	Dynamic act without limit

when the verb is joined to this consonant it expresses movement in a direction that is infinite (i.e., without limit or end). In other words, corresponding to the four grammatical uses of an action verb are four forms of process, which may be diagrammed as in Table 3.

To summarize, the Genesis narrative of creation is divided into seven atemporal units. Each one expresses a layer of the single event of creation. What that event is is stated in the first verse of the text – it is God's creation of the sky and the earth.[287] Hence, the seven days are seven components of this single act. A way to visualize the picture that our narrative draws is as follows: Imagine hearing an illustrated lecture about a single state of affairs. The lecturers begin by showing a single transparency of the event they are discussing. However, the event is complex and, if it is to be grasped, it must be seen as a construct of its different elements. The lecturers exhibit this feature with their transparency. In fact, while the event is unified, the transparency is not. What they have done is impose a number of transparencies on each other, and what you, the audience, see is this composite. Then they remove the transparencies and show them one at a time, beginning with the first, imposing the second on the first, etc. until the original picture is restored. This is in effect what happens in the Genesis narrative of creation.[288] Gen. 1:1 presents the composite picture; the event of the first day is the first transparency, the event of the second day is the second transparency imposed upon the first, etc., through the seventh and final transparency which restores the original picture.

The seven events that together constitute the single, unified, complex event of God's creation of earth and sky are the creation of (day 1) day and night (Gen. 1:2–5), (day 2) sky (Gen. 1:6–8), (day 3) earth and vegetation (Gen. 1:9–13), (day 4) celestial lighters (Gen. 1:14–19), (day 5) the swarm of the water and the flier in the sky (Gen. 1:20–23), (day 6) the life on earth and the human (Gen. 1:24–31), and (day 7) the Sabbath (Gen. 2:1–3).

These seven units can be categorized into three more general units. The first (days 1–3) is the differentiation of the initial space that receives God's intention into distinct regions. The second (days 4–6) is God's command to these created regions to generate prototypes that occupy them who are in turn commanded to generate an infinite number of individual, concrete, material exemplifications of themselves. And the third is the creation of the Sabbath which functions (a) to bring the act of creation to a close, (b) to introduce a ritual dimension into the conception of the picture's end, and (c) to introduce time into what had been a purely spatial picture.

In summary, what God does is perform a single act upon a single, undifferentiated space. This act can be expressed in three ways. One, he says (or names) something that separates the space into distinct domains. Second, he tells the spatial domains to produce their occupants. Third, he commands (or blesses) the occupants to reproduce their kind, and/or (in the case of the stars of the sky and humans of the earth) to govern the occupants of their domains.

The stages of the differentiation of space are the following: At first (i.e., on day zero) there exists a sphere of earth surrounded by a sphere of water over which God's wind[289] hovers. Just what are the dimensions of these spheres is not stated. For example, the region of earth could be no more than a point. In any case, because of their size and/or because the earth is encompassed by the water, they are not (in any specific sense) separate from each other. On day one they remain in the same state. What God does is create something called light that he uses to divide the region of the divine wind surrounding the earth and the water into a region of light (called "day") and a region of pre-existing dark (called "night"). It should be

Creation in Genesis 161

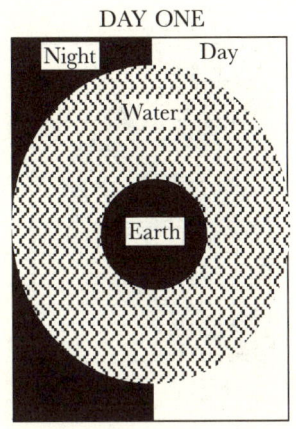

Fig. 2. Day Zero   Fig. 3. Day One

presumed that these created regions are not distinct from the water/earth space. Rather, the universe of undifferentiated space of water-earth, which originally was dark, becomes differentiated into day and night.

On day two God creates a second element viz., the RAQI'A.[290] It is a flexible kind of dimensionless transparency that God uses to divide the water surrounding the earth into three distinct regions – a ring of water surrounded by a ring of RAQI'A (which God names "sky"), surrounded by an outer ring of water. Then, on day three the sky is further extended so that part of the surface of the globe of earth protrudes through its surrounding water into the region of the sky. In so doing, God differentiates the surface of the globe of the earth into two regions – "seas" where the surface of the earth globe borders the ring of water, and "dry-land" where the surface of the globe borders the sky. Note that God produces two new elements – the light and the sky, in order to separate two pre-existent elements – the water and the earth. Also note that the term "element" here applies to regions of space, and not things that are locatable within space. All of the potential occupants are produced, not by God but, by the space.[291]

Fig. 4. Day Two

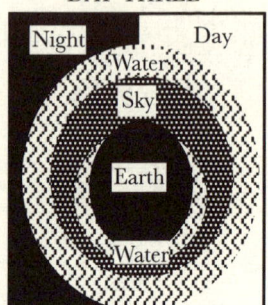
Fig. 5. Day Three

The parallels between space's generation of its occupants on days four through six and God's differentiation of space on days one through three are obvious. On day four God creates the celestial lighters who are to govern the regions of the day and the night created on day one. On day five the region of water produces the occupants of the water and the sky that were formed on day two. On day six the earth produces the occupants of its surface that was created on day three.[292] In other words, days one and four deal with the sky region, days two and five with the water region, and days three and six with the earth region.

Days three and six are parallel in yet another (less obvious) way: The concluding product of day six is the human and the concluding product of day three is the vegetation. Both vegetation and the human are parallel in that they differ significantly from the other products of their section of creation. In the first unit, what is created are spatial regions. However, vegetation is not space; rather it is an occupant of space, like the products of the second unit. Yet, as an occupant, it is significantly their inferior. For example, vegetation is the only occupant of space that cannot move in space. In this sense it is more like space than a spatial occupant. In fact, what it is is something in space intended to feed (i.e., preserve the life of) space's occupants. Similarly, the human is different from the other occupants. For example, while the lighters govern something but do not procreate, and every other living thing procreates but does not govern anything, the human alone both procreates and governs. Given this parallelism, it might also be the case that as the vegetation crowns the work of the first unit and points to the work of the second unit, so the human crowns the work of the second unit and points to the work of the third unit. In other words, by observing the Sabbath, man transcends the creation of this world. Furthermore, as the Sabbath concludes the creation of this world, the Sabbath may itself contain a hint to a world beyond this one that itself lies beyond the domain of the entire narrative of the Torah of which our picture is only the first chapter.

The conclusion of creation itself points to the fact that the first thirty-four verses of the Pentateuch are only a first tale that initiates a series of tales, the continuity of which constitutes the more general narrative of the Torah. Hence, what our story means has to be seen from the perspective of its whole context. That context again is a narrative whose events proceed in the following general order: The universe is created, the human enters the domain of the earth, the human is destroyed when God removes the created sky-barrier between the regions of earth and water, a remnant of the human (the Noachite) produces its kind until it is humbled at the Babel tower, and a member of the Noachites (Abraham) begins a new line of human life that governs Egypt and then is enslaved in Egypt. Finally, out of the tribes descended from Abraham with Sarah, (to Isaac with Rebecca, to Jacob-become-Israel with Leah and Rachel) arise a family of leaders (Moses, Aaron and Miriam) who lead their people through the desert of Sinai where they receive a law code from God that constitutes the people into a potential nation. The nation is only "potential" and not yet "actual" because, while it possesses the social-political-legal structure of a nation, it lacks any space on earth to occupy.[293] Finally, the people is brought to the edge of the land that it is destined to make its own. Moses briefly prophesies its rewards for obedience to its law code, and warns at great length of the punishments that its disobedience will entail. The narrative ends with Moses' death. What sense, then, are we to make out of the beginning of this story (viz., the creation of the universe) given its end in a detailed set of warnings about doom to a particular nation and the death of a specific individual leader?

If modern Bible scholars are right that the Torah was edited into a single text by the Israelite priesthood during the first national exile in Babylonia in the sixth century B.C.E., and that they themselves composed our creation story in that place at that time, then the following interpretation of the biblical concept of creation emerges: Creation is an epic myth that incorporates the best of scientific/philosophic opinion of that time to make sense out of the political-moral-religious situation in which Israel and (perhaps more importantly) its ruling religious priesthood found itself – viz., a nation Israel that (despite the fact that God separated it from among the nations) is in exile, and a priesthood of Levites that (despite the fact that God separated it from among the tribes of Israel) has lost (with the destruction of the Temple) its *raison d'être* (viz., its divinely ordained duty to administer the sacrificial cult).

They believed that the universe exists for one purpose – to serve God. The Hebrew verb here for serving is critical. It is 'AVAD, from which the nouns 'AVODAH (which means the sacrificial cult) and 'EVED (which means slave or servant) are formed. God created the universe to locate the earth; the earth was created to locate Jerusalem; Jerusalem was created to locate the Temple; and the Temple was built to offer sacrifices to God. Similarly, vegetation was created to sustain the living things of the regions of space to single out those earth creatures destined for sacrifice.[294] Similarly, humanity was created to govern the earth so that Israel could be singled out to produce priests who would offer the sacrifices. In other words, the sacrificial cult in Jerusalem is the final end of the universe.

However, now that cult no longer exists, because (in consequence of Judah's defeat at the hands of Babylon) the Temple no longer exists and its priesthood is in exile. How then could this have happened, or, given that it did, why does the universe continue to exist? The answer lies in the general structure of the particular stories that we summarized above that together constitute the narrative of the Pentateuch. Each individual story begins with a birth out of a previous destruction, and each birth proceeds to a new life that culminates in a death, out of which arises a new birth. In other words, the priestly composers of the Pentateuch viewed the universe as a living organism that, like all observed organisms, proceeds from birth towards maturity towards its end in death. Creation in the beginning of the Pentateuch is the beginning of our universe. Mosaic revelation – viz., the social-political-moral code presented in the middle sections of the Pentateuch – is its highest point of maturity. But what then is its end?

The most obvious answer is, the death of Moses, viz., the death of that human being whom the text judges to be the greatest of all recipients of divine revelation (i.e., prophets) who ever has or ever will exist.[295] (In this sense also the period of Mosaic revelation in the wilderness of Sinai is the high point of the universe.) His death marks a new birth, viz., the birth of the nation Israel. However, we, the readers of the narrative, know what its authors also know – that the nation born in the conquest of Canaan will reach its highest moment of maturity in the reign of David and will die in their own time.[296] This is why Moses' prophecy at the end of Deuteronomy is so short on rewards[297] and so long on punishments.[298] The statement of the rewards are brief because the authors knew that they are moot, i.e., because they knew that Israel will disobey the laws. Conversely, the statements of the punishments are lengthy, not only because they would come true but, because the authors could draw on their own lived experience in describing them.

In summation, the story of creation provides the macrocosmic perspective in which the exiled priestly authors of the Torah tell the microcosmic story of their now destroyed nation. Like all organic life, the story of nations and universes passes from birth to maturity to death. In the end there always is death. However, there is no bitterness in the epic Toraitic narrative. The Torah was not written to complain about Israel's disaster; it was written to explain it. Furthermore, the Torah is not (like Qohelet) a lament that there is no discernible purpose to existence. On the contrary, it is a work about cosmic value. It says that the universe does not exist for the sake of any of its creatures, no matter how noble or favored-by-God they may be. Rather, it exists for the service of God, and all events, including this national disaster, serve him. There is divine purpose to the tragedy. The destruction of the first Israelite nation will seed the birth of something new, something beyond the authors' own knowledge, that will restore world order and even enhance it. In other words, beyond the myth of creation and the historical knowledge of revelation, the entire epic of the Torah points implicitly to the hope for redemption.

CHAPTER 6

The account of the origin in Plato's Timaeus

Israelite priests, living in exile in Babylonia, wrote an epic myth that made coherent their lived experience of national destruction and their belief that what they did as priests was the most important activity in the universe. The myth itself placed their personal collective experience into a philosophical/scientific/theological causal framework that applied to absolutely everything – astronomy, physics, history, politics, ethics, etc. In forming their view of the cosmos they drew on what they knew as the best science of their day – that of sixth-century B.C.E. Babylonia. However, subsequent generations of Jews long lost any memory of the original sources that informed the author(s) of scripture. For many generations the revived nation had little interest in the scientific-philosophic component of their foundation text. However, as earlier generations sought through commentary to recapture the details of the legal parts of the Torah, post-ninth-century C.E. rabbis attempted to draw out of the words of Genesis the implicit philosophy and science of the text. In so doing they needed to reconstruct the text's underlying cosmology. As we have seen, they did so in terms of the cosmology and cosmogony in Plato's *Timaeus*. Clearly, this Greek text was unknown to the author(s) of Genesis. Similarly, we today know little if anything of the scientific framework of Genesis in sixth-century B.C.E. Babylonia. However, we can judge, solely on the evidence of what we do know – viz., the inherited texts of both Genesis and the *Timaeus* – how coherent the two cosmologies are. It should not be surprising if they are. After all, there are so many parallels between the Hebrew scriptures and the writings of ancient

Greece,²⁹⁹ that it is not at all unlikely that both works are unique products of a shared Mediterranean civilization. In any case, that judgment remains to be seen. First we will summarize the text of the *Timaeus* in much the same way that we looked at Genesis, after which we will compare and contrast these two foundational (for Jewish thought) accounts of the origin of the universe.³⁰⁰

The *Timaeus* is the first dialogue of a planned trilogy.³⁰¹ The second, the *Critias*, was not completed, and Plato never began the third, the *Hermocrates*. The dating of this dialogue relative to Plato's other works is open to scholarly debate. Some scholars consider it to be a middle dialogue from the same period in which Plato wrote the *Republic*. However, most scholars claim that the only finished works later than the *Timaeus* are the *Philebus* and the *Laws*. On this view the *Timaeus* was written later than 366 B.C.E., after Plato established the Academy and made his second journey to Syracuse.

The scene of the dialogue is set on the day of the Panathenaea. On the previous day Socrates instructed his guests – Timaeus of Locrus, Critias of Athens, and Hermocrates of Syracuse – about the structure of the ideal polity, i.e., on the content of the *Republic*. On this day they are supposed to return the favor and tell Socrates how such an ideal state could exist in the material world. Critias gives the first reply. The proof that such a state is possible is that in fact it once existed. Some 9,000 years earlier there was another Athens that saved the world from conquest by the nation of Atlantis, and the government of that Athens was very much like Socrates' ideal state. Critias heard this story in his youth from his grandfather, also named Critias, who heard it from Solon, who learned it in the city of Sais, Egypt from the priests of Neith, who is Athena. The report is reliable because, due to its ideal location, Egypt is able to escape the alteration of fire and flood that periodically destroy all human civilzation. Hence, they and they alone have reliable records that predate the last devastation by a flood of waters from below the earth³⁰² 8,000 years earlier.

Timaeus speaks next, and it is his story that concerns us. A proof that an ideal state can be made actual is that the universe itself is such a state. The deity (THEOS) formed a rational model of an ideal living creature and materially generated it. That living organism is the universe (KOSMOS). The details of this story are Timaeus' account of its creation.

It is of interest to note that both Critias' tradition of an ancient war between Atlantis and an earlier Athens as well as Timaeus' account of creation are stories that are used as a form of demonstration. Two arguments are given to prove that an ideal state such as Socrates described in the *Republic* can be actualized, and the form of both arguments consists in telling a tale. The story is a MYTHOS, which Timaeus describes as a probable or likely (EIKOS) account or story (LOGOS).[303] What the value of such an argument is with respect to other kinds of arguments, e.g., syllogisms, is briefly (but not extensively) discussed in this dialogue.[304] In any case, Timaeus' description of creation is, like its counterpart in Genesis, a "story," but not "merely" a story.

My summation of Timaeus' account of creation is divided into the following four parts: (1) The basic principles employed in the account. (2) What is produced by the deity through reason (NOUS). (3) What is produced in the receptacle through necessity (ANAGKE), and (4) What results from the influence of reason upon necessity. As in Genesis, the divisions express a kind of logical priority that has nothing to do with a temporal ordering. Timaeus' universe "is" created, but there is no time at which the universe "was" created. "Creation" expresses an eternal relationship that holds between the universe and the deity.

BASIC PRINCIPLES

There is no separate listing of the principles of creation in the *Timaeus* itself. The story is divided into the three categories of the products of deity through reason, the work of necessity in the receptacle, and the results of the deity's persuasion of the receptacle. However, it is clear from the text that the principles given in this section of this chapter are basic, and operate throughout the book. They consist of number and three pairs of opposites. The pairs are (1) being (TO ON) and becoming (GENESIS), (2) limit (PERAS) and the unlimited multitude (APEIRON PLETHOS), otherwise referred to as "the infinite dyad" (AORISTOS DUAS) and (3) reason (NOUS) and necessity (ANAGKE).

"Being"[305] is what is eternally. It is not subject to time, and, as such, never changes. It neither comes to be nor perishes. "Becoming"[306] is the complement of being. It is whatever is subject to time, changes, comes to be, and perishes. The story of creation told here is a tale of how one instance of being, the deity, uses other instances of being[307] to produce all of the instances of becoming.

Note that this way of explaining being and becoming is itself a story. If it were a literal explanation, then the dyad of being and becoming would be ultimate "classes," which they are not. They are ultimate "principles" (ARCHEI) under which the deity itself is subsumed. Timaeus' principles[308] are not classes. At the same time, they also are not "things." Just what they are is not clear. The same would be true of all of the principles discussed below.

The unclarity, in my judgment, is intentional. Again, Timaeus' account of creation is a story, and not what most people today would call a scientific explanation, precisely because the reality being described cannot be explained clearly within the limits of human discourse. At the same time, it should be said that Timaeus does not value obscurity as such. On the contrary, to the extent that this picture is unclear, it is "precisely" unclear, i.e., it is as precise but only as precise as the reality itself permits.

Timaeus' unclarity is that of a mathematician. Just as what it means to say, "π is 3.1416" is that the right-hand equation is as precise an approximation of that number which expresses the ratio between the circumference and diameter of a circle as is possible in a language of decimals limited by expressions of 10^{-4}, so saying "being and becoming, limit and unlimited multitude, reason and necessity, and number are principles of creation" is as precise an approximation of what is fundamental in a description of cosmology and cosmogony as is possible in ordinary language. As it is more accurate to say that π is "about" 3.1416 than to say that π "is" 3.1416, so it is more precise to speak about these principles in a "story" than to try, in vain, to describe the universe as such. The limitation in both cases is neither the knowledge nor the linguistic ability of the speaker. Rather, the limitation is a necessity dictated by the descriptive power of the language and the reality to be described.

Limit is equated in the text with the deity, unity (HEN) and good. The deity is often called the "DEMIOURGOS" and the "maker" (POIETES). As the God of Genesis makes ('ASAH) sky and earth, the deity of the *Timaeus* makes the cosmos. Again, this deity is intimately associated, if not identical, with the principle of limit (= unity = good). To the extent that what exists has limit, it has unity, and it is good. One is the case if and only if the other two also are true. All three are the work of Timaeus' deity.

The complement of limit is the unlimited multitude. It is equated with the receptacle (UPODOCHE) as well as with the dual principle of the great and the small (TO MEGA KAI TO MIKRON). More will be said about this second principle below in the section on the work of necessity. For now, let it suffice to note that insofar as the receptacle is the complement of the deity, as the deity is an active principle, the receptacle is passive. As the deity is what "makes," the receptacle is what is "made."

It is not possible to read Timaeus' distinction between limit and unlimited multitude without thinking of Pythagoras. Clearly this scientist and his school are an important influence on the author of this text. To what extent Timaeus' ideas are original is open to scholarly debate. In any case, it is worth noting that the division of limit and unlimit suggests a number of other pairs explicitly associated with the Pythagoreans that are appropriate to the *Timaeus* as well. The infinite dyad entails the following oppositions: odd and even, one and plurality, same and different, male and female, rest and motion, straight and crooked, light and dark, good and bad, right and left, and square and oblong. Four of these pairs of limit and unlimit are subsumed under the principle of number. They are odd and even, one and plurality, same and different, and square and oblong. Four of them are independent of number, but are explicit in this text and have particular relevance to the comparison of the *Timaeus* and the Genesis accounts of creation. They are male and female, rest and motion, light and dark, and good and bad.

Necessity is intimately associated with spontaneity (TO AUTOMATON), nature (FYSEI) and chance (TUCHE). Its complement, reason, is associated with design (TECHNE).[309] Timaeus' "chance" is not what we today would call "an accident." It is not something that "merely" happens. Chance events are caused as much as designed events. Note that it is chance and not design that is said to be necessitated. The difference has to do with the kind of cause involved in each case. The distinction agrees more or less with the division that some contemporary philosophers draw between causes and reasons. Closer to what Timaeus meant would be a distinction between effects that are or are not intended by the primary cause. Associated with reason is purpose or intention. Neither is connected with necessity. Consequently, since reason is associated with the deity in opposition to the necessity of the receptacle, if anything is contingent it is a product of the reason of deity. To the extent that a product is brought about by the rational, conscious choice of the producer, the product is a work of design by a designer. "Reason" (or "mind" in the sense of rational deliberation, i.e., the act of reasoning) is the distinguishing characteristic of the designer. In contrast, to the extent that a product is necessitated by its producer, independent of any rational plan, the product is called the work of necessity.

Note that it is not the case that some things happen by necessity while others happen by design, or that some events are planned while others occur by chance. Everything that exists in our universe is the joint product of both principles. Here, as with all of the principles discussed, Timaeus abstracted elements out of the generated world to provide an account

Plato's Timaeus

of the existence of this world. In other words, the "principles" of reality are not themselves "parts" of the generated world. Timaeus presents a "picture" of reality, and not the reality pictured.

The duality of reason and necessity can be read as a kind of dialectical synthesis of Plato's predecessors in physical philosophy, where the thesis and antithesis are pictures of reality generated through purposeless necessity or reasoned choice. On the one hand, the stories of Greek religion and theater tend to picture the universe as the product of the minds of deities. On the other hand, we find the attempts of most prior-to-Plato physical philosophers (FYSOLOGOI) to explain the universe entirely by non-mental, mechanical first principles.[310] In the latter case these scientists' first principles are first material entities out of which everything comes to be through the non-calculating physical nature of the elements (STOICHEIA). Thales decides (probably under the influence of Greek religious mythology) that water is this first principle. Anaximenes argues that it is air. Anaximander says it is the unlimited, and Heracleitus identifies it with fire.

The fragments of Heracleitus' writings give us a more detailed picture than either Thales or Anaximenes of what it means to say that there is a single material element that alone can account for all of physical reality. Fire as such is the element in a gaseous state. Water is liquified fire. Earth is solidified fire. In other words, it is not the case that first fire existed and then the other elements were generated out of it. Rather, all of the primary elements have always existed. Hence, the difference between these three early Greek scientists largely dissolves. Fire is not really fire, or water water, or air air. In each case there is a single kind of substance which becomes a different element, depending on the state of the primary material. The elements all have the same nature. They differ only with respect to the relative weight[311] of their composite forms. If the elements are distant from each other in the compound, then the composite is a gas. When the elements are most compact, the composite is a solid, and the in-between state is called a liquid. However, the elements in all of these compounds are the same kind of substance. In other words, long before there arose a group of physical philosophers who called themselves "atomists," the reigning voices of Greek scientists were in fact atomists in the sense that they maintained that everything in the universe is reducible to particles of the same nature that combine and separate from each other.

Different natural scientists explained this process of conjunction and disjunction in different ways. Only the atomists who came after Plato argued in effect that there are no principles beyond the physical makeup of the elementary particles themselves. These post-Plato philosophers pictured a universe of atoms in chaotic motion that "happen" to form from time to time a circular vortex, through a sifting process where like atoms combine to form objects. Leucippus argued that atoms differ in shape (RUSMOS), concatenation (DATHIGE), and position (TROPA). Democritus replaced concatenation and position with size and weight.[312] The motion of the vortex is circular simply because, since there is nothing to confine their motion except their collisions with each other, the particles spread out away from each other equally in all directions. Because collision is least at the circumference of the vortex, the particles at the extremity are most distant from each other. In contrast, freedom of motion is least at the center of the vortex, so that the particles here are most compact. Hence, the universe is formed, solely by necessity, into a series of spheres within spheres, each less dense than its contained spheres. Consequently, the world of earth is at the center of the universe. It is surrounded by a world of water, which in turn is surrounded by a world of air, and ultimately is encompassed by a shell of fire.

The atomist pictures of Leucippus and Democritus are the purest examples of universes ruled by necessity that Greek scientists developed. Plato's predecessors grudgingly moved beyond chance in order to explain why the elements combine and separate. According to Heracleitus the elements are, by their physical nature, in conflict with each other. The dominance of one entails the conquest of the other. Hence, their nature alone is not sufficient to account for the combination of particles into the mixed material objects of our world. He explained this generation in terms of an equilibrium of opposite forces governed by justice and harmony. According to Empedocles, fire and air are "active" opposing elements, in contrast to the opposing "passive" elements, water and earth. He claimed that they transform into each other in the process of combining into[313] and separating from[314] more complex substances. Like Heracleitus he too went outside of mere necessity in order to explain this transformation, and identified love and conflict as the ruling forces behind all change.

In this context the *Timaeus* can be read as an attempt to synthesize the explanations of the Greek philosophers with the stories of Greek religion. The former argued that everything that is exists by necessity. The latter proclaimed that everything that is exists through the design of deity. In contrast, Timaeus asserts that neither reason nor chance on their own could produce this world. Each group of thinkers has grasped one part of the story. The religious poets see the reasoned order, while the scientific philosophers recognize the world's chaotic nature. In reality the story of our world is a picture of a deity eternally using reason to persuade purposeless, natural chaos to a reasonable, purposeful, moral order.

The principle through which the chaos is ordered is number (ARITHMOS). Number is associated, if not identical, with form (EIDOS). It is a composite of the limit, in the sense of unity, with the unlimited, in the sense of the great and the small. Through number the deity makes the model or pattern (PARADEIGMA) for the living organism that is the entire universe (TO OURANOS). As in Genesis a living thing is a NEFESH CHAYAH, viz., a soul with a life, so all of Timaeus' living things, including the universe as a whole, is a soul (PSYCHE) embodied. The world-soul is a rational mind.

The parts of the universe are themselves living things. They are the celestial gods (OURANOU THEOI), the birds of the air, the fish of the sea, and the animals of dry land. All of their bodies are composed in geometric proportion (ANALOGIA) to each other from the four elements, so that (fire/air) = (air/water) = (water/earth). In other words, celestial fire (the realm of the lesser gods) is proportional to gaseous air (the realm of flying-life) as air is proportional to liquid water (the realm of sea-life) and water is proportional to solid earth (the realm of land-life). Note that the lesser gods, created by "the" deity, are the celestial objects. These creatures are the earth, the sun, the moon, five (then known) planets, and the fixed stars. Again, each god, like all other life, is an embodied soul.

DIVINE PRODUCTS OF REASON: THE ARITHMETIC OF SPACE

Timaeus' story begins with a body (SUMA) surrounded by a soul. At first the body is a sphere (SEFAIRA), uniformly lacking any qualities whatsoever, rotating on its axis. This body is the receptacle. The work of necessity will be the story of the receptacle's qualitative differentiation into distinct kinds of entities that inhabit the universe. Similarly, at first the soul, equally without differentiation, surrounds and penetrates all of the body. The work of the deity by means of reason is the differentiation of this universal-soul into a series of distinct souls that determine the different spheres into which the inhabitants of the universe reside. However, once again, this picture is intended to be an abstraction from the created reality. It is not the case that there are spaces chosen by reason and inhabitants caused by necessity. The different souls are not spaces without inhabitants and there are not different kinds of substances that can be conceived without some kind of spatial location. Spaces and spatial entities are not parts of the universe. Rather, they are "elements" in the sense that Franz Rosenzweig spoke of God, man and world as elements in *The Star*. In other words, they are principles in the same way that being, becoming, reason, necessity and number are principles, except that the former[315] are derived from the latter.[316] Hence, Timaeus' story is not a tale of the creation of the empirical world itself, but rather it is an account of the principles of its creation through different orders of principles.

The ordering of these principles is causal and not temporal.

Plato's Timaeus

It is not the case that at some particular moment in time the first principles produced the second principles, that produced the third principles, . . ., that produced the final principles of the material world. The creation spoken of here is eternal, i.e., it is not subject to time. What it means to say that the n + 1th principle was produced by the nth principle is that (to paraphrase Spinoza) the n + 1th principle could not be conceived and/or exist as a principle without reference to the conception and/or existence of the nth principle. More than two thousand years later Spinoza would attempt, in his *Ethics*, to give a philosophical explanation of what this kind of causal relation is. For whatever reason, Timaeus does not, which is one more reason why the *Timaeus* is MYTHOS and not science.

Timaeus begins with the differentiation of the all-penetrating soul's diversification through reason into distinct souls. First, reason creates the soul of the world, and then the individual souls of human beings. All of these souls, as soul, are the product of a two-step mixture of two sets of prior principles. The first set is the triad – existence (TO ON), sameness and difference. The second set is the pair – indivisibility and divisibility – that modify each member of the former triad. The mixture of divisible and indivisible existence produces "intermediate existence." Similarly, the mixture of the two kinds of each of the remaining members of the triad produces "intermediate sameness" and "intermediate difference." Soul itself is a product of the mixture of these first three products.

The final mixture is divided into "intervals" in a harmony based on mathematical and musical means. The first intervals are the numbers 1, 2, 2, 3, 4, 8, 9 and 27. The relationship between this sequence can be diagrammed as in Table 4. Each number is a power of the basic numbers – 2 and 3. In other words, $1 = 2^0 = 3^0$, $2 = 2^1$, $4 = 2^2$, $8 = 2^3$, $3 = 3^1$, $9 = 3^2$, and $27 = 3^3$. The sequence ends with 27, because 27 (3^3) is the solid number of the cube from which, as we shall see below, necessity generates the material inhabitants of the world.

Next, these initial intervals are filled in with other numbers generated by harmonic and arithmetic means to form a full musical scale. The complete list of arithmetic divisions of soul

178   The foundations

Table 4. *The mathematics of Plato's intervals*

Power/Term	2	3
0	1	1
1	2	3
2	4	9
3	8	27

are 1, $\frac{4}{3}$, $\frac{3}{2}$, 2, $\frac{8}{3}$, 3, 4, $\frac{9}{2}$, $\frac{16}{3}$, 6, 8, 9, $\frac{27}{2}$, 18 and 27.[317]

Next, reason differentiates soul – the proportional mixture of divisible and indivisible existence, sameness and difference – into the world-soul and individual human souls. The world-soul contains within it a series of principles of circular motion that influence the absolute motion of the inhabitants of the universe brought about through necessity. These inhabitants are the domain over which these motions have influence. Each motion is described in terms of a spin direction, a relative velocity, and a spherical domain. Their interaction with each other forms a three-dimensional geometric picture of spheres moving within spheres. However, the picture is pure geometry. They are no more or less "real" than any other geometric picture. It may be the case, as some critics have argued,[318] that Timaeus did not intend them to have any reality at all. It may be the case that Aristotle is the first astronomer to give existential import to what Plato's students, Eudoxus of Cnidus and Calliphus, only intended to be purely mathematical models for understanding the perceived circuit of celestial objects relative to perceivers on the planet earth. Or it may be the case that the existential import that Aristotle gave to his modification of Timaeus' geometry is more in keeping with Plato's intention in the *Timaeus*. It seems to me that the answer lies somewhere in between. Certainly the motions of Timaeus' world-soul do not exist in the sense that they are invisible globes (like Wonderwoman's airplane) in which bodies are located. As the work of reason, independent of the products of necessity, there is no body for these motions to have. Furthermore, as we have already noted, souls are principles and not substances. At the

Plato's Timaeus

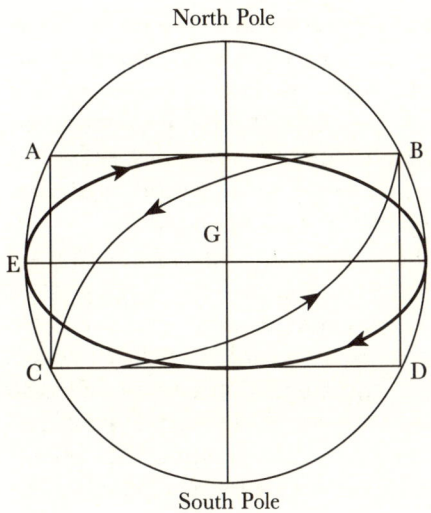

Fig. 6 Timaeus' celestial motions

Key
AB = Circle of Tropic of Cancer
CD = Circle of Tropic of Capricorn
EF = Circle of Equator
Vector EF = Movement of the same
Vector BC = Movement of the different = The Ecliptic
ABCD = Cube of the Zodiac
G = Globe of the planet, Earth

Table 5. *Timaeus' celestial motions*

Name of Motion (domain)	Length of circuit	Characteristics
1. same (universe)	24 hrs	d
2. different (Zodiac)	1 yr	u
Celestial deities		
3. Fixed stars	0	
4. Saturn	29 yrs, 166 days	is, id, d, c
5. Jupiter	11 yrs, 315 days	is, id, d, c
6. Mars	1 yr, 322 days	is, id, d, c
7. Mercury	1 yr, 322 days	is, id, d, s
8. Venus	1 yr, 322 days	is, id, d, s
9. Sun	1 yr	is, id, u, c
10. Moon	1 month	is, id, u, c
11. Earth	0	is, u

Key to the Abbreviations for the Characteristics
c = constant motion   d = down spin or circuit
id = influenced by motion of the different   s = sporadic motion
is = influenced by motion of the same   u = up spin or circuit

same time in some sense not explained in the *Timaeus* itself, they do exist, because "existence," together with sameness and difference, is part of a soul. Just as these motions exhibit sameness in that they are all three-dimensional circular spins, and difference in that they have different velocities, radii and spin direction, so they must in some sense exhibit existence. Still, the sense in which they exist is not explained.[319]

Timaeus differentiates the world-soul into eleven distinct motions that may be pictured as in Fig. 6.

Each motion is a circular spin that traverses the area of a sphere. They are distinguished by their radius,[320] the direction of their spin, and their speed. The domain of the motion of the same is the universe. Its spin is down,[321] and it takes twenty-four hours to complete one circuit. The domain of the motion of the different is the Zodiac, i.e., the region occupied by the so-called fixed stars. Its spin is up, and a complete circuit takes one year. The motion of the different is contained within the motion of the same at such an angle that the combination of the two produces a spiral twist. In other words, any object with no motion of its own under the influence of these two vectors would move in a spiral twist.

The remaining nine motions occur within the interconnected motions of the same and the different. Together these nine are called "celestial deities." Two of them occur within the influence of the motion of the same, and are so positioned that they are not effected by the motion of the different. One is the motion of the fixed stars at the circumference of the world-soul, which is subject to neither change of place nor position. The other is the soul of the planet earth, located at the center of the world-soul. It does not change place, but it does change position, viz., it rotates on its axis. Its spin is up.[322]

The remaining seven deities are all subject to the motion of the different as well as the motion of the same, since their domain resides in between the center and the circumference of the sphere of the world-soul. Moving from the center to the periphery of the sphere, these divine motions are the following: The movements of the domains of the moon and the sun are constant and up. The moon's orbit takes one month and the sun's one year. In contrast, the remaining motions are all down. The next two motions after the sun's are sporadic, i.e., at times along the circuit the motion ceases, and its body comes to rest. The total circuit of the motion of the domain of Venus,[323] including rest periods, is one year and 322 days. The same is true of the motion of the domain of Mercury.[324] The remaining three motions are all constant. The motion of the domain of Mars has the same spin direction and takes the same amount of time to complete its circuit, as do the motions of Venus and Mercury. The circuit of the domain of Jupiter takes eleven years and 315 days to complete. Saturn's takes twenty-nine years and 166 days.

The geometric notion of concentric spheres in order to analyze complex motions was developed by Plato's associate, Eudoxus. However, Plato's employment of this tool differs from that of his associate. Excluding the fixed stars and the rotation of the earth, whereas Timaeus posited nine motions, Eudoxus affirmed twenty-six. On Timaeus' model, the sun, the moon and the five known planets all have a single, distinct motion of their own, subject to the influence of the two motions of the same and the different. In contrast, Eudoxus requires three distinct motions for the sun and the moon, and four for the others. To these twenty-six motions Calliphus added seven – two for both the sun and moon, one for Venus, Mercury and Mars respectively, and one each for Jupiter and Saturn. Finally, Aristotle extended the list of basic vectors of celestial motion to fifty-five.

Note that we are not discussing the planets themselves, but their inherent motions, and these motions are identified both as souls and as deities. As such, they are existents, but they are not material substances. Again, it is true that Timaeus is telling a story, but that fact in no way entails that Timaeus did not believe that in some significant but unexplained sense of the term, these are real entities.[325] The same can be said, with an equal degree of ambiguity, about the deity's final creation through reason – the human souls.

Human souls are like the world-soul in that they are souls. As such they impose order on their characteristic bodies that are created by necessity. They differ from the world-soul in two respects. First, their characteristic motions are linear and not circular. Their six different motions, in contrast to the eleven concentric motions of the world-soul, are (1) forward, (2) backward, (3) right, (4) left, (5) up, and (6) down. The comparison between these two sets of motions in itself shows that as the six motions of a human soul belong to the same soul, so the eleven different motions of the world-soul are motions of a single soul. Hence, whatever it means to call these celestial vectors "gods," it is not that they are distinct entities. Second, human souls, unlike the world-soul, are influenced by their associated bodies. Sensations affect human souls, and, on Timaeus' theory of perception, sensations are forms of physical motion.

SPATIAL WORKS OF NECESSITY: THE GEOMETRY OF BODY

Timaeus continues his story of the cosmos with the diversification of the receptacle through necessity into distinct kinds of bodies. The receptacle has no qualities of its own. Without altering what it itself is, i.e., without affecting its own nature, it receives a quality (POIOTES) that is impressed upon it by the form of one of the elements. However, these qualities do not exist in the receptacle. On analogy with a mirror, the receptacle simply reflects them.

Timaeus' use of terminology is not as sharp as it could be. While he explicitly says that the qualities are images of the forms of elements, he also tells us that both the forms and the qualities are called "fire," "air," "water," and "earth." Furthermore, the terms, "quality" (POIOTES) and "form" (EIDOS) are often used interchangeably, and both are associated, if not interchangeably, with the terms "character" (IDEA), "shape" (MORFE), and "kind" (GENE).

The problem word is "form." At times it is used for a product of reason that is the model for the qualities, and at other times it is a synonym for a quality. In either case, Timaeus only distinguishes the mental models from their reflection in the receptacle. There is no third thing that is the form of the mental product that is reflected. The model, which we shall call "element," is an instance of being. It is "a perfectly real thing" (ONTOS ON), that is imperceptible and ungenerated. In contrast, the reflection, which is called a "quality," is an instance of becoming. It is "a sort of existent" (ON PUS),[326] that can be apprehended through the senses, and is generated.

That the elements are products of reason is in itself a problem. First, Timaeus never mentions them in the first section of his tale. There we are told only how the deity differentiates the world-soul into a set of distinct vectors that are the souls of the celestial objects. Nothing is said of any association between these motions and elements. Furthermore, this second section of the story is supposed to be the work of necessity in the receptacle, independent of the products of the deity through reason. Only in the third and final section are we to see how the two interact with each other. From this perspective Timaeus ought not to talk about qualities at all in section two, since they are the result of the interaction between the two different kinds of creation.

A second, more serious problem involves the metaphors that Timaeus uses to explain the relationship between elements, qualities and the receptacle. The primary analogy is that of a reflected image in a mirror. The second is that of the birth of a child. The two metaphors are incoherent. In the case of the mirror, there really exist only two things – the mirror and the object mirrored. The mirror is the receptacle, and the object mirrored is the mental element. In this case the qualities do not exist. They simply are reflections. Just as it would be possible[327] to say that the way an image appears in a mirror is a quality of the mirror, so, on this analogy, Timaeus can call everything that we would identify with a physical substance a "quality." On the other hand, the birth metaphor suggests a very different ontology. The quality, viz., what comes to be, is the offspring. The receptacle, viz., that in which what becomes comes to be, is the mother,[328] and the element (viz., what always is) is the father. Offspring produced by parents exist just as much and in the same way as their producers. In this case the world is made up of three kinds of entities – a rationally produced model ($\approx$ father), a necessary receptacle for the model's image ($\approx$ mother = nurse), and the copy or image itself ($\approx$ offspring). Therefore, while qualities[329] differ from their elements and their container[330] in that the qualities are generated[331] and causally dependent,[332] these bodies become independent existents with as much right to be counted among the entities of the universe as do souls and the receptacle. That the *Timaeus* describes the instances of becoming as a "sort of existence" indicates the ambiguity of the two metaphors.

In any case, at this stage of his tale it is clear that Timaeus posits the existence of model products of reason called "elements" whose reflections are the "qualities" fire, air, water and earth. What it means to say that they are generated is that they are reflected. We can distinguish between, on one hand, the products of reason that are ungenerated and "perfectly real," and, on the other hand, their reflected qualities that "sort of" exist but not "really." The former are said to have "being" and the latter "becoming." Outside of both of these categories is the receptacle. It neither is nor becomes. At the same time, it is not "nothing at all" (PANTELOS ME ON). If the qualities are the building blocks of the physical bodies of the universe, the receptacle is the "space" in which they reside. Like being, it is everlasting, because it is not generated. However, it also "is" not. Rather, it is what "receives" what comes to be. It cannot be known by reason, which only knows the product of reason. Also it cannot be perceived by the senses, which only apprehend the work of necessity. Yet, in some inexplicable sense, this nothing that is not absolutely nothing is known. In the language of Timaeus, this space (CHORA) for becoming is apprehended by a "bastard reasoning."[333] The *Timaeus* has nothing more to say about this distinct way of knowing.

At first there is being, becoming and space. From being, space[334] receives[335] qualities – fiery from fire, cool from air, moist from water, and dry from earth. Necessity[336] randomly shakes them, which, through a kind of winnowing effect, results in the division of the reflections in space into distinct regions. By what Timaeus calls "an innate impulse" (SIMTHITOS EPITHMIA), like qualities are attracted to their like, i.e., each quality is drawn to a quality of the same or similar density.[337] The movement follows all six kinds of linear motion listed above.

We have already seen that these motions define the last products of reason – the individual human souls. Hence, it is reasonable to conclude that whereas the creation of reason can proceed independent of necessity, the same cannot be said for the work of necessity. In other words, while the deity creates the universe from uncreated entities, its tools – reason and necessity – are not equal. In an important logical sense, reason is prior to necessity.

At this stage we can introduce an answer to the question: From where did the elements arise? They are the products of number, whose mental existence is the creation of reason's employment of the infinite dyad. This answer can be diagrammed as in Fig. 7.

Timaeus uses the geometry of Pythagorean physics to explain how the spatial qualities are differentiated into regions of the so-called four elements. The starting point is limit and the unlimited multitude. The former generates the odd, the latter produces the even, and number is the joint product of even and odd.

Each number has a counterpart in geometry. 1 = unit = point. 2 = line, and 3 = rectilinear plane. The higher numbers are distinct plane figures, solid planes and sensible bodies. These numbers in their relative order to each other are expressed geometrically as follows. Let n = any equivalent of[338] a plane figure; m = any equivalent of a solid plane; and p = any equivalent of a solid body. $3 < n < m < p$. For every number there is a geometric counterpart. Each geometric counterpart is a physical body, and each body is a compound of qualities mirrored by reason in space.

Note that there is no zero, and 1 is not a number.[339] Rather, 1 is a unit. By extension, the point is a unit for measurement in space, and is not itself an object. The first numbers are 2 and 3.

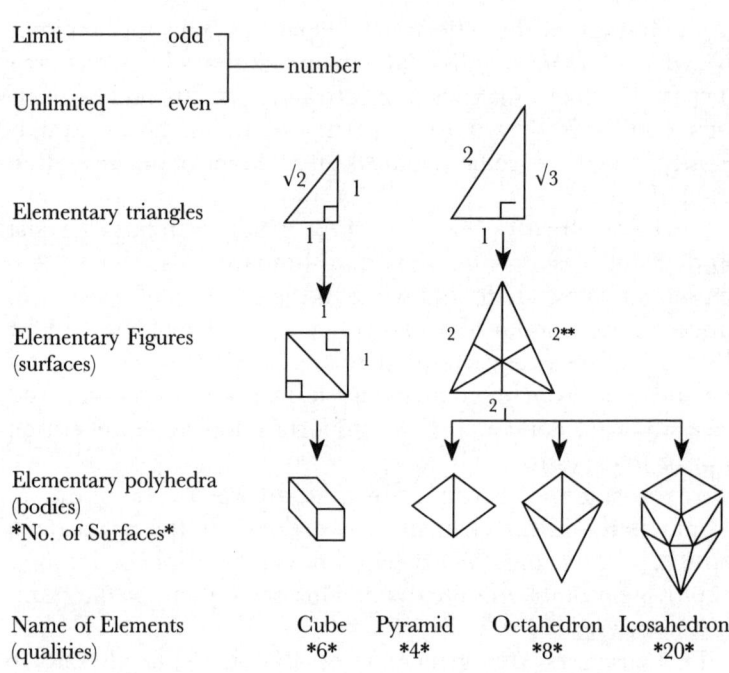

Fig. 7. Timaeus' elementary triangles

2 expresses the lines that form the triangles from which all solids are constructed. 3 expresses these triangles. Timaeus' physical world is constructed out of two distinct kinds of planes. These are (a) isosceles right and (b) half equilateral, scalene right triangles.

Note that Timaeus, like Pythagoras, could not avoid positing an element of the irrational in his construction of the physical world. Since "1" is the unit measure, the hypotenuse of the elementary isosceles right triangle has a length of $\sqrt{2}$. Similarly, one adjacent side of the elementary half equilateral scalene right triangle has a length of $\sqrt{3}$. Both are irrational

Plato's Timaeus

A. From isosceles right triangles

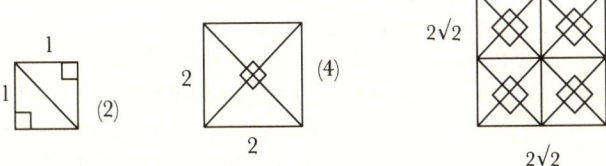

B. From half equilateral scalene triangles

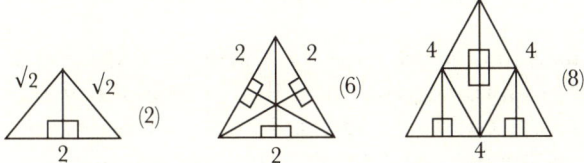

Fig. 8. Timaeus' elementary regular figues

numbers. These facts in themselves are sufficient motive for Timaeus to insist that the universe cannot be only a rational place generated by the reason of the deity. It would also serve as a motive for Plato's student, Aristotle, to insist that space is continuous.[340]

The elementary triangles are combined into the elementary regular figures. The relationship between these elements and their compounds may be diagrammed as in Fig. 8.

Four isosceles right triangles form a perfect square with four plane angles. Note that the length of each side of the square is the irrational, $\sqrt{2}$. Similarly, six half equilateral scalene right triangles combine to form an equilateral triangle, each of whose sides has a measured length of 2.

These two elementary regular figures form the surfaces or sides of the four elements. Earth is a six-sided cube. Fire is a four-sided pyramid. Air has eight sides,[341] and water has twenty sides.[342] The sides of the cube are formed from the perfect square, while the sides of the other polyhedra of the elements are formed from the equilateral triangle. Hence, the element earth has a fundamentally different nature than the other elements. Consequently, Timaeus excludes earth from his judgment that the elements can transform into each other.

Timaeus grades the elements according to size. Some plane surfaces in each of the four kinds of polyhedra are larger than others. In principle there is no limit on the number of different sizes possible, but Timaeus only lists three within each category. Each surface of a cube can be formed from 2, 4 or 16 elementary right triangles of uniform dimensions. Similarly, each surface of the other elementary kinds of polyhedra can be formed from 2, 6 or 8 elementary right triangles of uniform dimensions.

The elements combine to form the primary bodies that initially[343] are grouped together into distinct regions of the receptacle. The different sized cubes, pyramids, octahedra and icosahedra are attracted respectively to the regions of earth, fire, air and water. Although Plato does not explicitly say so in the *Timaeus*, based on the other ancient Greek creation accounts, it would be reasonable to picture each region as a sphere enclosed by other spheres. The sphere of earth bodies is at the center, surrounded by a sphere of water bodies, surrounded by a sphere of air bodies, surrounded by a sphere of fire bodies. Within each sphere there are subspheres of the different grades or sizes of the same elementary bodies, where the largest bodies are closest in the center and the smallest are closest to the circumference. The outer sphere is that of the cosmos itself. It is composed of twelve-sided bodies,[344] where each side or surface is composed of a pentagon. Timaeus does not describe the kind of triangles out of which this pentagon is formed. In any case, it clearly follows that in this picture, the sky is by nature at least as different as earth is from the other elements. Furthermore, as water, air and fire cannot be transformed into earth, so none of the standard four elements can be transformed into the celestial bodies.

Timaeus describes a sort of calculus for the transformation of the elements[345] fire, air and water into each other. The basis of the arithmetic is that the surface of each element is composed of a different number of the same kind of right triangle. Limiting ourselves solely to elements of the same size or grade, transformations can be expressed algebraically as follows: Water = W = 20. Air = A = 8. Fire = F = 4. The number in each case corresponds to the minimal number of elementary triangles needed to form the kind of polyhedron in question. The burning of a lamp, which is the transformation of a liquid into a heated gas, is expressed by (W = 5F), viz., (20 = 5 × 4). Water boiling, which is a liquid transformed into both a cold and a heated gas, is expressed by [W = (3F + A)], viz., [20 = (3 × 4) + 8]. Finally, water evaporating, which is a liquid transformed into a cold gas, is expressed by [W = (F + 2A)], viz., {20 = [4 + (2 × 8)]}.[346]

The limitation imposed at the beginning of the three examples was set only for the sake of simplification. In fact, Timaeus' picture of how elements change into each other is far more complex. Transformations occur between bodies that are mixtures of all four and not just one kind of element. Furthermore, the elements composing these bodies are of different grades. The calculus should still be applicable, but any real transformation would require a vastly more complex expression.

Note the following two consequences of these examples for Timaeus' theory of creation. First, none of the elements are what their names suggest. Fire is not fire, air is not air, water is not water, and earth is not earth. Rather, each term expresses the relative density of different kinds of physical objects, which is the difference between solids, liquids and gases. "Earth" means a solid, "water" a liquid, and air and fire are kinds of gases. "Air" means a cold gas, and "fire" a warm gas. Second, insofar as we may call the reflections of mental objects in space "matter," Timaeus is committed to a theory of the conservation of matter. What ultimately exist in the material world are right triangles. The triangles of this physics are the counterpart of "energy" in modern physics. As in modern physics in principle no energy is lost in the process of one kind of matter being transformed into another, so in Timaeus' physics the transformation of elementary bodies neither adds to nor detracts from the number of elementary triangles in the universe.

Timaeus' discussion of the transformation of elements leads directly to his account of perception and emotion. In fact, these two mental phenomena are understood simply as such transformations. For example, people feel pain when their normal state of equilibrium is suddenly disturbed and only gradually returns to normal. Conversely, they feel pleasure when the loss is gradual and the restoration is sudden. However, Timaeus' psychology lies outside of the topic of this book. The creation of these "perceived affections" (TA PATHEMATA AISTHETICHA) is the last work of necessity, and serves as the bridge into the last part of the *Timaeus* on the cooperation between reason and necessity.

THE CORRELATION OF REASON AND NECESSITY: LIFE IN SPACE

This final section of the *Timaeus* deals with the creation of the parts of the human body, their relation to the human soul, and how humans preserve and lose their lives through disease. Again, all of this lies outside of the domain of this book. All that is relevant to our concerns is the following: First, the number of individual bodies and of souls at creation is the same. Second, the only individuals initially created are male humans. When a man dies, his soul is united with another body. What kind of body depends on the moral nature of his life. The body exists for the purpose of housing the soul. Hence, the kind of body a soul assumes depends on the grade of the soul of death. Birds are born from humans who only viewed the cosmos in empirical terms. Land creatures are humans ruled by the mortal parts of their souls[347] to the exclusion of their immortal part.[348] Based on this doctrine of transmigration, Timaeus ranked the species of physical life as follows: male humans, female humans, birds, land-animals, reptiles and sea-life.

The *Timaeus* includes no picture of the planet earth. This part of the story can be filled in from other dialogues. Through the "innate principle" of the attraction of like matter to like matter, under the direction of the reason of the deity, the sphere of the planet earth is surrounded by air and divided at its surface by collections of water that form the seas. The *Phaedo* and the *Critias* map the inhabited surface of the earth in very different ways with respect to the interests of a geographer, but for our purposes the differences are not significant. Far more important is that in the *Phaedo* Plato asserts that below the surface land mass inhabited by the living is a region for the dead (OKEANOS) divided into four subregions. One is for philosophers, a second for the mediocre, a third for criminals, and a fourth for incurable criminals. Note that prima facie this doctrine of an underworld[349] directly contradicts the theory of transmigration at the end of the *Timaeus*.

It is misleading to say that according to Timaeus man was created first and only afterwards did the deity generate woman. The sense in which this statement is true is not temporal, because (again) the creation described in this story does not take place in time. There have always been people, birds, fish and animals of both genders. Rather, the terms "first" and "afterwards" express a moral order of species. In Timaeus' conception of the universe, events occur for moral as well as physical reasons, and the highest physical and moral end of material life is to engage in abstract mathematically based thought. For our purposes, this is all that is important about the fact that this story of creation concludes with psychology and transmigration. It tells us that the cosmology and cosmogony of this dialogue have at least as much to do with human moral instruction as they have to do with physics, or (more accurately) morality and physics are intimately tied together. Though one[350] has priority over the other,[351] neither is possible in isolation from the other.

We concluded our summary of Plato's *Timaeus* with a judgment about the relationship between divine reason (or purpose) and necessity (or chance) in drawing a picture of the universe and telling the story of its origin. That judgment is a key to relating Plato's doctrine of creation to alternative cosmologies, with specific reference to the text of Genesis and the way that the rabbis interpreted it. One way to view the history of physical theory is as a moving pendulum between two extremes. One tendency reduces all matter to discrete indivisible particles and explains physical motion in terms of chance principles of necessity. The other sees matter as part of an endlessly reducible continuum, and explains all physical motion in terms of choices made by agent(s) through purposeful design. The issues in this case have little to do with empirical data. Rather, in question is the kind of a priori schema best suited to order or conceptualize that data.

Physical philosophers ask the following questions about the complex phenomena they observe: First, is this complexity best understood by reducing objects to simpler entities, and then considering how the smaller elements combine and separate, or should we concentrate instead on the interaction between our elements, and consider the pattern of their interrelationship in motion? Second, are the elements in our schema discrete from or continuous with each other? Third, are the principles or patterns best understood on the mechanical metaphor of a machine, or would it be better to view them on analogy with living organisms that calculate their actions? Every schema pictures elements in an ordered array. Again, the issues are: (1) Which is more basic to the schema, the elements or their relationship, and (2) is the projected ordering mechanical or teleological? Those theorists who make elements basic tend to see them as discrete from each other, and employ mechanical principles of organization. Conversely, those theorists who make the relationships basic tend to consider the elements as continuous, and judge their interaction to be purposeful. These are the two extremes.

The writings of Homer and the other poets and dramatists of traditional Greek religion approximate the extreme of a universe governed by the reckoned choices of deities. Most Greek scientist-philosophers before Plato approximate the other extreme. Thales, Anaximenes, Heracleitus, Empedocles, and the Pythagoreans all posit basic, indivisible material substances as the building blocks of the universe. Anaximander may be an exception. Like the others, he speaks about a first matter (ARCHE), but he calls it the "unlimited" (APEIRON), which in itself suggests that matter ultimately is a continuum and not a set of discrete entities. The one clear exception is Anaxagoras.

Similarly, most of the early Greek scientists opted for purely mechanical principles to conceptualize the interaction of the elements of the physical world. However, there may be more exceptions in this case. The list of those who may have presented a teleological account include Heracleitus,[352] Empedocles[353] and Anaxagoras.[354]

In this context, Plato's *Timaeus* can be read as a movement toward the middle of the pendulum, away from his inherited two extremes. In his picture the universe is understood in terms of immaterial motions, intentionally generated by reason, brought into purposeful nexus by a single deity with discrete bodies, generated by chance necessity in continuous space. However, Plato's attempted synthesis does not last. Subsequent Greek science itself split into two extremes. On one hand, Aristotle presents a picture of the physical universe composed of infinitely divisible matter, whose primary intelligibility is found in immaterial, teleological principles that are themselves deities. On the other hand, the atomists Leucippus and Democritus posit discrete indivisible particles that combine and separate entirely by chance. This extreme atomism, with slight modification in the interpretations of Epicurus and Lucretius, dominates subsequent Roman science, and represents the consensus of scientific opinion in Asia Minor, when the inheritors of the world view of the Hebrew scriptures, viz., the early rabbis, come into contact with Hellenistic civilization.

As the model of the atomists is the most extreme swing of the pendulum in the direction of a universe pictured as chance combinations of discrete material elements, the way the early rabbis in the Roman and Sassanid worlds interpreted the Genesis account of creation is an equally radical picture of the universe in the opposite pendulum direction. In other words, the midrashic interpretations of the first chapter of Genesis form a more extreme picture of a continuous universe created by the will of a living deity than is literally found in the Hebrew scriptures themselves. Consequently, both the *Timaeus* and the Genesis accounts of creation are at least comparable as moderate pictures of the universe, in contrast to the more extreme opposing schemata in Greek and early rabbinic civilizations. In this sense, the move of the classical rabbinic commentators on Genesis to schematize Genesis in terms of the *Timaeus* reflects a return to the moderacy in the physical theory of both foundation texts.

Most biblical scholars claim that the Genesis account of creation was written in Babylonia in the middle of the sixth century B.C.E.[355] Most Plato scholars place the *Timaeus* about two hundred years later, some time after Plato's visit to the Near East. It should be noted that in Plato's time there already was significant trade between these two civilizations. I have no intention of claiming that the Hebrew story directly influenced Plato's tale. Rather, my only point is to establish that, with respect to history as well as philosophy, it is legitimate to compare the two accounts of creation. Against this background let us look in greater detail at how these two stories of creation compare and contrast.

There are obvious differences between these two works. We have already noted that (1) the creation of the planet earth is an important part of the Genesis account that is almost entirely ignored in the *Timaeus*. Furthermore, (2) Timaeus' celestial objects are alive, while in Genesis they are inorganic. Timaeus calls them "gods," and Genesis does not. (3) While the deity of the *Timaeus* acts to differentiate the universe, unity is so much more valued than diversity that God and reason are identified with both unity and good. In contrast, what the deity of the Torah seems to value as good is the very diversity he initiates in the pre-existent unity. (4) On Timaeus' ontology the only substance is the cosmos itself. It is an embodied soul. The different souls exist as modifications of the world-soul, and the different bodies are modifications of the world-body.[356] In contrast, in Genesis the celestial bodies, the earth, and the different species of celestial-, sea- and land-inhabitants are all separate substances.

Still, the similarities are noteworthy. (1) Both accounts are intentionally stories rather than scientific explanations of creation, because the authors judged story-telling to be a more precise way to present this picture than a literal description.

(2) The universe pictured in both cases contains aspects of reality that are, from the perspective of reason, negative. In the case of Genesis, there is the void[357] that is the primordial occupant of the deep.[358] In the case of the *Timaeus*, there is the receptacle. In both cases we are presented with something that cannot be known either by the senses or by reason. In both cases it is not something. However, neither is it nothing at all, since it is a necessary aspect in the intelligible order of the universe. In other words, there is more to the universe than what is knowable, without which the universe would not be intelligible. This additional component in and of itself would be a sufficient motive for the two authors to tell stories rather than to attempt to present literal explanations. Explanations can only be given for what is knowable by the senses and/or by reason. Stories enable their tellers in some way to express what is inherently inexpressible.

(3) While temporal language is employed in both stories, clearly neither tale purports to take place in time. The creation of the universe is supposed to be a reality, but the event is eternal. It is not the case that reason made souls, necessity made spatial objects, and then the deity made the cosmos. The order is in some sense causal or logical, and not temporal. Similarly, the divisions of the Genesis story into days also is not temporal. Days are not periods of twenty-four earth hours. They are in some sense causal or logical divisions in the presentation of an eternal event.

(4) Both stories picture the universe as being "made" on analogy with the way artists make artifacts out of unmolded materials. In both cases the given material is a real[359] nothing from which something is generated. In other words, both Timaeus and the author(s) of Genesis could say that God created the world out of nothing. However, "nothing" is something "real" that is not absolutely nothing at all.

(5) Both stories picture the universe as a harmony of opposites. In the case of the *Timaeus*, the opposites are deity/receptacle and active/passive. In the case of Genesis, they are

God/void and motion/rest. Furthermore, in both tales the three opposing pairs of male/female, light/dark, and good/bad play significant roles.

(6) Morality and physics are closely integrated in both stories. The world not only "is," it also "is good." In part this is a consequence of the employment of teleological principles of ordering. The question, "How does something come about?" is inseparable from "Why does it come about?" What is made by a deity is not "just" made. It is made to be something, and as such it can be judged. What "is" also "ought to be," and if it is not yet what it ought to be, then it ought to "become" it. In other words, both works of literature deal with "becoming" and not "being." Both authors understand the former word, in contrast to the latter, to entail moral values. In other (more precise) words, the created world that is good is a model for the empirical world that is becoming (and therefore [in principle] is not [yet]) good.

(7) In both accounts of creation the scientific/moral answer to the question, "Why is there a universe?" is political. Timaeus tells his story to support a political program, viz., to show that the political model presented in the *Republic* is possible, which implicitly entails a moral imperative to act towards its establishment. Similarly, the priestly account of the origin of the universe sets the scene for the statement of a political/religious constitution in the body of the Torah, which (again) entails a moral imperative to act towards its (in this case) re-establishment. Consequently, both accounts of the universe propose a view of physical/moral science that is inherently connected with politics.

In neither case are there grounds to make a radical separation (as did Kant) between the "is" and the "ought" and to consign the one to the domain of science and the other to the domain of religion. On the contrary, such a separation is a fundamental distortion of reality – empirical and moral, scientific and religious. However, this analysis is premature. Whether or not it is a correct judgment belongs to the final part of this book. Here our concern is limited solely to the question of the coherence between the two foundation texts – Genesis and the *Timaeus* – of the Jewish dogma of creation.

In conclusion, the judgment of the classical rabbinic commentators on the Hebrew scriptures that the *Timaeus* provides a sound schema to understand the story of Genesis is not without merit. Certainly the *Timaeus* is more useful for this purpose than any other schema available to the classic Jewish philosophers – viz., those of the Aristotelians and the atomists. However, the fit is not perfect. With respect to cosmology, Genesis and the *Timaeus* are significantly different. Although both give the nothing-in-itself of space a logical priority over either events in space or positive spatial objects, the ontology of the Hebrew scriptures is pluralistic, whereas Timaeus' universe is monistic. On the other hand, with respect to cosmogony, the situation is very different. The origin of the universe is to be explained by an active positive principle (associated with God) and a passive negative principle (associated with space). The origin itself is not part of the universe. Rather, it expresses an ideal beginning that points to an ideal end, between which resides the existing empirical universe(s) of movements (or events or processes) from and to asymptotes.[360] From this perspective, Timaeus' view of creation is more than consistent with Genesis; it amounts to a more detailed philosophic account of the literal meaning of the biblical text than are the tales of the early rabbis or of their disciples in the subsequent Jewish, Christian and Muslim theology that preceded Rosenzweig's *Star*.

This concludes our discussion of the two foundation texts of the Jewish concept of creation. We began by examining Rosenzweig's statement of the dogma, explored whether or not it could be called a "Jewish" position, and concluded that it could because of its coherence with what the classical rabbis taught about creation both in their philosophy and in their biblical commentaries. In the course of that discussion we discovered that the classical Jewish understanding of creation as a philosophical dogma was based on the use of Plato's *Timaeus* to schematize the text of Genesis. That discovery raised the question: is such a philosophical schematization of the Hebrew scriptures believable? Our answer is a conditional yes. There are serious differences between the two stories of the origin of the universe, but the similarities are sufficiently strong that, given the alternatives, the rabbis (and Rosenzweig) made a sound judgment about how to explicate this dogma. In other words, of all the then available schemata, the *Timaeus* provided the best fit. However, a best fit is not a perfect fit. Furthermore, the cosmology of the *Timaeus* is not contemporary scientific cosmology. It is reasonable, therefore, to look now at what kind of view of the origin and general nature of the universe arises out of contemporary empirically based and mathematically expressed astrophysics to see (1) is there a better schema for interpreting the Genesis text, and (2) given the currently best available philosophical/scientific schemata, is the Jewish dogma (i.e., a belief rooted in the biblical text as interpreted through Jewish tradition) believable as an account of the origin of the universe?

PART 4
A believable view of creation

Introduction

To understand the universe is a positive religious obligation, but it is one that can never be perfectly fulfilled. If knowledge means to understand in detail with certainty, then there can be no knowledge. The available sources for judgments are the evidence of sense experience, the written records of what others (past and present) have experienced, rational/logical reflection on that evidence, the words of revealed scriptures, and the written records of how others (past and present) have interpreted these scriptures. None of these sources in themselves are perfectly reliable. Not all experience is veridical; there are illusions and hallucinations. Not all rational reflection is demonstrative; arguments are no better than their premises and reason itself cannot provide true premises; furthermore, there is more than one kind of logic, and it is not always clear what kind is appropriate to any particular kind of question. Clearly there is no way to prove that any particular scripture in any religious tradition really is revelation, and if it is not, its value as a basis for truth judgments is questionable. Finally, even if we grant that a particular text does in fact record divine revelation, there is no way to be certain that our, or anyone else's, interpretation of the text is what its purported divine author intended to communicate. Furthermore, there is always new sense experience, some of which is novel, and we often discover new ways of thinking that change the way we understand both our past experience and our revealed texts. This novelty often has in the past, and certainly can in the future, change what we think we believe to be true about almost anything.

These judgments about the limits of knowledge are especially the case when the subject for understanding is the origin of the universe – a topic concerning which there is no relevant direct experience and there is a very large variety of significantly different purported revealed scriptures and interpretations of them. Hence, beyond any general skepticism about knowledge, the ideal of understanding the origin of the universe is among the most worthy candidates for doubt. Clearly in this case the best that can be hoped for is to approximate the ideal but never to reach it, to do the best we can as human beings to understand creation but never truly to succeed. The best we can do in fulfilling this religious duty is to explore what we have learned from the past as it is preserved in our religious traditions and to correlate that with the best data and judgments of contemporary empirical science. The result will be a belief that is the best possible one, which in this case means a belief that is least likely of all presently conceived alternatives to be false. (A best possible belief cannot even be one that is most likely to be true.)

Towards this end, in the fulfillment of a positive religious duty, we have explored in this book what the sources of Judaism say about creation. We began with what is the fullest account of creation in modern Jewish philosophy (viz., that of Franz Rosenzweig), and related it to what the sources of classical rabbinic Judaism (particularly in rabbinic philosophy and biblical commentaries) teach about creation. The quest has led to the following conclusions.

The present universe is to be pictured primarily as a set of interdependent processes that arise from an origin and are directed towards an end. In religious language, the end is called "redemption" and the origin is called "creation." These processes are asymptotic in both directions. No matter how far we look into the future, the end will never be reached; at best it can only be approximated. Similarly, no matter how far we look into the past, the origin can never be discovered; it too can only be approximated. As such both extremes are not themselves part of the actual processes of the universe. They are ideals that define actual experienced directions. As such, while the processes themselves are in space and time, their two terminal points are not. Consequently, both creation and redemption are atemporal.

The ultimate principles that the models of creation and redemption provide for understanding the present dynamic universe are God and space. God functions primarily to establish the ends towards which the universe moves. In this sense, creation teaches the identity of morality with the will of God and asserts that the universe is moral. That is not to say that it is good. It says the opposite. What is is always to be understood as a movement towards an end. It is the end that is good. Because the universe never is at the end, the reality of God entails that whatever is only is to be made better, but in itself it is not good. In other words, God functions on the model of creation as the teleological principle of everything. Conversely, space functions primarily to establish the chance-necessary-mechanical-mathematical causes from which the universe arose. On this view it is space that has ontological priority over every other candidate for existence. At first there is universally uniform space. Then space is differentiated by an act of divine will. Out of this space erupt events. Concrete substantive objects are the least primary of all candidates for existence. Objects exist only within dynamic events that define them, and the events themselves receive their nature from their spatial domains.

As space is prior to events which are prior to objects, i.e., as the non-substantive is prior to the substantive, so the indefinite is prior to the definite. What is at first is not definitely anything. It is simply undifferentiated space. What occurs through time is that this real vagueness becomes increasingly clear. Being clear is not something true of the experienced universe. Rather, it is a moral ideal towards which the space of the universe, and every process that it generates, tends but does not achieve. Similarly, as the indefinite is prior to the definite, so the negative is prior to the positive.

The universe and everything in it is not yet something. Rather, it is to be understood as a movement from what was absolutely nothing towards what will be absolutely something, i.e., from an infinitely remote endless quantity of total nothings towards an ideal single total something. Again, whereas the principle of the end is an ideal, thoroughly positive, thoroughly moral and thoroughly volitional God, the principle of the origin is a primordial, thoroughly negative, thoroughly amoral, and thoroughly necessary space.

What these two models – God and space, redemption and creation – generate is a general picture of a universe that is dynamic more than static, infinite more than finite, moral more than amoral, and negative more than positive. Every individual thing and person has meaning only in the context of its participation in continuously changing events in time and space. Taken in themselves, they are nothings that arose from nothing trying to become something in response to the (known or unknown) will of God. This is what emerges as the meaning of what Genesis entails when it says that at first space became something (on days 1–3), then space produced objects that strive to become something (on days 4–6), where the something that moves them is the end of Sabbath rest (day 7).

Furthermore, this world is not unique. There were worlds before ours and there will be worlds after ours. Our world arises out of the end of what came before. This is what Genesis calls TOHU and BOHU. That past end is the space of our present origin. And our world contains within it the seeds of the world that will come after ours. For example, the initial OR and the upper regions of the waters created at the beginning of our world are intended solely to function in the next world. In other words, OR and the upper waters relate our world to the world to come as TOHU and BOHU relate the universe that was to our universe. All of this says that there can be no perfection (no divinity if you will) in the lived world. Our world is neither good nor divine. Rather, it is the space that defines the constant struggle to attain the unattainable – God and the good. In other words, the universe in principle is neither good nor divine; its sole (moral and religious) worth is that it allows movement towards worth – moral value in politics (familial, national and global), and religious value in worship.

So far the focus has only been on the past record of revealed scriptures and their interpretation in Jewish tradition. The result is a view that properly can be called "Jewish." Now the question becomes, is this one of a possibly infinite number of possible views in itself believable. The answer turns on looking at the second set of data discussed above, viz., the present record of the conclusions of theoretical physicists about the origin of the universe, based on their mathematical schematization of what experience reports about the dynamics of the heavens. Furthermore, to ask if what the traditional texts teach is believable and to explore an answer to this question by comparing Jewish texts with contemporary relevant scientific and/or philosophical texts whose source of authority is external to the Jewish tradition is, as we have seen, itself an inherently Jewish way to pursue the fulfillment of this religious conceptual obligation. As the classical rabbis looked to the best scientific view of their time (viz., that presented in Plato's *Timaeus*) to understand what they could believe about creation as this belief is informed by the beginning narrative of the book of Genesis, so we will look at the best of current scientific speculation about the origin of the universe to examine how we can understand and believe in our revealed story of creation.

First we will survey the variety of different theories in contemporary physical cosmology, and then we will explore how these findings in contemporary empirical science correlate with what we found to be the classical Jewish dogma of creation. This correlation will be the basis for some final judgments about what one ought to believe about the origin and general nature of the universe.

CHAPTER 7

Creation from the perspective of contemporary physics

This chapter[361] summarizes the general contemporary scientific consensus about cosmology and cosmogony. The data considered in this case come from physics. We will examine the published conclusions that theoretical scientists have drawn from empirical data about astro- and particle-physics, and the attempt of philosophers of science to apply those findings to their own interests in ontology and epistemology.

The views summarized here are taken from secondary rather than primary works, which (as such) requires a word of explanation. In general my emphasis throughout this book has opposed this kind of approach. When dealing with the Hebrew scriptures, I did not try to present a general view of the scriptures on creation. Rather, I focused on a single text, viz., Genesis 1–2:3. Nor did I base my judgments primarily on what secondary sources had to say about this material. Rather, I went directly to the text itself and formed my own opinions, guided by rules of coherency and consistency, and based on my own reading of primary material. Secondary works were consulted and the Genesis text was considered within a context of the Hebrew scriptures in general. However, these more synthetic considerations were always submissive to my own reading of the chosen primary material. I followed this approach in order to avoid two basic errors all too evident in many works in intellectual history. First, by synthesizing significantly different works one is always in danger of presenting mere fiction, since the synthesis may be foreign to what the synthesized authors ever said or anyone else for that matter ever believed. Second, by relying on secondary sources to interpret primary texts one compounds the danger of misreading the original sources, since one may not only pass on one's own inevitable misinterpretations, but one will pass on the mistakes of the secondary source as well as misread the secondary source itself.

Of course, choosing a single text has problems of its own. For an enterprise such as this one, the goal is to examine the wisdom of different periods (P1, P2, ..., Pn) of intellectual/spiritual history. It is not just to study a single text from that period ($P1_1$, $P1_2$, ... $P1_n$, $P2_1$, $P2_2$, ... $P2_n$, ... Pn_1, Pn_2, ... Pn_m). The text chosen (Pn_m) must be in a significant respect representative of its period (Pn). The criteria for selection include the following factors: (1) The internal clarity and completeness of the text itself, (2) the extent to which the view presented in the text is coherent with other texts of the same period, and (3) the extent to which this text influenced subsequent discussions of (in this case the subject of) creation in later periods of (in this case) Jewish religious thought. Based on these criteria I focused my attention on Plato's *Timaeus* for science in the Hellenistic period, the MIQRAOT GEDOLOT for medieval biblical commentators, Gersonides' *Wars* in the so-called medieval period of Jewish philosophy, and Rosenzweig's *The Star* in the modern period of Jewish theology.

There are only two instances when I have not used this approach. The first was my discussion of midrash in chapter 4, and the second is now in my discussion of contemporary physical science. The reasons for this departure in these two cases are significantly different. The proper text for discussing creation in the midrash, specifically with respect to the quantity of material that deals directly with the subject, is *Genesis Rabbah*. However, this text exhibits little internal consistency. It consists for the most part of a set of one-liners by different authors without any apparent attempt to set any of their judgments in a coherent order. In fact the internal structure of the text dictates that each independent author-statement should be considered on its own. However, in this case there is not a sufficient quantity of material within *Genesis Rabbah* itself to determine what any of the quotes mean. The situation of text reading is further complicated by the fact that the current level of scholarship into this genre of literature cannot make any precise statements about the historical time and place of either the individual rabbis quoted or the editors of their quotations. My first temptation was to ignore this material altogether – not because it is unimportant, but because the time is not yet right in the history of research in this field to attempt to generalize with any degree of probability about what individual texts of midrash mean. However, for one simple reason, I had to resist this temptation. Midrash had a profound influence on the subsequent medieval biblical commentators and Jewish philosophers. Hence, for that reason alone, some attempt had to be made to deal with the literature. In this case, while no historical context could be applied to the individual texts, they could be read within the framework of the full corpus of midrash in general terms. This work had been done by Louis Ginsberg in his *Legends of the Bible*[362] and by George Foote Moore in his *Judaism in the First Centuries of the Common Era*.[363] Hence, without in any way denying the above noted criticism of such an approach, I based my discussion of midrash on these synthesizing secondary texts rather than on specific primary works. All of this is to say that while I believe that my use of this early rabbinic material is the best possible reading of it for my purposes in a work that examines the doctrine of creation out of the sources of Judaism, the reading is far from definitive. It is only the best possible reading for this purpose at this particular time and place. I have no doubt that subsequent scholarship will require a reassessment of my present conclusions about this literature.

The use of secondary, general literature for assessing what contemporary science has to say about creation dictates the same qualifications on my summary in this chapter. However, in this case the necessity of basing judgments on this kind of literature is entirely different. Here there are no problems of authorship, historical setting or sufficient quantities of published statements about the subject by representative scientific authorities. However, this genre of literature has problems of its own for the goal of this enterprise, viz., to judge the truth value of the concept of creation in relation to Jewish tradition. The first problem is the transient nature of conclusions in contemporary science. The second is the dependency of scientific thinking on mathematical models.

Let me begin with an example of the first problem. In 1984, I conducted a conference on creation that brought together a group of leading Jewish philosophers[364] in conversation with a group of leading astrophysicists.[365] The papers finally appeared in print in 1986.[366] From the perspective of the Jewish philosophers, the two-year span between presentation and publication seemed reasonably short, whereas the physicists were somewhat disturbed by the delay. At first the physicists' reaction surprised me, but now I understand it. Statements that in 1984 were at the cutting edge of the field, by 1986 were already somewhat obsolete. In 1984, everyone assumed more or less the kind of standard Big Bang Theory that Steven Weinberg summarized in *The First Three Minutes* (New York, Basic Books, 1977), and people were just beginning to discuss how this model for the origin of our universe relates to the then new "inflationary" universe models of Alan H. Guth, J. Richard Gott and others. Furthermore, the physicists could talk about an initial singularity of infinite density and temperature at zero time without any uneasiness. Similarly, they could discuss the transition stages of the early universe without any reference to quantum mechanics. However, such a discussion would have to be radically different today as I am composing this chapter in 1988, only four years later, particularly in the light of the fact that information has reached the general public both about string theories, (that call into question much that is said about fundamental particles) as well as about Stephen Hawking's speculation on a grand unified theory of the universe (GUT, which calls into question much that is said about singularities in four dimensional space-time).

In and of itself, the recognition that all scholarship in time becomes obsolete is an optimistic prediction. It means that our knowledge improves. However, the rate of change in the physical sciences as compared with the humanities is alarming. Changes that occur in physics in months occur in the humanities in decades. Furthermore, at least some works in the humanities remain classics, i.e., relevant to contemporary discussions irrespective of current developments in the academy, whereas, at least in the physical sciences, development seems to condemn most past speculation to irrelevancy. Hence, I can speak about thinkers such as Rashi, Ibn Ezra, Nachmanides, Sforno, Ibn Daud, Maimonides, Gersonides, Spinoza and even Rosenzweig with confidence that their writings will remain relevant for centuries and my interpretation of their texts can be of value for at least decades. However, in the case of the texts of the relevant contemporary theoretical scientists, there can be no comparable confidence. The speculation of scientists such as Carl D. Anderson, David Bohm, Niels Bohr, Paul Dirac, Murray Gell-Mann, Stephen Hawking, Werner Heisenberg, Edwin Hubble, John von Neumann, Ilya Prigogine, Erwin Schrödinger, John Schwarz, John Wheller, Eugene Wigner, and even Albert Einstein, James Clerk Maxwell or Isaac Newton could lose all relevance to a particular question of science that has bearing on religious belief in as little as a few decades.

I wrote my first draft of this chapter in 1988. In writing it I drew on information I had been gathering since 1981 about contemporary physics. I cannot be sure when this chapter will be in print, but I do know that by the time it appears the scientific cosmology summarized here will be significantly different. It will not help to start the research over, since the problem will continue to hold no matter what I read now. In other words, of necessity the science to be considered will be significantly obsolete in ways I cannot predict as I write about it no matter when I write about it, and in this case, dealing with the thought of a single physicist rather than a general summary of a momentary consensus of contemporary physics is not a solution. Hence, I can have no more certainty that what I say about contemporary physics is true than I could when I summarized early rabbinic commentaries on the Hebrew scriptures.

The second problem is even more serious. Here the issue is the very mode of language appropriate to discussing contemporary scientific models of creation. In general, the pictures that physicists draw to schematize their empirical studies are mathematical and the mathematics cannot adequately be translated into discursive speech. Consider the following example: According to Hawking, string theories tend to presuppose a picture of the universe in anywhere from ten to twenty-six dimensions.[367] Now an n-dimensional universe is not difficult to discuss in the language of linear algebra. Equations with two variables can be pictured on a two-dimensional plane by objects such as rectangles and circles. Similarly, equations with three variables can be pictured in a three-dimensional space by objects such as boxes and spheres. Furthermore, even though it involves some imaginative difficulty, it is possible to picture Einstein's four-dimensional space[368] by projections into planes or three-dimensional geometric space. However, such pictures in drawings or even ordinary descriptive language become mind boggling when the space to be drawn has more than four dimensions. Even the name "string" for the fundamental building blocks of the universe defies speech when we consider the fact that this is an object extending in ten to twenty-six dimensions. However, this is only the bare surface of the problem.

If the discussion of modern physics were limited to interpreting simple algebra, the task would be immensely simpler than it in fact is. As such it would be no harder in principle to give a realistic interpretation of modern physics than it is to discuss the possible claims about reality that Plato's use of Pythagorean geometry entailed for his theory of creation. However, the reality is that modern models for both particle physics and astrophysics presuppose a level of mathematical sophistication far beyond that presupposed by the *Timaeus* – e.g., infinite series, complex numbers, hyperbolic functions, partial differentiation, multiple integrals, vector analysis, Fourier series, coordinate transformations, asymptotic series, and probability series – none of which can be explained to the majority of informed readers who are interested in the religious dimensions of questions like, how was the world created and what does the universe look like. Nor is it the case that the mathematics itself presents an answer to these questions.

Rather, the mathematics are a pure, as yet uninterpreted, formalism that physicists use for applications in the sense world in the form of objects like hydrogen bombs and electron microscopes. The positive as well as negative practical uses in engineering of this realm of rarified speculation and experimentation suggests that the models should say something about the nature of reality, but the tendency of the majority of modern physicists themselves is to leave that kind of work to "lesser minds," viz., philosophers and theologians.

This widespread lack of interest on the part of modern scientists in the significance of their work for "doxis" as opposed to "praxis" is in itself one of the "problems of our times," where intelligent people often mistake the judgment "it works" for "it is true." For example, according to the Heisenberg uncertainty principle,[369] at any instant of time a particle may occupy a determinate position or it may have a definite momentum, but it cannot have both, and in fact the more definite one is the less definite the other is. The most straightforward interpretation of this equation would be that in reality particles, the proposed building blocks of the physical universe, have indefinite positions at indefinite rates of speed, and this judgment is not the result of any lack of knowledge. In other words, it is not the case that we simply cannot determine their definite position and momentum; in fact, they do not have one. Now, given that this interpretation of the mathematical equation is correct, what does that say about reality, since the complex physical entities in our experienced world are constructs from these particles? That this equation "works" for solving problems in physics is not in itself reassuring. The issue is, is the equation true, i.e., does it express what reality is like, and if it does, how, beyond mathematical equations, are we to understand the nature of reality?[370] At this level, physicists tend to lose interest in the question and turn it over to humanists.

Consequently, the discussion of contemporary physics with respect to Jewish doctrine has seemingly insurmountable obstacles unlike those encountered in any other period of Jewish intellectual history. The scientific doctrine is expressed in a set of symbols, viz., mathematical language, whose complexity transcends what can readily be translated into ordinary speech, and is posited by authorities who rapidly (from the perspective of the history of ideas) change their theories and exhibit limited interest in applying their skills to understanding the universe, whereas the people who are most interested in such understanding have insufficient mathematical sophistication to apply the insights of contemporary theoretical physics to their legitimate religious and intellectual concerns. In short, there is a radical separation today between the interests of scientists and religious humanists unlike that ever experienced before in Western history, and there is no easy way to bridge the gap between the two. In this respect, this chapter on the doctrine of creation as it emerges out of current theory in astrophysics is an attempt at forming a bridge. However, "bridge-work" in this case requires a price. No judgments are ever absolutely precise. In this case, a priori, precision is practically impossible on purely linguistic grounds. Similarly, no judgments that purport to give information are ever absolutely certain. In this case, under the best of circumstances they are barely probable. They are only slightly more than reasonably informed, imaginative guesses.

As we have seen, Timaeus called his "myth" of creation a kind of "bastard thinking." It is equally an appropriate way to characterize the following discussion. It is the proper mode of reasoning for what you know to be at the very horizons of your knowledge. It is the best method of expression for what exactly is vague and precisely is uncertain. In short, given the limitations of human language, any theory of creation must be, as it always has been, a myth, viz., a story that pictures all that is in general, that is informed, however inadequately, by everything that we know in particular. It is a mode of thinking as appropriate to the contemporary period as it was to the worlds of the Hebrew scriptures and the classical rabbis.

What follows is a summary of a contemporary myth about creation out of the sources of astronomy and particle physics. The story is primarily about how our universe came about and what it looks like. For practical reasons, the model ignores both string theory in elementary physics and inflationary models in astrophysics. What effect they would have on modifying the myth presented below is at this particular time an open question.

Our universe originates at time zero with the explosion of a single, infinitely small amount of pure energy at an infinite temperature with infinite density. Energy at this level is a kind of amorphous soup in which no distinction can persevere. Diverse parts may interact for an instant, but on account of the great temperature and density of the soup, these potentially diverse components continually annihilate each other. Each annihilation produces radiation. Hence, at this stage of the univere, energy is primarily radiation, and the radiation is in equilibrium.[371]

As this energy disperses through empty space, in time energy becomes less dense and its temperature decreases. At the end of 10^{-41} of a second[372] the temperature of the energy is about 10^{32} K.[373] At this stage the universe is sufficiently cold to end the thermal equilibrium of the gravitational radiation. When the universe has cooled to 1.3×10^{15} K, the weak[374] and the electromagnetic forces[375] also begin to dominate the radiation. These forces enable the energy to divide into diverse, separate conglomerates. These groupings of units of energy that are attracted to each other constitute the different elementary particles.

The distinguishable particles at this stage of the universe include photons (gamma rays), leptons and antileptons,[376] and quarks and antiquarks.[377] While some of these particles have no electric charge, others of them have a charge that is either positive or negative.[378]

At 10^{-9} of a second the temperature of the universe has cooled to 1.5×10^{12} K, at which time the particles become dominated by the fourth and final force, the so-called strong force.[379] At this stage of interaction nuclear particles are formed, including pi mesons, protons and neutrons. By 10^{-2} of a second the temperature of the universe is 10^{11} K, and there is approximately one nuclear particle[380] for every 10^{10} elemen-

Contemporary physics 215

tary particles. By the time that the universe is three minutes and two seconds old, its temperature is 10^9 K, 14 percent of the nuclear particles are neutrons, and 18 percent of the energy in the universe is in the form of nuclear particles. When the temperature of the universe is 4×10^3 K, the energy of matter exceeds the energy of radiation, so that we have what for the first time can be called our physical[381] universe.

When the universe is 7×10^5 years old, its temperature is 1.5×10^3 K and the nuclear particles form into atoms. The radiation from the continuous mutual destruction of particles ceases to be the dominant occupant of the universe. In its place is an increasing number of clumps of matter that come together to form the complex nuclei of atoms. In other words, the universe has become sufficiently cool for attraction to dominate over annihilation, so that nuclear particles can form into composites that become stable atoms, no longer subject to rapid annihilation. Out of this attraction, after more than 10^{10} years have passed since its origin, our present universe of extended bodies composed of molecules that are composed of atoms has evolved.

Exactly what the particle constituents of these atoms look like is itself somewhat mysterious. First, relative to their size, the distances that separate them are enormous. For example, the nucleus of an atom consists of a group of positively and neutrally charged particles, protons and neutrons surrounded by an area(s) occupied by associated negative particles (electrons). The volume of a proton in a nucleus of an atom is 2.1×10^{-45} of a meter cubed. The volume of a typical nucleus, that contains from ten to twenty protons, is 1.1×10^{-43} of a meter cubed. The volume of a typical atom is 1.1×10^{-28} of a meter cubed. That means that a proton occupies less than 2 percent of a typical nucleus.[382] Furthermore, the nucleus itself occupies only 10^{-15}th of the volume of the atom. The consequence is that the vast majority of the space occupied by an atom is empty.[383] For example, given these relative proportions, if a nucleus were the size of a bowling ball (about 1 foot across), the atom would have a radius of 19 miles.[384] Second, particles move in waves in which their position and velocity at any particular moment is probable but never definite. In other words, the physical objects that constitute our universe, whose reality seems to be tied to the apparent fact that they are the kinds of things that occupy specific places at specific times, i.e., things whose reality is inherently definite, are nothing but conglomerates of more basic entities in relatively great isolation from each other, whose real position and momentum are in principle indefinite. Hence, the picture presented above of the development of the universe has to be modified in at least one major respect. The above tale depicts an initially thick soup of energy that evolves into a material universe of separate discrete material entities that occupy distinct space at distinct time. However, in reality the universe seems instead to become increasingly less dense and more discrete but never in fact actually discrete. Being definite in every physical respect seems to be as much an ideal limit of an asymptote as is the origin of the universe itself. In fact, the entire history of the physical universe, from beginning to end, is conceived in ideal rather than actual terms. The limit origin is an infinitely dense, hot and small, amorphous, singular unit. Similarly, its possible end[385] also is a limit. It is an absolutely cold, infinite expanse of real space in which static objects are so remote from each other that it can be said that the universe has ceased to exist.

In general, independent of any allusions to quantum mechanics, it is clear that the model drawn above is far too simple. The general picture of an exploding universe subject to continuous expansion can be pictured in one respect as an expanding balloon. Just as the expansion of every part of the surface of the balloon relative to every other part of the surface can be described through a set of curves, so the increase in relative distance between every galaxy (viz., every discrete aggregate of stars) in the universe can be diagrammed through a set of curves, all of which have their origin at a single point. However, in other respects the balloon metaphor is entirely misleading. The balloon, for example, has interior as well as exterior space, and it is by no means apparent that the physicists' model of the universe presupposes a greater universe that encompasses our own. Furthermore, as a balloon expands, so do the points on the balloon's surface. However, as the space between galaxies expands, the size of the galaxies themselves do not change. Furthermore, what is happening on the surface of the balloon is caused primarily by forces external to that surface, viz., someone or something pushing air or some other gas beneath the balloon's surface. However, whatever forces are pictured to have initiated the expansion of our universe are contained within that universe itself. Furthermore, the balloon analogy presupposes definite spaces in space and definite moments of time that are in themselves discrete from each other. However, the physicists' model of the universe presupposes a single space-time in which time and space are no more independent of each other than length has any physical reality independent of width and/or depth. Properly speaking, the growth of the universe ought not to be pictured as something expanding in space in a time sequence. Rather, the universe itself is expanding in a single real space-time, which in some as yet uninterpreted sense, has a significantly different kind of reality than the things it contains.

The above problems have to do with the attempt to picture what contemporary physicists think about the origin of the universe without using their mathematical tools. In this sense, the problems have to do with the limits of our ability to form non-mathematical images of what experience enables us to infer about how our universe began and what it looks like. However, irrespective of our limitations, there are aspects of the above model that are in themselves perplexing. What, for example, it means to say that our universe began at time zero is not clear. It could mean, the first time in a real sense, i.e., the moment of time before which there were no other moments of time. On the other hand, the "zero" in this case can be nothing more than an arbitrary number which stands for the origin of the process, viz., creation, under discussion. For example, an equation that measures the distance of the flight of an arrow through space can treat the time at which the arrow is fired as time zero, and the place from which it was fired as place zero, without any implication that this is a time before which there is no time or a space before which there is no other space.

However, just what exactly it means to say that the energy from which our universe originates is infinitely small and has infinite density and temperature is vastly more perplexing.

What is "infinitely small" could be nothing, or it could be something that for all practical purposes is nothing.[386] The two interpretations need not amount to the same thing. The same two kinds of interpretation also apply to calling the density and the temperature infinite. The problem is that an infinite number is not the same kind of number as a rational number. The latter designates something definite while the former does not. The adjective "infinite" is easiest to understand when it modifies a process. For example, to say that positive integers[387] are infinitely augmentable means that you can never reach an end to increasing them.[388] Yet, what it means to say that the number of positive integers is infinite is less easy to grasp. Normally, when we posit the number of a given set, we assert something definite that actually exists. For example, the number of members of my family is four, which means that in actuality there are exactly four persons who either exist or have existed that belong to my family. However, this cannot be what it means to say that the number of members of a set is infinite. For example, the number of integers is not any actual group of numbers that anyone at any time in all of history may have counted. That number of numbers would be very large, but it would still be a finite, i.e., definite, number. Now, there is a way to talk about infinite numbers of members of sets. Such numbers are dealt with in the theory of transfinite numbers. However, it is not at all clear how such a theory can apply to the numerical designation of an actual density, an actual temperature, and an actual size of something that existed in the past. In general, as Georg Cantor developed his theory of transfinite numbers, they apply to sets, but sets are not things that occupy space and time. Yet, presumably the initial density and temperature of the universe describes an actual spatial-temporal thing.

It could be argued, as suggested above, that the stuff of the origin of our universe is not real. Given how this singularity is derived, it could be claimed that it is nothing more than an asymptote.[389] We human beings reach our judgments about the development of the physical universe without any direct knowledge of either its beginning or its end. Rather, we observe how the physical objects in our universe are moving relative to each other from somewhere in the middle of the process. Then we form an equation that expresses this observed change, and we apply the equation to both an origin and an end of the process. Since the end, being in the future, does not exist now, there is no serious conceptual difficulty in making assertions about infinity. For example, to say that the universe will become infinitely large need mean nothing more than the claim that there will never be a final time at which the universe would not become larger. However, statements about infinite points in the past cannot be interpreted in the same way unless we give up the notion that time is one directional. In general, we assume that what is in the past was actual and definite, and if that is the case, it is not at all clear what it could mean to say that in the past there existed an actual, definite thing that was infinite. We must refine what we mean by infinity and/or we must change what we ordinarily mean by time. In any case, the physicists themselves have little to say about this critical question of understanding what their model of the origin of the universe means.

In short, the above model, even if it is to be judged as myth, needs further clarification of a sort that lies beyond either our inadequate grasp of the tools of modern mathematics or the issues that physicists who are only physicists consider. In particular, three questions need to be addressed: First, is zero time the beginning of time? Second, in what sense is our universe infinite? Is it one infinity among many infinities, all of which are part of a larger infinity, or are all infinities the same size? Third, is space and/or time and/or space-time something or nothing? Are the horizons of our universe themselves within a larger space that is in some sense absolute or does the term "space" itself have meaning only relative to its occupants? Again, all of these are open questions that either physicists cannot answer or have not answered. Yet, all of these issues are relevant to answering how the universe began and what it looks like.

In this connection it is of interest to ask, can the Jewish concept of creation studied in this book be used to fill in the gaps in the physicists' account of creation, or is it incompatible with contemporary physics? Furthermore, if the two stories are complementary, how would contemporary physics affect a theological reading of Genesis?

Note that there is no issue of one discipline serving as an authority for the other. Nor is one source invoked as a standard for judging the truth of the other. These are modern thought-games that were largely foreign to the classical commentators and Jewish philosophers studied in this book, and they are equally foreign to the approach used here to do philosophic theology.

Consider a series of pictures of a single event, drawn from a number of different perspectives. There need be no question of which pictures are true to their object. None or any of them may be, and this judgment is quite independent of comparing the different pictures. On the contrary, the value of each stands or falls on its own. Yet, there is still value in looking at all of them together. Given, that they are all, in some significant sense, "true" to their shared object, they can together constitute a richer comprehension of the event, each capable of filling in gaps in the other's description. This is all the more the case when each picture uses a different kind of medium with a different set of rules. For example, when one artist is a poet while another is a sculptor, or, similarly, when one composer is a romantic while another is a minimalist, even though they all use their talent to represent a single event, they cannot be judged relative to each other. One poem may be better than another, but it is meaningless to say that a poem is better or worse than a statue. Furthermore, it is equally meaningless to judge a minimalist piece of music by the rules of composition of a romantic symphony. In a parallel way, the insights of geniuses such as the authors of Genesis, *Genesis Rabbah*, *The Timaeus*, *The Exalted Faith*, *The Guide of the Perplexed*, *The Wars of the Lord* and *The Star of Redemption* ought not to be viewed as competitive. Knowing is not that kind of game.

The classical rabbinic commentators read Plato and scripture not to decide who was better, but to know what is true. Similarly, we can look at the radically different perspectives of classical theologians, philosophers and contemporary theoretical physicists to help us form our own best judgments about "what is true" without any primary concern for whose vision of reality is better. We look to all of their judgments because we believe that all of them have something to teach, that what they teach is significantly different, and that our own comparison of them can yield a richer approximation of a true answer than would be the case if we limited our search to only one source.

This concludes the summary of the current positions in physics on the origin of the universe. Now we will examine whether or not the views of contemporary science and Jewish philosophy correlate. We will do so by investigating how (if at all) the current consensus in theoretical physics can inform us about the meaning of Genesis. Can scripture be read, without distortion, to be coherent with these findings? Do they agree or disagree with the way the rabbis interpreted the text? Do they suggest a better (i.e., more coherent with the actual words of Genesis and/or a reading that is more likely to be true) schema for understanding scripture than the picture Plato provided in the *Timaeus*? The above summary suggests a minimum of four aspects of the Jewish concept of creation that interface with contemporary scientific cosmology. They have to do with (1) the nature of the origin of the universe in relation to time, (2) the role of space in the story of creation, (3) the relation of the physical universe that creation produces to other possible universes, and (4) the relationship between the domains of science and ethics from a religious perspective. Each of these aspects of creation bears directly on how we should interpret specific texts in the Genesis narrative. These are (respectively) that (1) "in the beginning," God created (2) "out of nothing" (3) "heaven and earth," and (4) "it was good." The question of the correlation between the Jewish and the contemporary scientific stories of creation will be examined from the perspective of the biblical text.

IN THE BEGINNING

Rosenzweig claimed that creation is an atemporal model for the origin of the universe. It is not itself part of actual human or physical history. The present universe is to be understood as a set of interrelated processes that are best understood as movements in space and time. These motions point towards both an origin and an end. However, both extremes are endlessly remote, i.e., they are asymptotes. Hence, as the end in redemption is an ideal, so the origin in creation also is an ideal.

What does modern science say about this analysis? Is the beginning of the universe an event in time? There are different ways to answer this question depending on what is meant by "time." The minimal answer is mathematical. Time is one of the dimensions in geometric space which corresponds to one of the variables in equations that describe motion in the physical universe. Assuming that these equations say something about reality, then time is necessarily interconnected with, and in principle no different than, any other dimension in which the inhabitants are located, be the dimensions four or twenty-six. However, there does seem to be one important difference between time and any other dimension, viz., everything in it moves in only one direction. With respect to length, width and height, things can move positively and negatively, i.e., they can become larger or smaller. However, our current common sense would suggest that time can only be positive, i.e., things get older but not younger, i.e., things move towards the future but not towards the past. There is at least one interpretation of quantum mechanics, viz., that of John Wheeler, that would recommend that this so-called common sense is not sensible. His interpretation arises as one solution to the so-called measurement problem.

It is generally acknowledged that light is made up of photons in a way that is analogous to bodies of water being made up of drops. As water moves in waves, so individual photons oscillate as their conjunction forms waves of light. Furthermore, as waves of water that begin at different points adjacent to each other and/or at different rates of speed come into contact with each other and produce a distinct pattern identified as "interference," so different waves of light in the same way exhibit an interference pattern. This in fact is what happens when we perform the so-called "interference experiment."[390] What is surprising is that the same pattern will emerge even if the photons are emitted one at a time in such a way that they cannot have come into contact with each other. At this point the analogy between waves of water drops and light waves breaks down. Presumably the water waves formed interference patterns because the waves intersected with each other. However, this is not the case with light. Their pattern is independent of any contact. When the experiment is conducted, the photons will exhibit interference even when no individual photon could have come into contact with another photon. In this case we would have expected the photons to exhibit a radically different effect as they were detected, namely, the same kind of pattern that bullets would exhibit when they are fired at random in rapid succession from a gun towards an object with a single slit, towards an object with two other slits, towards a target-wall. In fact this is the pattern that photons do exhibit in the so-called photoelectric effect.[391] In this case light acts like a stream of particles rather than a wave. Furthermore, if in addition we measure which slit the photons pass through on the second screen, then, when they hit the third screen, they exhibit the pattern of particles rather than a wave. It would seem that the mere act of measuring the exact position of the photons in and of itself changes the way that they behave. If they are measured, then they behave like particles, but if they are not measured, then they behave like waves.

A solution to this seemingly surprising result is suggested by the Heisenberg uncertainty principle. The precise determination of the momentum and the location of any particle is inversely related to each other by a fixed constant [viz., $h/(2\pi)$]. The consequence of the location becoming precise, i.e., without any uncertainty, maximizes the uncertainty of the momentum of the particle, so that a specific determination of the particle's location as it passes through the second slit radically changes when it will strike the third screen. However, the same effect occurs on the third screen even if the measurement of the photon's location on the second screen is not determined until after the experiment has been concluded. It is this surprising result that gives rise to the by-now-infamous paradox of Schrödinger's cat.

As a light wave moves through a field, it vibrates like a wave at every angle to the plane along which it is moving. However, the wave can be sent through an electric field where the light's vibration becomes polarized at a fixed angle, in which case the oscillation itself is positive and negative along a single line, e.g., up and down along a line that is perpendicular to the light's direction forward.[392] Suppose we now pass the light through a box with two different exits. In one (A) the only light emitted is perpendicular to the direction forward and in the other (B) the only light emitted is horizontal to the direction forward. There is an equal possibility for any individual photon to pass through either A or B, so that it can be expected that about half of the photons emitted will pass through each channel. Now, suppose we attach a box to one of the channels. The box contains a live cat, a photon detector, and a vial of poison. The vial and the detector are connected in such a way that if the detector measures a photon, it will release a switch that will break the vial, release the poison, and thereby kill the cat. Now, we direct towards the series of boxes a single photon. The odds are fifty-fifty that the cat will live or the cat will die. We will not know which is the case until we open the box, i.e., measure what was the position of the photon. If it is true that the act of measurement itself affects what was the position of the photon, even if the measurement occurs after the event, then the earlier event itself was altered by the measurement. Presumably before the measurement was made, the cat was neither alive nor dead. Rather it had an equal possibility either way. However, the measurement itself determines what, with precision, was the fate of the cat, not only now when the measurement is made but in the past as well. In other words, it would seem that a simple present act of measurement can alter past facts.

What ultimately seems absurd about this consequence is the presupposition that time can only have one direction. However, if that is not the case and time as a dimension is no different than any other dimension, then the cat paradox is not at all paradoxical. In this case the otherwise radical disparity between the past and the future is dissolved. Past and future are directions beyond the horizon of an experienced present. Both are nothing except a limit on a process of lived life in the now. As such, both the beginning and the end of absolutely everything are something ideal rather than actual.[393]

It is in this sense that the term "at first" (BERESHIT) has been understood in all of the cosmologies considered in this book. In all probability the authors of the biblical text had no time reference for the story of creation at all. "At first" simply marks the beginning of the story. The sages of *Genesis Rabbah* speak of time, but the time of this world is not the only time. The world created is part of a larger universe that includes some things that are eternal and other created entities who existed prior to the creation of Genesis. The medieval commentators speak of the moment of creation as a first moment in the history of this world. However, that moment is itself part of God's creation. Hence, since time is created, the act of creation itself does not occur in the time of this world. If creation in fact takes place within time, then it is an entirely different notion of time than our conception of a unidirectional plane in which physical objects and events are located. The clearest assertion that the story of creation is not a temporal event is found in the philosophy of Gersonides. That God is one entails that God has a single act with which God himself is identical. Since God is eternal, i.e., not subject to time, so creation is eternal. Rosenzweig, for his part, locates creation in time, but once again the time is not the time of our physical world. It is not a time that is divisible into moments. Rather, it is a pure, otherwise indistinguishable past, in which pre-existent elements – God and world – come out of their isolation into a relationship that makes the lived present of revelation secure. The initial universe from which ours originates is not our universe. It is a pre-existent, chaotic mass which admits neither distinction nor intelligibility, including the minimal order and structure of measurement that time itself makes possible. In short, the singularity from which the universe originates is the origin of time, but, as such, it is not itself within time.

OUT OF NOTHING

Both Genesis and Timaeus claimed that the sense in which the universe was created out of nothing is that the universe originates as a pure undifferentiated space. It is space that is the first principle of dynamic existence, and space (as space) is not anything at all. Rather, it is a purely negative ideal. In other words, the dynamic universe is to be understood as a set of interrelated processes whose direction is from an origin in nothing to an end in something. As the end is thoroughly something (viz., God), so the origin is thoroughly nothing (viz., space). What is in between these ends (viz., the empirical world) is a correlation to the two ideal limits.

What does modern science say about this analysis? In what sense is the universe created out of nothing? There are different ways to answer this question as well, depending on what is meant by "nothing." If it is what is not a "thing," then, again, there are different ways to answer this question, depending on what is meant by a "thing." If things are material entities that have mass, then photons are not things. If the term "thing" encompasses all particles, then the pure energy that existed at the origin of our universe is not a thing. In both of these senses it follows that the world was created out of nothing.

The classical rabbinical commentators also posited that the nothing out of which God created the universe was nothing only in the sense that it was not a thing. All of them identify at least the biblical "TOHU" and "BOHU" in this kind of role. Both the midrash and Rashi add the referents of the biblical term, "TEHOM," the element fire, wisdom, and the throne of Glory. Similarly, Nachmanides adds the biblical terms for earth and sky to his list of pre-existent no-things that are the stuff of cosmic creativity. However, nowhere is the parallel between classical rabbinic and modern cosmogony clearer than it is in the commentary of Gersonides, who affirmed that to be something is to be in some sense intelligible, but the initial material out of which God formed the universe was utterly unformed, unintelligible. Clearly, Gersonides' "first matter" (which he identified with scripture's BOHU) and his "body that does not preserve its shape" (which he identified with the MAYIM) function in his model of the early universe as pure energy and photons respectively function in the standard Big-Bang picture of our universe's first three minutes.

It is interesting to note from this perspective, that the affirmation of creation out of nothing in this modern scientific theory is in fact more radical than it is found in classical Jewish thought. While the rabbis affirmed creation out of nothing as creation out of no-thing, there is little evidence to suggest that by "nothing" they meant anything more extreme than this. The stuff from which God informed our universe was a positive entity. Yet, in contrast, some contemporary cosmologists will affirm that beyond the horizon of the present expanding universe, there is in fact nothing, not even space. Furthermore, many contemporary cosmologists will speak of the original so-called singularity from which our universe arose as something actual (i.e., something not just ideal) that was at a first moment in the past a real nothing, viz., something that occupied zero space and had infinite temperature and density. No major Jewish philosopher before Rosenzweig meant anything so extreme by his claim that the universe arose from nothing.

HEAVEN AND EARTH

The fact that Genesis posits a number of negative entities whose creation preexists the story of creation (viz., TOHU, BOHU and the dark) suggests that our universe is not the first one. Furthermore, the fact that the rabbinic commentaries on Genesis suggest that a number of things created in this universe exist solely to be used in a subsequent universe (viz., the light of the first day, the upper waters of the second day, and the rest on the seventh day) suggests that our universe will not be the last one.

What does modern science say about this analysis? Is our universe the only universe? There are different ways to answer this question as well, depending on what is meant by "universe." If what we mean by this term is "everything there is," then there is only one. However, this is not the only way that the term is used. Sometimes it functions as a synonym for a domain. In this sense a universe is everything subject to the same set of rules. So, for example, when we speak about a "universe of discourse" or even "the physical universe," we tend to use the term in this more restricted sense. Medieval Muslim, Jewish and Christian philosophers called the different celestial spheres different universes precisely because they were regions subject to different sets of natural laws. It is in this more restricted use of the term, "universe," that the above question becomes interesting.

Specifically with reference to contemporary astrophysics, the different versions of what are now called collectively "inflationary universe models" all claim that our universe is only one among many. Imagine the "Big Bang" universe described above as a single bubble in a caldron of bubbles. The caldron is "the" universe, while each bubble is a single universe, one of which is our own. Each of these universes may or may not be subject to different natural laws. In either case that is not what makes them separate universes. They are separate because they do not interact with each other. In other words, in this sense, a universe is a continuous spatial-temporal domain that includes everything that could possibly interact with each other.[394] In this case the limit on interaction, given that these are all "Big-Bang" universes, is the assumption that no signal travels faster than the speed of light. As a universe expands from its originating explosion, no thing can travel faster than the photons. Hence, the motion of the photons sets the outer-limits, i.e., the horizon, of a universe.

At least one conceptual advantage of this view over the simpler single Big-Bang model is that it provides a sensible answer to the question: What is there beyond our presently finite universe? On the simple Big-Bang picture given above, the answer would be – nothing, not even space and time. The inflationary modification of the picture changes the answer to a more comforting (with respect to current common sense) answer – viz., other infinitely expanding universes, all of which are infinitely distant from each other. Furthermore, it enables us to say that the first moment of our universe is in fact an actual moment in time that is not a time before which there are no other times. However, this picture will not solve any of the other perplexities that we discussed above in connection with time. It does not make any greater sense than we already have out of what it means to speak about actual past and present infinities, and it in no way solves any of the measurement perplexities that emerge from quantum mechanics.

There is another sense of universe that I want to consider, based on Leibniz's analysis of the meaning of modal predicates. Assume that every fact or state of affairs that ever has happened, is happening, or will happen anywhere can be expressed through an endlessly long set of simple declarative sentences. For example, this extremely lengthy conjunction would include the following elements: "At t_n (i.e., at the moment that I am writing this sentence), Norbert Samuelson's right leg is crossed over his left leg, and Norbert Samuelson is wearing a blue sweatshirt, and Norbert Samuelson is in his study, and . . ." The conjunction would contain an extremely large (possibly endless) number of elements that describe any particular moment, as well as an extremely large (possibly endless) number of sets of elements about different moments in the actual history of our universe. Let us call this sentence an accurate statement of our actual universe. Now repeat this sentence, change a single element in it to one of its complements, and then change every other element that is logically inconsistent or incompatible with this one new element. For example, suppose we make Samuelson's sweatshirt red, or green, or white or any other color that is not blue. Presumably for every (or at least most) elements in the original sentence there are an extremely large (possibly endless) number of complementary elements. Let us call each of these a possible universe. Presumably the number of such possible universes is infinite, i.e., there would be no end to listing them. Now, current common sense would suggest that, of this infinite number of logically possible universes, one and only one of them is the actual universe. However, paradoxes that arise from the measurement problem could suggest, as it did to Hugh Everett, that this common sense is not sensible.

Consider again the Schrödinger's cat paradox. What was perplexing was the conclusion that after the photon did or did not pass through the channel connected to the detector, before the measurement was observed, the cat was neither definitely alive nor definitely dead, but was instead possibly alive and possibly dead, with a .5 probability in both cases. Everett proposed the following solution to the seemingly disturbing conclusion that in reality[395] things can be probably one way or another, without actually being either way. Both possibilities exist in actuality; only they exist in different universes. In other words, in one universe the cat is alive, and in another universe the cat is dead. We do not know in which of these two universes we live until we observe the measurement. However, we also exist in both of these universes. In one of them we examine the box and find the cat to be dead, while in the other one we find the cat to be alive. Hence, for every possible position of a particle at any given moment, and for every possible momentum for a particle at any given position, there is a different actual universe.

Everett's interpretation of quantum mechanics is called the "parallel universe" theory. Consider a universe limited to three coins that are independently flipped in a single period of time, each having an equal probability of landing as heads (H) or tails (T). In this case there are eight equally possible universes,[396] viz., HHH, HHT, HTH, HTT, THH, THT, TTH, TTT. By analogy, in the parallel universe theory, each of these eight possible universes exists simultaneously.

In fact *Genesis Rabbah* presented us with a similar model of cosmology. The sages of the midrash described a region of the universe called "The Great Ocean" in which there are an endless number of islands whose inhabitants are all of the possible species of entities that do not exist in our world. In other words the midrash also presented us with a picture of a universe composed of parallel universes.

In the case of the sages, the motive for presenting such a model was theological rather than physical. They did not give reasons for what they said, but the motive is not difficult to deduce. If God is perfect, then there is nothing possible that he cannot do. Furthermore, given that he is perfect, there is nothing that he can do that he does not do. Consequently, if multiple universes are possible, then God must create all of them.

From this theological perspective, there could only be two reasons why God would create only one universe. The first is that there is in reality only one possible universe, i.e., the actual universe is in every respect a necessary world. This was the alternative that most medieval Jewish (and Muslim) philosophers adopted. The second is that this is the best possible universe. This is the alternative that most modern Jewish (and Christian) philosophers adopted.

According to the first reason, everything that was, is, or will be exists necessarily either in virtue of itself (as in the case of God) or in virtue of a cause (as in the case of everything other than God). Some events appear to be chance occurrences only because we human beings do not have sufficient knowledge to grasp the total chain of interlinked events in the past that necessitated what exists in the present. Human beings have options or choices in the sense that in some instances an act of self-determination is the proximate cause of what an agent does. However, the agent's volition also is necessitated by prior causes. Spinoza also adopted this position, and in doing so he followed in the footsteps of his teachers – Maimonides, Gersonides and Chasdai Crescas.

Yet, quantum mechanics calls this first line of traditional Jewish philosophy into question. If the so-called "Copenhagen Interpretation" of quantum mechanics, advocated by Niels Bohr, is correct, then everything ultimately is indeterminate, because it is the nature of the most fundamental units of the universe to be themselves indeterminate. If then there is only one universe and God created it, he did so not for the first reason (viz., that he had to do so), but for the second reason. God desired (i.e., chose) to do so. Given that God is good, his choice would not have been arbitrary. Rather, in the words of Leibniz, he chose this universe because it is the best possible one. In other words, the reason why this universe exists is because it is, in some as yet unexplained sense, "good." This conclusion leads us into the final (and most important) aspect of the Genesis account of creation that we will consider in correlation with contemporary physics.

IT WAS GOOD

Genesis presents a negative ideal as the origin of the universe that hints at a positive ideal of its end. These two ideals function as models to explain the dynamic process of the dynamic universe itself. This movement is understood in moral terms. The universe moves from being indefinite to becoming definite, from being undifferentiated to being differentiated, from being negative to becoming positive, and from being evil to becoming good. All of these sets of terms are correlated. Being indefinite, undifferentiated, and negative are bad; similarly, being definite, differentiated and positive are good. Hence, the physics and the ethics of the Torah are inseparable. Furthermore, both Genesis and the *Timaeus* picture the moral ideal that renders intelligible the physical world in political and religious terms. In the latter case, the moral ideal is the polity described in the *Republic*. This national ideal functions microcosmically as a moral imperative for how all individuals should govern their lives and macrocosmically for the structure of the universe itself. In the former case the moral ideal is the polity whose laws are spelled out in the body of the Torah. This theocratic ideal functions microcosmically to govern the lives of every individual and macrocosmically as the TELOS of the universe itself. Here the universe exists so that Israel may exist so that its priesthood may offer its sacrifices to God in accordance with his revealed word in the Torah. In other words, both scripture and Plato present a scientific story whose principles of intelligibility apply to absolutely everything – ethics as well as physics, political theory as well as biology, etc.

What does modern science say about this analysis? At best it has nothing to say about it; at worst it judges it to be scientifically absurd. In other words, in terms of relationship between science and ethics, there is a radical disparity between the two pictures of the universe under consideration.

While the emphasis in this chapter has been synthetic, the intention has not been to deny that there are significant differences between all of these pictures. The single most important way that the model of the physicists conflicts with the other pictures of the universe discussed in this book is that the category of ethics has no relevance to modern physics. The claim of modern physics is in fact stronger than the oft-time expressed statement that modern science is morally neutral. The truth is that modern science takes a definite stance on morality. It is that knowledge is morally neutral. All of the sources previously considered asserted that in some sense the universe is good, and that to understand this judgment is an essential part of understanding the universe. In contrast, all of modern physical science makes a radical separation between determining what is true and judging what is good. In other words, the biblical statement that creation "is good" (TOV) cannot in principle have an interpretation from the perspective of astrophysics.

In part the issue has to do with whether or not the universe is intelligible if we assume that what happens occurs either by chance and/or exclusively through mechanical causes. God's nature and will are not admissible terms in the schemata of contemporary science. Most discussions of the relationship between science and religion today seem to come down to this issue. Certainly every Jewish account of creation posits God in the central role of creator. However, in the light of the major texts of classical Jewish cosmology and the current theories in physical science, this dispute appears to be less critical than one would suppose.

On the scientific side, statements about forces and the employment of mathematical constants and equations seem to have more to do with finding ways to organize observed data than to assign causes. The more mathematical theoretical science appears to be, the less conceptualization in terms of causes seems relevant. Consider the following example: Given the sequence of terms $(\frac{1}{2})$, $(\frac{2}{3})$, $(\frac{3}{4})$, $(\frac{4}{5})$, . . ., we may be able to deduce that its defining rule is $[1 - (1/n)]$, with the index n increasing in each term of the sequence from 2 to infinity. The defining rule makes the sequence intelligible in the sense that it enables one to explain by a single formula each term that has appeared and each term that will appear. Now, it so happens that in this case I thought first of the defining rule and then used it to generate the sequence. Here I am the cause of the sequence, and the defining rule is what I thought. However, the defining rule is not a cause; I am the cause. Furthermore, it need not be the case that if a sequence has a defining rule, there is any cause. It could be the case that a mere random sequence of numbers also would have, by chance, a defining rule. Now, the laws of physics are more like defining rules of sequences than they are like causes.

It is in this sense that the current discussion of entropy in relationship to the so-called Game of Life has particular interest. The second law of thermodynamics states that any process that changes a system will increase the quantity of its entropy (or, disorder). In other words, whatever order there is in an independent system, its increased disorder is irreversible. Increased order in any system must be taken from outside of the system, but the total balance between increased order in one system and increased disorder in the other system will always tip in favor of disorder, i.e., the joint system will have undergone an increase in disorder. Now, the ultimate independent system is the universe itself. Hence, the second law of thermodynamics suggests that whatever order there was in the universe had to be there from the beginning. With the passage of time the disorder will irreversibly increase.[397]

Given that what happens in the universe happens by chance, we can picture what this law of entropy means with the space modeled in Fig. 9.[398] Consider a square divided into 25 equal internal squares with two particles free to move at random anywhere within this rectangle. Label the columns 1 through 5, and the rows a through e.

Contemporary physics

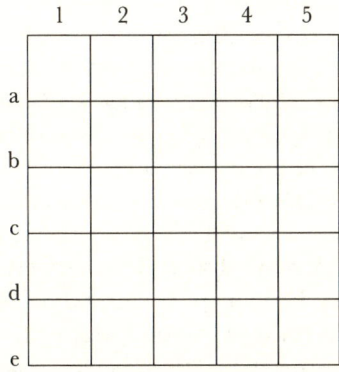

Fig. 9. A space model to illustrate entropy

Assume that you have order when the two particles are next to each, but disorder in every other case. Locate the first particle at 1a. The probability that there is order in this case, viz., that the second particle occupies the internal squares 2a, 1b or 2b, is $(\frac{3}{25})$, while the probability that there is disorder is $(\frac{22}{25})$. Now calculate the probabilities of order and disorder for every other position within the square, and your conclusion will be that the probability of order is $(\frac{144}{625})$ or about .23, while the probability of disorder is $(\frac{481}{625})$ or about .77. In other words, within this relatively confined space, the probability of disorder over order is roughly 3 to 1. Remember that in a typical atom, the "empty" space between a nucleus and an electron is so large that an electron would be nineteen miles away from a nucleus the size of bowling ball. In other words, given the few subatomic particles in the universe relative to all the spaces that they can occupy, the improbability of any set of two or more particles coming together to form an ordered whole is extremely high.[399] Hence, given that order occurs by chance, it is extremely unlikely that there would be any order at all. In fact, the existence of our present universe is so unlikely, that the assumption that it merely happens to exist borders on being unreasonable.

This argument is in fact one kind of version of a modern argument from design for the existence of God. It is a basis for claiming that the existence of any order whatsoever, let alone organic life, let alone intelligent life, is such an enormous blessing, that any amount of evil from a human perspective pales by comparison. However, it may not be as unreasonable as it would seem for there to arise an ordered universe by purely mechanical, chance principles. One possible reply to the above argument from design is what is called the "anthropic principle." There are at least two versions of the principle (a strong and a weak one). The difference between them need not concern us here. For our purposes, what both state is that the reason that we know that our highly unlikely universe exists is that if any other kind of universe existed, there would not be anyone around to know about it. This response is most compatible with the many worlds interpretation considered above. In this case the anthropic principle states that every possible world exists, but we are only conscious of those few possible ones that are compatible with our own existence. Hence, what is in reality very rare appears to us, from our perspective, to be disproportionately likely.

There is a second reply to the above version of an argument from design that is less theoretical and more experimental. It involves using a computer program called the Game of Life. The game is played in the following way: You are given a screen divided into a matrix of squares of equal size. Each square is either occupied or unoccupied. If it is occupied, then it is said to contain a life form. You begin the game with an arbitrary number of life forms organized in any particular way that you want. The game has the following two rules: (1) If a life form has more than 3 or less than 2 life forms in the eight spaces immediately adjacent to it, then in the next generation (i.e., the next configuration of life forms on your screen), the previous life form dies. In other words, an occupied space surrounded by zero, one, four or more occupied spaces will become unoccupied in the next picture on your screen. On the other hand, if exactly two or three life forms are adjacent to a given life form, then the life form in question will survive or persist into the next generation. (2) If a space is empty and there are exactly three life forms next to that space (i.e., three occupied spaces out of the eight adjacent spaces), then in the next generation that formerly empty space will be occupied by a newly born life form.[400] In this way, operating solely with these two mechanical rules, where the only variable is the initial configuration of life forms (which, after all, can be selected at random), many life forms will persist and even undergo a process of growth and decay in which some complex forms give birth to entirely new life forms that undergo their own generation and corruption. In other words, what the Game of Life illustrates is that it is not unimaginable that a universe of living things very much like our own can be generated by chance and purely mechanical-mathematical rules.

On the Jewish side as well, the popular dispute over the role of God in creation is less critical than it would appear to be on the surface. Statements about God are not as different from scientific non-theistic claims as they appear to be at first glance. With respect to classical Jewish philosophy, a consequence of God's unity is that God has a single act, and there is no distinction between God as actor and his activity. Hence, the judgment is not so much that God creates but that the act of creation and the creator are one and the same thing. Consequently, to understand creation – i.e., from and to where everything is moving and why – is the same thing as understanding God. Another way to say the same thing is that the formula(s) that make intelligible the physical universe are not "in" the mind of God; they are the mind of God. Presumably this is what it means to say that God knows only himself and in knowing himself he knows everything in the universe. In this context, Einstein's belief that to discern the laws of nature is to discover how God thinks is essentially a classical Jewish philosophical stance. This is all the more apparent in Maimonides' case where all that we know about God is what he does, but any inference from what he does to what he is is invalid. However, more to the point is Gersonides' analysis of divine attributes, in which what we know about God ultimately is expressible in moral rather than descriptive language.

With specific reference to creation, Gersonides' claim is that the ideal limits that determine, either as a first or as a final cause, everything that occurs in the world of generation and corruption are in fact what we do and can know about God. Hence, it is not knowledge of God that informs us about creation; rather, it is knowledge of creation that informs us about God.

It is at this level that the major issue between classical-Jewish and contemporary-scientific cosmologies is most apparent. For the former, but not for the latter, statements of scientific laws ultimately are ideal-limit claims and not descriptive generalizations. In this respect Rosenzweig was true to his teacher Hermann Cohen, whose own teachers in turn included the classical rabbinic philosophers. Rosenzweig posited the element God as the limit of the asymptotic function of creation. It is God who is "the" good, and the world is characterized by an infinite number of unique nothings striving to become something by endlessly moving in the direction of the divine element. In one sense it would be more accurate to say, not that the world is good but, that it is becoming good. The only sense in which "is" is appropriate is that only the ideal in Rosenzweig's ontology is positive, is "something." Our real physical world is primarily nothing. In this sense, to use Trefil's example given above, the 19 miles of empty space between the 1 foot in diameter nucleus and its related electron says more about the physical world than do these two positive, physical somethings. In other words – in agreement with the ontologies of both Genesis and the *Timaeus* – space is prior ontologically to both spatially located objects and events. In other words – in the language of contemporary physics – the different kinds of fields to which particles are subject have ontological priority over the particles themselves. In other words, it is the field that determines how particles behave and not particles that define their fields.

Note that the sense in which it is here claimed that the universe is good is radically different from any sense of this moral term as it functions in any current version of the problem of theodicy. The "good" spoken of here is a good appropriate to the mind of a deity whose laws will turn our sun into a red giant, at which point all that remains of inorganic and organic life (including human beings), will evaporate. It is also a "good" appropriate to a mind whose laws of thermodynamics make no distinction between human saints and sinners. In this connection, consider the example above of the sequence whose defining rule is $[1 - (1/n)]$. That sequence also has a limit. As n becomes larger, $(1/n)$ increasingly comes closer to being 0, in which case the rule increasingly comes closer to having the value of 1. Here "1" is the limit of the sequence. It is that number which the sequence endlessly approximates but never exceeds.

It is in the sense described above that Jewish philosophy, both classical and modern, says that the universe has value. The present universe is expressible through a set of equations whose infinitely remote origin and end is expressed by a limit, and that limit, in both cases, is God. In other words, the "good" of the universe is a divine good, and a divine good (as Maimonides emphasized), need not be (and probably is not) a human good. In other words, the purpose of the universe is for the good of God and not ultimately for the good of any of its inhabitants.

What this purpose(s) is is not itself part of the doctrine of creation. Creation holds out the promise that the universe is good, but does not explain the goodness. The moral value of the universe is a function of its limit-end, not its limit-origin, and the content of the end is the doctrine of redemption. Hence, the consistent Jewish conception of creation, in opposition to contemporary physics, is that a most rational cosmology connects creation essentially to a doctrine of redemption, and redemption deals not with what was or is, but with what will be. Rosenzweig understood creation in these terms. So does Jewish philosophy and theology.

In conclusion, the classical understanding of the Jewish dogma of creation is coherent with contemporary scientific cosmology in most respects. In fact, in many ways it provides a far better schematic fit for interpreting the text of Genesis than does the *Timaeus*. There is, however, one critical difference. The ways that contemporary astrophysics interprets its empirical data stands at that extreme end of the pendulum of scientific interpretation that favors chance and necessity over purpose and design. While the Jewish dogma of creation admits the validity of a mechanistic, quantitative conception of the universe, it insists that such a schema is never more than a partial account. Reality is (at least) as much qualitative as it is quantitative. The universe cannot be understood solely in terms of moral principles; but neither can it be adequately comprehended by amoral, mechanical ones. Hence, while contemporary astrophysics provides a far better mechanism than does Plato's *Timaeus* to understand the origin of the universe (viz., the sense in which the universe is primarily a nothing whose origin is in nothing), it fails to capture the sense in which this nothing is a motion towards something. In other words, the universe originates in the nothing of creation as a process towards the something of redemption, and redemption is an ethical utopian concept that as such falls totally beyond the scope of modern science. Contemporary physical cosmology provides us with the best tools yet developed for understanding creation, but the price to be paid for this advantage is that there is no way to understand redemption. In this sense the *Timaeus* remains schematically superior.

However, is it true that the universe has an end? Is it in fact the case that there are moral ideals towards which the events in our universe point with as much (or, as little) certainty as they point to an ideal origin in nothing? In other words, is there a moral dimension to the universe, as the Jewish dogma of creation affirms, or, are the only principles that guide the physical universe principles of mechanical chance, as the mathematical formalisms of contemporary physics implicitly entail? This question transcends the domain of contemporary physical science and brings us into the range of contemporary philosophy of religion. In the next, and final, chapter we will correlate the Jewish concept of creation with relevant positions in contemporary philosophy of religion. The critical question that we have raised here – viz., is the universe moral? – will conclude both that chapter and this book.

CHAPTER 8

Creation from the perspective of contemporary philosophy

Our study of the doctrine of creation as it arises out of the textual sources of Judaism is now complete. We began by explicating the doctrine in the single most important text of contemporary Jewish philosophy to deal with creation – Rosenzweig's *Star of Redemption*. Next we discussed whether or not this view could be called Jewish. The answer was a demonstration that Rosenzweig's concept is coherent with the general discussion of this concept in the relevant classical rabbinic texts – viz., the works of medieval Jewish philosophy and the rabbinic philosophical commentaries on the first chapter of Genesis. Here, rather than presenting a summary of all of the literature that (because of its general character) would lose the richness of the details of each individual religious philosopher and theologian, we chose specific texts upon which to focus as representative of the best of Jewish thought in each period of Jewish intellectual history. This approach led us to examine the creation concepts of Ibn Daud, Gersonides, Rashi, Ibn Ezra, Nachmanides and Sforno. Our conclusion was that Rosenzweig's concept of creation is in fact coherent with the major writings of these central rabbinic voices. Furthermore, their accounts enabled us to spell out in richer detail what Rosenzweig's own words tell us about creation. This study exhibited the dependence of the Jewish concept on the application of a philosophic-scientific schema to render intelligible the words of the Hebrew scriptures, and we discovered that the primary text that traditional Judaism used for this purpose was Plato's *Timaeus*. That discovery led us to examine how good a fit the cosmology and cosmogony of Timaeus was for reading the narrative of Genesis. In other words, we turned our attention to the question: Is the model of the *Timaeus* a useful philosophic tool for understanding what scripture says about the origin of the universe? Our conclusion was a qualified yes, viz., of all the philosophical-scientific systems available for the classical rabbis to use, that of Timaeus was the best. This qualification served as a transition to the final question about Rosenzweig's account of creation, viz., can we reasonably conclude that it is true? Or, more precisely, we asked, using Rosenzweig's concept to guide us (in the light of how that concept fits into the classical Jewish understanding of the question), what can we judge to be the case about the origin of the universe.

The answer to this final question turned on two methodological premises that emerged as conclusions from our historical survey of Rosenzweig's Jewish sources. First, to give an answer to this question is to tell a story. Questions like these (probably like most questions of religious interest) are not subject to precise explanation or strict demonstration, for they are neither simple empirical reports nor analytic-mathematical deductions. In other words, religious concepts like creation are neither (in Kant's terms) synthetic a posteriori nor analytic a priori judgments. They are analytic as well as synthetic and a priori as well as a posteriori. There is no precise way to separate out these elements in the concept and it is by no means clear that such a separation would be informative.[401] Second, the story should be informed by the relevant findings in contemporary science as well as by relevant traditional religious texts. Neither scientific evidence nor revealed scriptures in themselves are sufficient to fulfill the religious duty to know the origin of the universe, and the same is probably true of most religious dogmas of belief. On final analysis, the issue is not, is revelation to be judged by the canons of reason or reason by the dogmas of revelation. Rather, each is to be pursued on their own terms by their own standards of evidence, but neither independently is sufficient to do the best that one can to attain wisdom. The "best" in this context involves trying to correlate the independent judgments of each source of belief.

Both premises are epistemological. The first makes a

judgment about the epistemic status of religious dogmas like creation and the appropriate way to reach conclusions about statements of this status. The second part of this judgment further entails a statement about how contemporary religious thinkers ought to understand the relationship between religion and science. The entailed epistemological conclusions for the philosophy of religion in general are in themselves controversial. They will be our first consideration in this concluding chapter on the doctrine of creation from the perspective of philosophy.

The results of our comparison of the previously determined Jewish concept of creation with the current consensus in theoretical physics about the origin of the universe led us to conclude that the former is believable. Furthermore, in most respects, the findings of contemporary physics help religious people to fulfill their obligation to understand creation by filling in details of the dogma that by no other means would be accessible, which in turn enables us to achieve a far more coherent understanding of what it means to say that at the beginning God created this universe out of nothing than would have been possible on any alternative schema, including that of Timaeus.

This determined correlation between the rabbinic understanding of Genesis and the philosophic understanding of the mathematical formalisms of astrophysics led us to a number of conclusions about ontology that are in themselves controversial. They are controversial because they are incoherent with the wide spread judgment, by both technical philosophers and ordinary people, that the philosophy and science are atomistic in the sense that to understand everything that is is to understand rules that govern interactions between definite objects that happen to occupy space and time. According to this current "common sense," events are to be understood in terms of their agents who in themselves occupy specific places at specific times, complex agents are reducible to events between simpler spatially-temporally determinate agents, and these events also are to be understood ultimately in terms of their most simple agents. In contrast, the correlation of the Jewish dogma of creation and contemporary physical theory is far more coherent with a process ontology in which there are no such ultimate reductions. These reductions fail because (1) there are no ultimate determinate entities upon which to base a foundationalist view of reality, i.e., no agent is so determinate that in principle it cannot be reduced to some other more simple agent ad infinitum, (2) the world that we experience is never definitely anything, i.e., dynamic reality is in principle indefinite and indeterminate, and (3) agents as such are not ultimate. The last point is the most controversial. It says that agents exist only as constituents of the events in which they act, and the acts themselves are ultimately to be understood as vectors in time-space. In other words, the model for reality is not, as it has been since the time of Aristotle, a static picture of substances with determinate characteristics. Rather, the model for reality is a dynamic picture of amorphous space that reflects its occupants in motion. These explicit ontological conclusions will be our second consideration in this concluding chapter on creation as a philosophical doctrine.

Finally, our quest for what is and is not believable in the Jewish concept of creation revealed one major disparity with contemporary physics. Whereas the Jewish concept insists that an adequate understanding of the universe must be in some significant sense teleological, modern science admits only mechanistic explanations. Furthermore, from the perspective of creation, both the universe and our knowledge of it is subject, in some important sense, to moral judgments. In contrast, modern philosophy makes a radical separation between judgments of science and judgments of ethics. This disparity between our two ultimate sources for belief in creation will be the third and final consideration of this book.

EPISTEMOLOGICAL REFLECTIONS ON CREATION

At the beginning of this book we described how Rosenzweig made a radical separation between philosophical-scientific thinking (*Denken*) and religious belief (*Glaube*). His distinction served as the basis for our distinction between philosophy (viz., reasoning from empirically derived premises) and theology (viz., reasoning from inherited texts that claim to be scripture, i.e., to be in some sense divinely revealed). Until now we have not questioned this separation. However, our conclusions about the way to interpret religious dogmas – viz., by correlating results from thinking about the implications of both scientific judgments and religious texts – entail that Rosenzweig separates what should not be separate. Science is relevant to religious belief and religious texts are relevant to scientific speculation.

The source of Rosenzweig's distinction is the Protestant division between the secular and the religious, where all of science falls within the domain of the secular and is completely outside of the subject matter of religion. This distinction in part explains why Rosenzweig is almost unique as a modern Jewish thinker in his concern with creation. For everyone else the issue of creation is a question of modern physics, and physics belongs exclusively to the domain of the secular sciences. As we have seen, for Rosenzweig creation is neither part of the physics of the old philosophy nor a subject for revealed belief. Rather, it is the central topic of the metaphilosophy of the new thinking. By a dialectic of thought rooted in negation, the new theology of negative philosophy transcends the Hegelian dialect of positive thinking, demonstrates the inadequacy of the old thought, and points beyond itself to the need for belief through which revelation is grasped. As such creation is itself not part of revealed belief, because it is the basis for discovering belief. But neither is it part of science, for it transcends objective analysis. It belongs properly to the realm of scientific story-telling which as such bridges science and religion.

As we have seen, Rosenzweig's new thought of creation is a revival of Plato's use in the *Timaeus* of MYTHOS, where "myth" is to be understood as a kind of "bastard reasoning" where you can give a probable account of something like the space of the universe that sort of exists, but not really. It is the appropriate method to discuss something that on one hand is not "something" but on the other hand is not "nothing at all." Hence, Rosenzweig, in good Protestant tradition, reverses the direction of classical religious thought from both Plato and the

Hebrew scriptures, to attempt to make the content of the inherently equivocal story-pictures determinate in religious science.

However, Rosenzweig calls this new thought a "rationalism." Like his medieval predecessors, he places no value on unclarity. Quite the opposite is the case. A language ought to be as precise as the object discussed, and, in this case, where the object itself is in reality equivocal, the appropriate language of expression ought itself to be precisely imprecise.

From the perspective of Rosenzweig's place within the tradition of modern Jewish philosophy, his argument is with Spinoza. We have not paid any attention to him before because Spinoza had no theory of creation. However, it is important to consider him now in connection with the present question, viz., the relationship between science and religion. In this case the critical texts to be considered are Spinoza's *Ethics* and his *Tractatus Theologico-Politicus*.

Spinoza also made a radical distinction between the religious and the secular, but it is not identical with how contemporary Protestant society and theology make that separation. Rosenzweig in at least one sense accepts Spinoza's distinction. For both of them, to know is the activity of the individual scientist whose source is his experience and his reason, whereas to foster political morality is the responsibility of a community's clergy whose source is their imagination. However, Rosenzweig turns Spinoza's judgment upside down. What in Spinoza is an inferior role becomes for Rosenzweig a superior one.

Spinoza's true religion is a handmaiden of science. Religion is a political repository of myths created from the images of artists. It cannot judge the value of its stories, because it has no access to truth. A good myth is one that pictures for the sake of moral inspiration for the ignorant masses what the scientist knows to be true. Conversely a bad myth is one that pictures what the scientist knows to be false, and as such is a poor heuristic device to educate those who are intellectually deprived.

In contrast, Rosenzweig claims that the objective results of scientific inquiry only point to the inadequacy of the scientific method. Real knowledge transcends science. It requires the spoken images of poets inspired by divine revelation. In other words, whereas for Spinoza the value of the honest work of prophets rests on the objective findings of scientists/philosophers, for Rosenzweig the value of the sincere efforts of these latter thinkers rests on the subjective inspiration of prophets.

The contrast between Spinoza and Rosenzweig in this respect is informative of Rosenzweig's enterprise in yet another way. Maimonides concluded that statements about God and the universe really inform us about human morality. The heart of Spinoza's critique of Maimonides in both *The Ethics* and *The Tractatus Theologico-Politicus* is a rejection of this claim. For Spinoza and his disciples in Western civilization, science tells us the truth, and the truth is what is rather than what ought to be. In other words, modern science is morally neutral. Rosenzweig grants Spinoza and his descendants through Hegel their science, and then returns to Maimonides. Science has its truth, but there is a higher truth. The truth of metascience is what ought to be, and not what merely is.

However, Rosenzweig's reading of Maimonides' theological cosmology is itself not above suspicion. For example, Rosenzweig says[402] that from Maimonides' correct judgment that divine creativity (*Schöpfertum*) is divine power (which, in turn, is God's essence), he mistakenly concluded that divine creativity is the result of divine need (*Bedürftigkeit*), which entails a denial of both divine freedom (*Freiheit*) and the judgment that the world in itself has internal meaning (*Sinn*). It is not at all obvious that Maimonides emphasizes the revealed inner necessity of God to the point of exaggeration or that he did so because he ignored the caprice dictated by the self-structure of the conceived deity of philosophy. That Rosenzweig's own theory of divine providence so closely parallels Maimonides' theory[403] in itself raises doubt about the correctness of Rosenzweig's evaluation.

Again, at least with respect to Jewish thought, Rosenzweig's real argument was with Spinoza, and not with Hegel. However, even if we grant (at least from the perspective of classical Jewish thought) that a story about a created universe that inherently has moral value is preferable to a picture of an eternal cosmos where moral categories are inapplicable, is it in fact the case that Rosenzweig's fairly simple sets of oppositions and identities are the correct way to pose the problem?

Whereas the medieval Jewish commentators and philosophers either equated the generation of Greek philosophy with the creation of the Hebrew scriptures, or at least saw them to be compatible, Rosenzweig accentuates their difference. This sharp separation stands in marked opposition to the direction of classical Jewish thought. However, it is not without precedence. It is implied in Abraham Ibn Daud's radical rejection of the emanationist cosmology of Solomon Ibn Gabirol[404] and in Maimonides' critique of any doctrine of necessity that an Aristotelian cosmogeny entails in marked opposition to any alternative creation schema.[405]

Still, it is appropriate to ask, is scientific thinking (which Rosenzweig judged to be inherently emanationist and amoral) opposed to religious belief (which Rosenzweig judged to be inherently creationist and moral)? Certainly there are in principle more than these two alternatives. In fact there are as many possibilities as there are ways to combine these three sets of opposites (reason/revelation, emanation/creation, and amoral/moral).

The *Timaeus* is a study of creation, not emanation. However, its cosmology is based on pure mathematics, not revelation. Conversely, the theory of Ibn Daud's cosmogony is emanation, not creation, and his authority for it is scripture, not mathematics. According to Rosenzweig's dialectic these combinations could not be made by careful thinkers, but in fact they have been made by first-rate critical scientific and religious minds. It occurred in the past. Furthermore, it is occurring in our own day where, for at least the moment, the dominant view of the mathematically oriented astrophysicists is that we live in a universe created out of nothing.

Rosenzweig employed two analogies to illustrate the need for thinking to move beyond static elements to their dynamic relations. The second is for our purposes particularly informative. He said that a point x_n is determined in a geometric object only through a differential equation that expresses that object. In other words, a geometry limited solely to algebra can only present static forms that, as such, are in principle unreal. Only when it employs calculus can its models apply to the real world of concrete objects in a continuously changing nexus of space and time.[406] This example points to the strength of Rosenzweig's epistemology. In some sense Rosenzweig's understanding of the use of mathematical thinking is more sophisticated than that of his more recognized contemporaries such as Bertrand Russell. Whereas they think that thinking is limited to positive thought and their only model for such thinking is simple algebra, Rosenzweig, under the influence of Hermann Cohen, was able to incorporate negative thinking on the vastly more sophisticated model of calculus. Still, his understanding of mathematics remains too narrow in that his mathematical expressions are arbitrarily restricted to at most two variables.

To summarize, I want to argue now against Rosenzweig's radical separation between what is knowable by science and philosophy on one hand and by the study of revealed texts on the other. In one sense this book is an argument for my claim. It is an illustration of how the two are not so radically separate and how the two should be interrelated. Hence, my positive argument for a form of unity of reason and faith, science and religion, is the example of this book. Negatively, my argument against Rosenzweig's radical separation of the two, despite its long history in contemporary Protestant and Jewish thought, is that it presupposes a mistaken view of the limits of reason. Specifically with reference to Rosenzweig's argument, his analysis of the domain and character of logical thinking within empirical science and analytic philosophy presupposes its dependence on the perceived nature of mathematics. I accept this judgment. I too agree that developments in the understanding of mathematics play a major (if not dominant) role in directing the way that scientists and philosophers both carry on their respective enterprises. However, I want to argue that Rosenzweig's understanding of the nature of mathematics is distorted, that distortion leads him to present an inaccurate view of the domain of rational thinking, and it is the latter inaccuracy that underlies his radical separation between reason and faith. It is this specific form of critical argument that will be our concern in the remainder of this section on epistemology.[407]

There are three passages in *The Star* where Rosenzweig discusses mathematics. The first text is his statement of his indebtedness to Hermann Cohen's application of his infinitesimal calculus to philosophy. The second is Rosenzweig's critique of the limitation of Cohen's method. The third deals with Rosenzweig's alternative method.

Rosenzweig's first reference to mathematics comes at the end of his introduction to the work as a whole.[408] It is an explicit statement of the influence of Hermann Cohen and his use of mathematical thinking on Rosenzweig's philosophy. The conclusion of Rosenzweig's reasoning is a geometric model, viz., a Star of David. As we have seen, it is constructed from the intersection of two triangles. One is formed by connecting the elements (God, man, world), and the other is formed by connecting the relations/courses (creation, revelation, redemption). However, it would be more accurate to say that his model of reality is a single triangle, where the sides are vectors from points of origin to limit ends. As we discussed above, the points are the elements – God, man, world; the vectors are the relations – creation, revelation and redemption. Rosenzweig explicitly said that his model for this kind of thinking is mathematical.[409] However, the math itself is new. Again it is not the simple geometric model of Platonic thought based on a correspondingly simple notion of numbers and harmonics, limited exclusively to the domain of positive rational numbers.

The next reference to mathematics[410] occurs as a central part of Rosenzweig's discussion of creation. It is an explicit statement of why he limits the value of reason in epistemology. This passage is the central datum for my critique of Rosenzweig's critique of rationalism.

According to Rosenzweig, creation is a relational concept that points beyond the elements God and world to God's revelation and the world's redemption. By implication, the third element, man, must emerge as a critical link between God and the world. Hence, the symbolic, i.e., scientific, equation of creation out of nothing points beyond its expression of the relation between God and the world to a grammatical, i.e., formal linguistic, expression of a relation between all three elements, which entails the three relations. The last claim (about scientific equations and linguistic expressions) forms the core of Rosenzweig's argument for the need of a spoken language to express the three-dimensional nature of reality over and above the silent language of science and mathematics whose grammatical structure limits its range to two-dimensional conceptual space.

Rosenzweig's argument can be summarized as follows: To express something symbolically and to think scientifically solely out of the sources of human reason is the same thing. The model for all such thinking is the algebraic form "$y = x$." While this form is adequate to express a two-term relation, viz., the relation between the variables y and x, it is limited in that it cannot express any relation that is more complex. For every algebraic expression there is a comparable geometric figure mappable in Cartesian coordinates, and the dimensions of that figure correspond to the number of variables in the algebraic expression. Hence, two variable equations are represented in two-dimensional space, three variable equations in three-dimensional space, etc. However, scientific thinking is limited to two variable expressions which, as such, cannot adequately express three-dimensional reality. That reality is three-dimensional is expressed in Rosenzweig's symbolism by designating the number of elements precisely as three. That he judges scientific reasoning to be inadequate is represented by the fact that well-formed propositions in his formal logic can only admit two variables.

In this form the fallacy of Rosenzweig's argumentation should be apparent. There is no such thing as logic. Rather, there are logics, i.e., multiple sets of rules to structure formal ways of speech. Rosenzweig's elemental logic is only "a" logic, which is inadequate for the very reasons that Rosenzweig himself states. The reality that he wants to express is three dimensional, and, as such, requires a three-dimensional rather than a two-dimensional calculus.

It is of interest to note that while Rosenzweig restricts the use of algebra to thinking about elements, he does not so restrict geometry, since his ultimate picture of reality is a solid Star of David constructed out of two pyramids – one joined at points representing the elements (God, man, world), and the other joined at points representing their interrelationships (creation [between God and world], revelation [between God and man], and redemption [between man and world]). This device in itself illustrates that something is wrong with Rosenzweig's argument, since, without doubt, this geometric shape in principle has an algebraic counterpart.[411]

My main point is that Rosenzweig made a radical separation between logic and language that has no justification either in Rosenzweig's arguments or in fact. To take but one small example, the standard equation in Newtonian mechanics that a quantity of kinetic energy equals one-half the mass of an object multiplied by the square of its velocity ($E = \frac{1}{2}mv^2$) is an algebraic equation whose three-dimensional geometric counterpart looks like a pyramid.[412]

In fact algebraic expressions can be well formed with any number of variables/dimensions, and three is no more reasonable for expressing reality than two. (Einstein used four.) Rosenzweig's error was to equate Hegel's philosophy with philosophy and Hegelian logic with logic.[413] In fact logic(s) is (are) a kind of language and the grammar(s) of ordinary spoken language is (are) a kind of logic. Hence, Rosenzweig's radical separation between the kind of thinking involved in picturing elements and describing relations, founded on his radical separation between silent and spoken language, is without justification.

Rosenzweig recognized that thinking about reality would require a language that could encompass depth, i.e., a set of signs that could express more than two dimensions. He mistakenly inferred that such thought must transcend reason, so he concluded that theology, from the perspective of mathematical thinking, must be a "mystery."[414] However, merely to think multidimensionally in terms of algebra in no sense is mysterious. Here once again Rosenzweig drew lines where no separation is called for.

Similarly, Rosenzweig argued[415] that algebraic symbols (i.e., philosophic thinking) are inadequate to express creation on heuristic and philosophic grounds. As a teaching device it is inadequate because the movement from Yes to No to And in the elements now becomes a movement from the No in unity with the Yes in the direction of a new No. Rosenzweig claimed that this packing-unpacking movement would be too complicated to illustrate through algebraic symbolism. However, he also said that it could be expressed geometrically,[416] and, in so doing, Rosenzweig failed to recognize the complete symmetry between algebra and geometry, i.e., there is in principle an algebraic formula or set of formulas that could express any geometric shape. Once again, Rosenzweig made distinctions when none are called for.

The third and final reference to mathematics[417] points to one especially interesting alternative because it can also serve as a viable defense of Rosenzweig's position in the light of the above criticism of his second reference to mathematics. This third text occurs as a central part of Rosenzweig's argument for the role of communal ritual activity as a/the way of transcending the limitations of what is knowable both by the logic of science and philosophy (Part 1 of *The Star*) and the grammar of prophecy and poetry (Part 2 of *The Star*).

In Part 2 of *The Star* Rosenzweig used midrash (in the sense of a literary/theological explanation of the meaning/implications of specific verses in the Hebrew scriptures) to spell out the nature of the three kinds of movements between the elements. These directions from nothing to something constitute what Rosenzweig understood to be new kinds of relations (*"neue Beziehung"*). In fact they are not as "new" as Rosenzweig believed them to be. In mathematical terms they are vectors that originate in a point and move towards an asymptote.[418]

Rosenzweig claims that his picture model of the dynamic universe as a Star of David is beyond geometry. However, he

254   *A believable view of creation*

overstates his claim. It is beyond a certain kind of geometry, viz., Euclidean geometry whose range is limited to static pictures. However, a dynamic geometry of vectors from origins to limits is no less geometry than a static one of lines between points. Just what kind of geometry would this be? As Rosenzweig argues, the critical differences between his and Euclid's turn on three issues. The first we have already mentioned, viz., his is a geometry of directions whereas Euclid's is one of static states. The other two are the following: In Euclidean geometry (1) the lines between points exist within their domain necessarily (viz., they are the shortest distance between two points),[419] and (2) they are grounded (*begründeten*) in (i.e., formally deduced by) pure logic. In other words, given two points and the definition of a line as the shortest distance between them, there necessarily is one and only one connection between the points that can count as the line that they define. In contrast, (1) the directions that link Rosenzweig's elements are contingent,[420] and (2) they are grounded in history,[421] i.e., they are inferred from experience.

The question is, what kind of geometry is there (or, can there be) whose domain is defined by contingent vectors where choices for pictures/models are made by empirical inference? The obvious answer is, a geometry that is coordinated with non-linear equations, i.e., what is popularly called "chaos theory," viz., models for scientific reasoning (e.g., in weather forecasting) whose basis in mathematics consists of fractals.[422]

To summarize: The "lines" of Rosenzweig's new "relations" or "interconnections" are asymptotic functions, i.e., two-dimensional curves from an infinite number of origins infinitesimally distant that extend towards an endlessly expanding limit.[423] Rosenzweig seems to have believed that an Euclidean line (viz., the shortest possible connection between two points) is a different kind of geometric structure than a curve. In other words, he did not seem to be aware that a straight line is merely a kind of curve that differs from others only in that its slope is zero. Nevertheless, in spite of the limitations on his knowledge of mathematics, Rosenzweig was able to construct a model of what he thought was a new kind of geometry on which to model his "new thinking." We know that the geometry he had in mind is an instance of fractals. In fact the kind of fractal relevant to his new thinking is the Koch curve.[424]

Rosenzweig presents the new kind of interconnection (*Zusammenhang*) formed from his vectors of relation as a triangle of elements and/or relations. Each connection is to be pictured as a vector-segment[425] formed from a third point external to a basis in a particular relational line. His two kinds of connections intersect to form a six-pointed star. Rosenzweig claimed that these intersecting connections form a structure (*Gestalt*) rather than a figure (*Figur*), and therefore are something beyond mathematical conception (i.e., something that is "über-mathematischen"). However, they are only beyond Euclidean geometry and linear algebra; they are not beyond mathematical conception as such.

In conclusion, to interpret Rosenzweig's model of the Star of David in terms of fractals overcomes one major part of the objection in the second section of this discussion, viz., why – despite the obvious evidence from Cartesian coordinates that there is no difference in domain between geometry and algebra – Rosenzweig could introduce a geometric model in the third sector of *The Star* and limit mathematical (i.e., algebraic) thinking to the first section? While fractals are a certain kind of algebra, viz., non-linear equations, they are best understood in terms of geometry. However, the force of the initial objection for Rosenzweig's critique of reason remains. Rosenzweig's simplistic identity of reason with Idealism has at its foundation an unjustified simplistic restriction of mathematics to linear equations in algebra and static Euclidean forms in geometry. No Jewish philosopher since Spinoza[426] has had a better grasp of the relationship between mathematics, science and philosophy than Rosenzweig. No other philosopher to my knowledge[427] had a better intuitive grasp on how modern math can impact on both philosophic and religious thinking. Yet, Rosenzweig's deficiencies in this respect misled him to a kind of conceptual relativism. It is unfortunate (in my judgment) that the deficiencies in knowledge in Rosenzweig's thinking have had greater historical influence on the best of his successors[428] than the strengths of his theological intuitions.

Fortunately there are many alternatives, one of which has emerged in the body of this book. There are many ways to be rational, i.e., logical. Which are proper depends on the kind of question being asked and the sources for answering the question. No one way can be called "scientific" or "religious" to the exclusion of the other. Rather what differentiates the domains of science and religion is the source of the initial texts to which reasoning is applied. In the case of science, the fundamental texts are reports of human sense experience; in the case of religion, they are transmitted records of revelation. Sometimes each deals with subjects excluded from the domain of the other.[429] However, often the two domains intersect, as they do when the question becomes: What ought we to believe about the origin and general nature of the universe? In such cases what each discipline has to say should be viewed rigorously and authentically by its own rules of inference within its own syntax, and, at the same time, each should be viewed with an eye towards honest correlation. To achieve such a correlation should enrich our understanding of the question at hand (e.g., creation) from either perspective – scientific or religious.

ONTOLOGICAL REFLECTIONS ON CREATION

The implications of our study of creation for ontology provide a notable example of enriching both our scientific and our Jewish understanding through correlation. Here we have found that both domains of thinking suggest that the substances or objects or spatial/temporal hosts of this universe are inherently indefinite and indeterminate. However, this conclusion is incoherent with the most fundamental assumption in ontology since the time of Aristotle, viz., that reality consists fundamentally of determinate, defined things. Different philosophers have analyzed them in different ways and called them by different names, from Aristotle's "substances" to the referents of Russell's "proper names." However, all of them would agree that what these terms name are concrete things that together constitute a complete list of the primary constituents of existence. Undoubtedly this philosophical prejudice lies behind the impulse of popularizers of physics to speak more about charged particles than about the fields that determine the charges.

However, not all philosophers have systematized their universe in this way. A notable exception is Alfred North Whitehead, in whose ontology processes or events (not things or objects) are primary. In this respect it is of interest to note that probably no contemporary philosopher had a deeper understanding of twentieth-century physics and mathematics than Whitehead, and that the development of his ontology was intended primarily to make sense out of the "new" twentieth-century science and mathematics.

Furthermore, Rosenzweig's friend and colleague, Martin Buber, also developed an ontology in which things are not primary. Rather, in Buber's case, relations are primary (viz., I-Thou and I-It) and define their terms (viz., their objects – I, you, it, we and they). In Buber's case, the motive for breaking with "conventional philosophical wisdom" was neither scientific or mathematical. Rather, it was theological and ethical. In terms of theology, Buber determined that the language of revelation (viz., of the Hebrew scriptures) was first and foremost a language of verbs, where what pass as nouns most often are participles. Given that the language of scripture is itself a reflection of reality, what verbs express (viz., actions) are more fundamental to understanding reality than are the referents of nouns.

In terms of ethics, Buber determined that the domain of the moral is the domain of interpersonal relations. If there are no real relations then there can be no actual morality. However, if our philosophy begins (as it has at least since the writings of Descartes began to influence philosophy) with the isolated individual (be it a mind or a body), and there is little likelihood that we can infer from this starting point other individuals (let alone relations with them), then there seems to be no way to ground ethics in philosophy.[430]

In other words, modern science and mathematics, Jewish

theology and biblical language, and philosophical ethics combine (or, correlate) to suggest that a picture of the universe that makes events primary and objects secondary is a better model of reality than a picture in which objects are primary and events are secondary. Furthermore, our discussion of creation from the perspective of Jewish texts and contemporary physics has suggested an even more radical move in ontology. It suggests that a picture of the universe that makes space determinate and objects indeterminate is a better model of reality than a picture in which objects are determinate and space indeterminate. In between the indeterminacy of objects and the determinacy of space are the events of this world that are motions that are more determinate than their objects but less determinate than their space. In fact all that is definite about them is from where they came (viz., from the nothing of undifferentiated space), to where they are going (viz., to the total something of a self-fulfilled God), and that everything has moral value – objects are judged in the context of their activities that in turn are judged either negatively as regressions in the direction of their origin (creation) or as progress towards their end (redemption). In other words, ontology ultimately is to be understood in terms of ethics. In other words, the ought determines the character of the is, and not the other way around. These final two sentences bring us to the last, and most difficult, reflection on the consequences of this book's study of the doctrine of creation out of the sources of Judaism – the issue of ethics.

ETHICAL REFLECTIONS ON CREATION

As the standard account of creation in modern physics begins the universe with a bang that ends in the whimper of a cold death of endless expansion, so this study of the Jewish dogma of creation that began with the linguistic bang of Rosenzweig's extreme linguistic expression of creation will end in a whimper about ethics and redemption. We will say relatively little here about ethics or redemption precisely because (given what we said at the end of the section on ontology) it is redemption and the "ought" that determine the "is" of this created world and not the reverse. What our study of creation has shown is that any account of the origin and nature of this created world is not adequate unless it is related to a conception of its redeemed end in the ideal of the world-to-come. In other words, creation points to the end, but it does not determine its nature. Furthermore, since this world is defined as a movement from its atemporal origin in creation to its atemporal end in redemption, all that we can know now about redemption is that it is not creation. In other words, whatever else redemption is, it is the antithesis of creation. As creation is negative, redemption is positive; as creation is indeterminate, redemption is completely determinate; as creation presents a universe unified as homogeneous space, redemption projects a universe segregated into determinate spatial objects, etc. Beyond these conclusions, all of which amount to the judgment that creation points through revelation to redemption, there is little else to conclude about redemption solely from a study of creation.

My hope is that I will be able to write a book on the concept of redemption out of the sources of Judaism whose format will resemble the structure of this book. However, that study lies in the future, largely beyond the scope of this book. All that remains for me to do here is point to some of the problems in ethics that the concept of creation raises for a concept of redemption to solve. The first set of problems revolve around criteria for individual moral judgments. The second set focuses on political theory.

How are we as creatures in a created world to decide what is good and what is bad, what we should do and what we should avoid? The simple answer is, do what moves the created universe in the direction of its redemption away from a regression in the direction of its origin. The origin is amorphous indeterminate space; the end is self-determination in unity with God. In other words, what is right is to strive to make oneself more like God. In other words, the guiding principle of moral behavior that the concept of creation suggests is the classical principle of imitatio dei.

The most obvious expression in the Torah of this moral

principle is the phrase, "You shall be holy because I the Lord your God am holy."[431] However, there are important differences between the way the scripture states the principle and its generalization into an all-inclusive moral guideline in subsequent religious thought as imitatio dei. First, the Torah's statement restricts the way God is to be imitated, viz., to being holy. Second, the Torah's statement is associated with specific laws; it does not function as a general statement to imitate God in all his ways.

In both respects the Torah statement is preferable to the general principle of imitatio dei. First, the notion of "holiness" (KEDUSHAH) in the Torah accords with our understanding of the way that God functions in creation. To be holy is to be separate, i.e., to become holy in some significant way means to separate yourself from everyone else. Similarly, God functions in the Genesis account of creation by making what is otherwise unified separate, and what Genesis associates with the judgment, "it is good" is the initiation of division.

Second, it makes no sense to say that it is good for humans to try to be like God in all ways. The deity of creation is not just a force that causes life; he is also a force that causes death. For example, when a woman is giving birth to a child, and a struggle occurs such that if the woman is to survive the child must die and if the child is to survive the woman must die, the rabbis identify the "pursuer," i.e., the one who threatens both lives, as God. In some true sense, every natural mishap in creation is as attributable to God's creative influence as is every blessing, for creation is entirely the work of God. Furthermore, as the editors of the Torah observed, normal life is not birth and growth. These are only the first division of life; its second and final division is decline and death.[432] In other words, whatever it means to say that the created world reflects the will of God, that will creates entities universally intended to reach their end in death. Now, in some sense all of this human evil may be for some divine good, so that perhaps from a divine perspective it can be seen to be good. However, human death (particularly when it is innocent [e.g., the death of a child at birth]) can never be judged morally good from a human perspective. Hence, an unrestricted application of the principle of imitatio dei would be evil, for the principle in itself is not able to distinguish between proper and improper applications. For example, it seems to be right to say, "Introduce order and meaning into human society as God introduced order and meaning into the world." However, that same logic should lead one to the moral absurdity of saying, "Destroy human society as God destroys whole galaxies." Again, both imperatives seem to be logical applications of the general principle of imitatio dei if it means what it says, viz., whatever God does you ought to try to do.

The critical point is that right moral behavior cannot be determined from empirical nature, because what the concept of creation teaches is, not that the world "is" good but, that the world "ought to become" good. What the world itself is is the setting from which moral challenges arise; it is not itself a model for correct moral choice.

How then are we to know how to guide our behavior if we cannot learn it from created nature? The traditional Jewish answer is revelation, viz., obey the Torah as it has been interpreted by the chain of rabbinic tradition. In other words, what bridges the gap between creation and redemption is revelation. How intelligible this notion is – viz., that ethics must be grounded in revelation – is a debate that has a long history of its own, at least as long as the history of Jewish, Muslim and Christian discussions of creation.

In summary, creation guides our thought about the world through revelation to redemption. Creation and redemption are absolute poles that point respectively to standards for what is bad and what is good. However, these values are not human values; they are divine, and what is good/bad for God and the world in general need not be good/bad for human beings as participants in creation. For example, when the first Temple was destroyed, the lesson to be learned was not that the priests had wasted their lives. On the contrary, what the concept of creation entailed for them was a promise that out of their collective death would arise a new, as of then unknown, form of worship in the service of the creator deity. Similarly, failure and success teach us nothing in and of themselves about the way we as creatures are to serve our creator. (Divine providence does not mean that there is a direct and immediate correlation between doing what is right and receiving rewards.) Creation teaches that the universe exists, not for the sake of its occupants but, for the sake of its creator. Our place is to serve him, but there is no way to know (from nature) how to serve him except through revelation.

For now this is all that I can say about individual morality out of what the concept of creation in and of itself entails. However, there is much more to say about political ethics, and in this case the consequences are even more unsettling. There is a tendency in political theory to see the ideal polity as an imitation of the structure of the universe. This was certainly what we found in reading both Genesis and the *Timaeus*. In both cases the state proposed is an imitation of the picture of the universe, and the claim that the social organism of the state ought to be modeled on the nature of the territorial organization of the universe functioned as an argument in favor of the proposed political structure. In the case of the *Timaeus*, as the universe exists under the governance of a single, absolutely rational agent, so the state should be governed by a single person whose source of authority is his wisdom. In the case of Genesis, as the universe exists to serve God, so the primary duty of a state is to serve God, which entails that every other institution within the state should exist in order to promote the work of the institution of the priesthood.

This notion that the political structure of the state ought to mirror the physical structure of the universe is not unique to ancient Greece and Israel. It operates in modern political thought as well. Briefly, the modern European liberal political theory that asserts the primacy of the individual citizen over the collective of the state is coherent with the kind of atomism that has characterized modern science. In Newtonian physics the universe ultimately consists of a plurality of individual things, called particles. Ultimately only they are real, since all complex or compound entities are reducible to configurations of these most elemental things. In a sense, all collections and compounds are merely illusions, since all that really is there are the particles. Similarly, the modern theory of the state presupposes real individuals who come together to form a collective; the collective has no existence of its own beyond the individuals who create it. Ultimately individuals are not only the cause; they are its only reality. In other words, any list of what exists ontologically will include the citizens of all states, but not the states themselves. This ontological notion, in conformity with Newtonian physics, underlies the "common sense" quality of the judgments of political liberals like Thomas Jefferson who granted to every individual human being inherent rights, but none to the state itself. In other words, whether or not individuals are good or bad, positive contributors to the welfare of the polity or drones or (even worse) leeches, they retain their inalienable rights to live and pursue their individual happiness. However, should the state prove to function to the disadvantage of its citizens, they have every right to destroy it (through revolution) and replace it with another. Since states have no reality beyond the reality of their citizens, they have no rights. In other words, states deserve loyalty only insofar as loyalty is in the interest of their citizens; otherwise, there are no positive obligations to the welfare of the state.

This very conception of political ethics is incoherent from what we have discovered about creation in this book from the perspectives of both scripture and contemporary science. In terms of the latter, the mathematical formalisms of contemporary physics break down into absurdities when we are too strict in applying them to fundamental particles. An example of this point is Schrödinger's cat-in-a-box. The critical move behind the paradox is that the equations that describe the universe in quantum mechanics express probability values, and probabilities apply only to collections of individuals, but not to individuals themselves.[433] Now, what does it mean to say that physics reflects reality when the equations that physics employs are probability equations? Minimally it suggests that collections (e.g., polities) are more real than their individual constituents (e.g., citizens).

The most immediate consequence of this kind of thinking

would be to deny the moral validity of individual rights. In fact this was the political position in both the Hebrew scriptures and Plato's dialogues. There individuals have meaning only as members of nations, while nations have relatively unrestricted authority over the life and death of individuals who in turn owe relatively absolute allegiance to their states.

It is not my intention here to argue in favor of political totalitarianism. On the contrary, my own political sensitivity suggests that the notion of individual rights is morally correct. Rather, I wish solely to point to a consequence of taking seriously modern science and biblical faith in moral and political theory. The problem is real, and the solution is not (it seems to me) to make ethics independent of either science or religion.

I have no answer to any of these problems. I only have a strategy for finding an answer. It is to conduct the kind of integrated study of science and religion, philosophy and theology, into ethics in terms of the religious concepts of revelation and redemption, that guided this present study into ontology and cosmology in terms of the religious and scientific concepts of creation in Jewish philosophy.

Notes

NOTES TO INTRODUCTION

1 I.e., that form of exegesis of the Hebrew scriptures most commonly associated with the earliest collections of rabbinic linear commentaries (i.e., by the "Sages") in marked distinction from the approaches developed by medieval rabbinic commentaries (viz., grammatical, philosophic and mystical commentaries) and the methods used by contemporary biblical critics (viz., historical and literary commentaries). Cf. Part 2, chapter 4, pp. 00–00.
2 Cf. Martin Golding, "Constructive Interpretation and Maimonides' Theory of Juristic Reasoning" (unpublished paper, 1988 meeting of the Academy for Jewish Philosophy). A similar point could be made about the way that scientists use mathematical models to form postulates.
3 In N. Samuelson (ed.), *Studies in Jewish Philosophy*. Lanham, University Press of America, 1987, pp. 143–160.
4 David R. Blumenthal's translation in his *The Commentary of R. Hoter ben Shelomoh to the Thirteen Principles of Maimonides*. Leiden, E. J. Brill, 1974.
5 Cf. Menachem Kellner, *Dogma in Medieval Jewish Thought*. Oxford, Oxford University Press, 1986.

NOTES TO PART 1: INTRODUCTION

6 *Mensch* in the original German. I considered translating this term as "human" rather as "man," but rejected it. The reasons for preferring "human" are obvious. First, the general use of the German term itself permits a gender neutral interpretation. Second, as the singular word "world" in some sense is intended to be associated with the multiplicity of things in the world, so Rosenzweig intended the singular word *Mensch* to be associated with the multiplicity of persons in the human species of both

genders. Third, "human" would make Rosenzweig's philosophy less gender biased than it otherwise appears to be.

It is this third reason that made me decide not to translate *Mensch* as "human." To make Rosenzweig not gender-biased in this way would be, in my judgment, a distortion of what he in fact intended. For example (as we shall see in chapter 1) Rosenzweig distinguishes between two levels at which the Hebrew scriptures speak of an actor acting. The lower level of action is passive and the higher level of action is active. The text expresses this difference by the verb's gender – feminine, which is inferior, is passive; masculine, which is superior, is active. However, when I am not merely describing what Rosenzweig says, but am reconstructing his thought (as I do in chapter 2), then I translate *Mensch* as "human." Conversely, when considering the biblical text of Genesis, the Hebrew term "ADAM" is translated as "human." In this case I believe that the gender-neutral translation of the noun is a proper interpretation of the biblical text itself. (For example, HA-ADAM is said to be created male and female [Gen. 1:27].)

NOTES TO CHAPTER 1

7 Cf. Menachem Kellner, *Dogma in Medieval Jewish Thought*. Oxford, Oxford University Press, 1986.
8 B. Spinoza, *The Theological-Political Treatise*, chapter 14.
9 Henceforth referred to as *The Star*.
10 An earlier version of the discussion presented below appeared as "The Concept of 'Nichts' in Rosenzweig's 'Star of Redemption'," *Der Philosoph Franz Rosenzweig (1886–1929)*. Band II. Das neue Denken und seine Dimensionen. Wolfdietrich Schmied-Kowarzik (ed.). Freiburg, Verlag Karl Alber, 1988, pp. 643–656.
11 What follows below is a summary of what Rosenzweig actually says. Here I attempt to be as true to Rosenzweig's language as I can. Consequently the style of my writing in this section differs significantly from what has been said so far. Undoubtedly, some readers will find the transition difficult. However, I hope the difficulty will not be insurmountable. In my judgment it is critical to allow Rosenzweig first to speak for himself before assimilating his thought into my own conceptual constructs. A similar change of style for the same methodological reasons will occur in the subsequent chapters in this book when other thinkers are given the same detailed consideration that I here give to Rosenzweig.
12 *The Star*, Part 1, Introduction, "Vom Tode."
13 I.e., dies.

14 *The Star*, Part 1, Introduction, "Die Philosophie des All."
15 *The Star*, Part 1, Introduction, "Hegel."
16 *The Star*, Part 1, Introduction, "Die Welt."
17 *The Star*, Part 1, Introduction, "Das Metalogische."
18 *The Star*, Part 1, Introduction, "Das Metaphysische."
19 *Ibid.*
20 *The Star*, Part 1, Introduction, "Der Ursprung."
21 *The Star*, Part 1, Introduction, "Mathematik und Zeichen."
22 *Ibid.*
23 Cf. Plato, *Timaeus* 31B–32c, 35B–36b.
24 In fact Plato's right triangle itself as a model for a rational construction of the material universe points beyond itself to the need for a cosmology that can encompass the irrational. In the right triangle given below, let the length of each adjacent side be the basic unit length 1. According to the Pythagorean theorem, the length of the hypotenuse must be the square root of 2, which itself is an irrational number. See Fig. A.

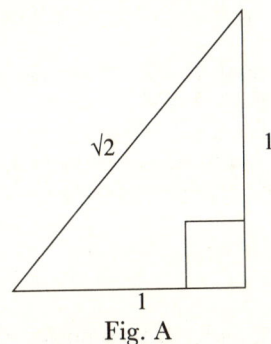

Fig. A

25 Rosenzweig could have reached the same conclusion exclusively through medieval Christian theology without any reference to Maimonides' theory of attributes. For example, the definition of God in Anselm's demonstration of God's existence as "a being no greater than which can be conceived" does not mean, contrary to what Descartes thought, a most perfect being. Rather it means that anything anyone possibly can think of necessarily is less perfect than God. In other words, God is in principle beyond thought.
26 *The Star*, Part 1, Book 1, "Negative Theologie."
27 *The Star*, Part 1, Book 1, "Göttliche Natur," "Zeichen."
28 I.e., "y," which means "the nothing of the knowledge of God."
29 I.e., "x," which means "what is not merely nothing."
30 *The Star*, Part 1, Book 1, "Göttliche Freiheit."

31 *The Star*, Part 1, Book 1, "Lebendigkeit des Gottes."
32 Viz., man. E.g., Descartes.
33 E.g., Spinoza.
34 *The Star*, Part 1, Book 2, "Negative Kosmologie."
35 *The Star*, Part 1, Book 2, "Zur Methode."
36 Rosenzweig seems to be thinking of what Kant called a priori synthetic judgments.
37 *The Star*, Part 1, Book 2, "Weltliche Ordnung."
38 *The Star*, Part 1, Book 2, "Weltliche Ordung, Urwort, Zeichen."
39 *The Star*, Part 1, Book 2, "Weltliche Fülle."
40 *The Star*, Part 1, Book 2, "Weltliche Fülle, Zeichen."
41 Much like Aristotelian ὕλε.
42 Much like Aristotelian μορφε.
43 *The Star*, Part 1, Book 2, "Wirklichkeit der Welt."
44 *Ibid.*
45 *Ibid.*
46 *Ibid.*
47 *The Star*, Part 1, Book 2, "Wirklichkeit der Welt, Zeichen."
48 *Ibid.*
49 *The Star*, Part 1, Book 2, "Der Plastische Kosmos."
50 *The Star*, Part 1, Book 3, "Menschliche Eigenheit."
51 *The Star*, Part 1, Book 3, "Menschliche Eigenheit, Urwort."
52 In other words, in the language of Jean-Paul Sartre, he knows himself in pre-reflective consciousness as a *pour soi*.
53 *Ibid.*
54 *The Star*, Part 1, Book 3, "Zeichen."
55 *Ibid.*
56 *Ibid.*
57 *The Star*, Part 1, Book 3, "Menschlicher Wille."
58 *The Star*, Part 1, Book 3, "Menschlicher Wille, Zeichen."
59 *The Star*, Part 1, Book 3, "Unabhängigkeit des Menschen."
60 *Ibid.*
61 Rosenzweig's division of his basic metascientific formula into two forms, a self-contained "y = y," and a relational "y = x," in some ways suggests Buber's *Ich-Du* and *Ich-Es* relations, the former applying completely to God, the latter applying completely to the objects of the world, and both applying to man. However, this is only a matter of similarity; the two formulations of source-words are not identical. In marked contrast to Rosenzweig's formulation, neither *Ich-Du* nor *Ich-Es* are self-contained, i.e., they both express relations.
62 *The Star*, Part 1, Book 3, "Unabhängigkeit des Menschen, Zeichen."

63 *Gestalt*, whose meaning, as it is employed by Rosenzweig, coincides with the standard use of the Hebrew term "SEDER" in medieval Jewish philosophy.
64 See *The Star*, Part 1, Übergang, "Rückblick: Das Chaos der Elemente."
65 See *The Star*, Part 1, Übergang, "Das offenkundige 'Vielleicht'."
66 Struktur, whose meaning, as it is employed by Rosenzweig, coincides with the standard use of the Hebrew term "'EREKH" in medieval Jewish philosophy.
67 See *The Star*, Part 1, Übergang, "Das herrschende 'Wer weiss'."
68 *The Star*, Part 1, Übergang, "Ausblick: Der Weltag des Herrn, Ordnung."
69 *The Star*, Part 1, Übergang, "Ausblick: Der Weltag des Herrn, Bewegung."
70 *Ibid.*
71 *The Star*, Part 2, Introduction, "Die Historische Weltanschauung, Schleiermacher."
72 *The Star*, Part 2, Introduction, "Die Historische Weltanschauung, Historische Theologie."
73 *The Star*, Part 2, Introduction, "Die Historische Weltanschauung, Aufgage." We will not deal here with Rosenzweig's reasons for this judgment. It is found in *The Star*, Part 2, Introduction, "Vom Glauben" and "Die Theologie des Wunders," and I have discussed this section elsewhere with some detail, viz., in "Halevi and Rosenzweig on Miracles," in David R. Blumenthal (ed.), *Approaches to Judaism in Medieval Times*. Brown Judaic Studies #54, Chico, CA, Scholars Press, 1984, pp. 157–172.
74 *The Star*, Part 2, Introduction, "Neuer Rationalismus."
75 *The Star*, Part 2, Introduction, "Philosophie und Theologie" und "Theologie und Philosophie."
76 Henceforth referred to simply as "theology."
77 Henceforth referred to simply as "philosophy."
78 *The Star*, Part 2, Introduction, "Der neue Theologie." Heinz-Jürgen Görtz defended this thesis in the paper that he delivered to the 1986 international conference on Franz Rosenzweig in Kassel, Germany.
79 *Ibid.*
80 *The Star*, Part 2, Introduction, "Grammatik und Wort."
81 *The Star*, Part 2, Introduction, "Der Augenblick."
82 In at least this one respect, Rosenzweig accepted Hermann Cohen's analysis of the philosophic significance of the fact that the mathematical formalism for modern science is calculus.
83 Gen. 1:4, 10, 12, 18, 21, 25 and 31.

84 Cf. Gen. 2:2–3.
85 Rosenzweig seems to ignore the fact that at least one constituent of the world, viz., the water (vs. 2, "HA-MAYIM"), is expressed with a definite article prior to the creation of the seas (vs. 10, "YAMIM").
86 In this sense the charge could be raised that Rosenzweig's interpretation of creation is not Jewish.
87 As will be discussed below in more detail, Rosenzweig's use of the term "Eigenschaft" parallels Plato's interchangeable use of the terms IDEA, MORPHE and GENEI in the *Timaeus*. Note that in the *Timaeus* it is the UPODOCHEN and not the THEOS who creates the qualities that Plato literally calls ideas, forms and/or kinds.
88 Gen. 1:31.
89 *The Star*, Part 2, Book 1, "Die Weissagung des Wunders."
90 Gen. 1:31. Note however that it also occurs in Gen. 1:7, 9, 11, 15 and 24.
91 I.e., he is at this level the UPODOCHEN, and not the THEOS, of the *Timaeus*.
92 Or, similarly, the THEOS of the *Timaeus*.
93 *The Star*, Part 2, Book 2, "Die Liebe."

NOTES TO CHAPTER 2

94 Some friends urged me to write a prolegomena to this book on epistemology and methodology. I have resisted their suggestions. In part this is because I am more interested in ontology than epistemology and in trying to reach some theological conclusions than in discussing how it is possible to reach them. Also, since my way of doing philosophy of religion and Jewish philosophy differs significantly from the kinds of approaches that are currently fashionable, I want first to illustrate my approach and only then discuss its viability.

NOTES TO PART 2: INTRODUCTION

95 It seems to me that in one way or another Rosenzweig was familiar with and influenced by these commentaries, which were written in the margins of the standard edition of the Torah that Rosenzweig would have studied. However, my concern in this book is not to demonstrate what more specifically are his historical sources. (For example, it is obvious that he studied Judah Halevi's *Khuzari* and was familiar with Maimonides' *Guide of the Perplexed*.) Here the interest in Jewish philosophy and theology is

sufficiently served by showing that both the substance and style of Rosenzweig's interpretation of scripture is consistent and coherent with the way previous rabbinic philosophic commentators read scripture.

NOTES TO CHAPTER 3

96 Saadia Ibn Joseph Al-Fayyumi. Born in Egypt in 882 C.E., a controversial gaon of Sura from 928 C.E. to 930 C.E., and again from 937 C.E. until his death in 942 C.E..

97 Saadia Ibn Joseph Al-Fayyumi. KETAB AMANAH UL'ATKADAT. Leiden, Brill, 1880. Translated into Hebrew by Judah Ibn Tibbon. SEFER EMUNOT VE-DE'OT. Leipzig, C. W. Vollrath, 1864. Translated into English by Samuel Rosenblatt. New Haven, Yale University Press, 1948. Selections translated into English by Alexander Altmann. *The Book of Doctrines and Beliefs*. New York, Atheneum, 1969.

98 "Aristotelian science" is rooted in but not identical with Aristotle's science. It consists of the corpus of the ancient Greek works attributed to Aristotle as they were interpreted first by Greek commentators, most notably Alexander of Aphrodisias and Themistius, and later by Muslim commentators, most notably Al-Farabi, Ibn Sina (Avicenna), and Ibn Rushd (Averroes). No attempt will be made here to distinguish which, if any, of the Aristotelian views properly are attributable to Aristotle.

99 As the style and language of presentation in summarizing Rosenzweig's theory of creation changed to that of Rosenzweig himself, so the style and language here changes to that of the medieval Jewish Aristotelians. The motive in both cases is the same, viz., to let the summary of what these philosophers said be as accurate as a summary can be. As in the previous part, the discussion of the summary will revert to the predominant style of the author of this book.

100 AL-AKADAH AL-REFAYA'AH. Composed in Judeo-Arabic in 1160 C.E. Survived through two Hebrew translations. One by Samuel Motot, entitled HA-EMUNAH HA-NISAAH and a second, better known translation by Solomon Ibn Labi, entitled HA-EMUNAH HA-RAMAH. It is divided into three books. In the first he explains the presuppositions of Aristotelianism to his intended audience of cultured Jews who know of but little about his new science. In the second he explains the six basic principles of Judaism in the light of the new science, and in the third he applies it to ethics.

272   *Notes to pages 83–86*

101 MILCHAMOT ADONAI. Completed in 1329 C.E., probably in Avignon. Henceforth referred to as *Wars*.
102 While Maimonides is critical of Aristotelian astronomy, he does not believe that it is possible to propose a better one. His interest is to defend the rabbinic dogma of creation against doubt based on science by demonstrating the limits of science. He does not present a summary of the astronomy that his discussion presupposes, but it is presented only to show what its limitations are and not to propose a cosmology of his own. His summary is fundamentally in agreement with Ibn Daud's picture of the universe, but provides far less detail.
103 For our purposes Ibn Daud's cosmology is a preferable paradigm to Gersonides' for precisely the opposite reason. Whereas the other rabbinic philosophers provide inadequate detail, Gersonides offers too much. He wrote for a sophisticated audience that was familiar with Aristotelian astronomy as it had progressed and developed over the two hundred years since Ibn Daud introduced it into Jewish thought. There are some points at which Gersonides' more developed astronomy conflicts with that of Ibn Daud, but none of them are important for the purposes of this chapter. They do not affect their shared general conception of cosmology. Hence, the greater conciseness and generality of Ibn Daud's cosmology makes his text more useful for our purposes than Gersonides' deeper, more rigorous exploration of astronomy.
104 If God were defined by two or more properties, then God would be complex. Similarly, if God possessed a non-essential property, then it would be possible to say that with respect to this property God is diverse, since we could distinguish at least two properties of God, viz., one that defines him and another that merely is true of him. Consequently, given any property F that can be predicted of God, it follows that God is (in the sense of identity) F.
105 SHEFA' in Hebrew. Commonly called "emanation."
106 Ibn Daud, in conscious opposition to Ibn Gabirol, believed that only immaterial objects reside in the supralunar world.
107 Aristotle's distinction between formal and material causes is an obvious source for Ibn Daud's two chains. However, given the general similarity between his cosmology and that of Plato in the *Timaeus*, one cannot help but see some similarity between Ibn Daud's formal and material causal principles and Plato's works of mind and necessity.
108 Ibn Daud, HA-EMUNAH HA-RAHAM, 159a1.
109 The ultimate source of Ibn Daud's first spheres is Plato's

Timaeus. However, there are notable differences between their respective cosmologies. Plato posits a first, all-encompassing sphere whose one natural motion is called "the Motion of the Same." Immediately within it is a second sphere whose one natural motion is called "the Motion of Diversity," which is identified with the Zodiac. These different spheres with their distinct motions are simply givens. No attempt is made to generate one from the other. In Plato's cosmogony, they are caused independently by contrary powers. The Motion of the Same follows from the principle of limit (PERAS), while the Motion of Diversity follows from the opposing principle of the Unlimited Multitude (APEIRON PLETHOS). The situation is significantly different in Ibn Daud's description of the heavens.

Ibn Daud's First Sphere parallels Plato's Motion of the Same. However, Ibn Daud, in contrast to Plato, distinguishes between the sphere and its motion. Consequently, Plato's different motions are associated with different deities, but not different spaces. Each spirit determines a distinct region of a single space, viz., the receptacle. In contrast, Ibn Daud's different motions are caused by souls/angels/intellects that are associated with different living bodies. Now, for our purposes there are no important differences between Plato's celestial deities and Ibn Daud's separate intellects. However, there is an important difference between their respective understandings of heavenly space. In Plato's case celestial space is nothing more than a host for different forces that "sort of" but not "really" exists. In contrast, Ibn Daud's spheres are as real as anything else that is not God.

Furthermore, Ibn Daud's "Sphere of Diversity" differs significantly from Plato's corresponding motion. First, Ibn Daud's sphere is not an independent force. Rather, it is a secondary power brought about ultimately by the Motion of the Same. Hence, whereas Plato's first two motions are equal with respect to an ordering of causes, Ibn Daud's Motion of Diversity is subordinate. Second, Ibn Daud's Motion of Diversity ranks fourth and not second in terms of causal priority. It is the product of the interaction of three distinct motions and the Epicycle. Consequently, Ibn Daud's cosmology exhibits considerably more unity than Plato's. Whereas Plato's universe is generated from at least two contrary, independent principles, Ibn Daud's cosmos ultimately is monistic with respect to causal order. In general, beyond the similarities between the two cosmologies, while Plato pictures a single space differentiated through a plurality of independent forces, Ibn Daud posits a unity of forces that determine a real pluralistic physical universe.

110 In the chapter under consideration Ibn Daud distinguishes between souls that govern spheres and intellects that cause spheres, whereas in I:8 the terms "soul" and "intellect" are used interchangeably. The suggestion that the soul of sphere n is the intellect of sphere n − 1 reconciles this apparent contradiction.

111 Moved or unmoved, i.e., necessary with respect to a cause or necessary by nature.

112 As noted above, heavenly bodies are not literally material entities. Nor are they literally forms. They are bodies that resemble matter in the sense that they are entities whose essence does not entail their existence.

113 This is not to say that there are not others in this tradition who come after Gersonides. However, none of them are as accomplished as a philosopher as he is. The one exception to this generalization (of course) is Chasdai Crescas. However, Crescas' intent was not to contribute to the development of this enterprise, viz., to the symbiosis of the best in philosophy and religion (which at this time meant Aristotelian science and rabbinic Judaism). Rather, his goal (to which he successfully contributed, as did Al-Ghazali for [or against] Muslim philosophy) was its destruction.

114 Principles 2 (God is one) and 3 (Statements about God are equivocal).

115 Principles 1 (The primary purpose of the Torah is to teach that God is incorporeal) and 5 (Mosaic prophecy as interpreted through rabbinic tradition is true).

116 Principles 4 (God governs the world through his angels) and 6 (God's rule of the world is both good and compatible with human choice).

117 Viz., that God governs the world through his angels (Basic Principle 4).

118 Book 2, chapters 2–28.

119 In the tenth chapter of his commentary on Sanhedrin, PEREK CHELEK.

120 It can be argued that by referring to God as "the creator" in the first principle, which entails belief in creation, Maimonides certainly must be said to have believed in creation and considered it to be a basic doctrine. Furthermore, Maimonides explicitly called belief in creation a "principle of the Torah" in his "Letter on Astrology," and he himself later interpreted his fourth principle that deals with God's eternity to entail belief in

creation. My point is not that Maimonides did not believe in creation as a basic Jewish belief. Rather, my claim is that even for Maimonides, in spite of its importance for his formulation of Jewish faith, creation was a far more problematic belief that any other fundamental rabbinic doctrine.

121 Chapters 1, 5, 6 and 29 deal with his methodology. He states the alternative views in chapter 2, and analyzes them in chapters 3 and 4. He demonstrates the general thesis against the Aristotelians in chapters 7 through 15. His reasoned conclusion is that the universe was created, but, once created, it is indestructible. His own more detailed account of how the universe was generated is laid out in chapter 17, demonstrated in chapters 16, 18 and 19, and defended against the Aristotelians in chapters 20 through 29. Having developed in Part 1 of the *Wars* the above cosmological schema out of his Aristotelian sources, he turns in the first eight chapters of Part 2 to determine the correct meaning of the first chapter of Genesis.

122 As was the case in summarizing the thought of Rosenzweig and Ibn Daud, the language and style in the following summary changes to that of the philosopher under discussion (viz., Gersonides). As in the previous cases, the change is intended to maximize accuracy. I will return to my original form of presentation at the conclusion of this chapter when what Gersonides said is scrutinized for its contribution to determining a contemporary Jewish doctrine of creation.

123 The following summary of Gersonides' interpretation of Genesis 1 is based almost entirely on *Wars* 6:2:1–8.

124 According to Gersonides, that God "made" ('ASAH) the universe means that its production is an act of will.

125 Which is what scripture means by saying that God "finished" (YAKHAL) it.

126 In other words, in Gersonides' judgment, Maimonides' distinction between necessity and intention with respect to God's act of creation is empty. In God's case it has no substantive meaning, a point of criticism that will later be made explicit by Spinoza in *The Ethics*. Jacob Staub attempts to interpret Gersonides as saying that with respect to his will, God's action is necessary, whereas with respect to the effects of his will, God's action is not. (J. J. Staub, *The Creation of the World According to Gersonides*. Chico, Scholars Press, 1982, p. 56.) In so doing Staub reads Gersonides in a way that is consistent with Maimonides' radical separation of acts of necessity and intention. However, I do not believe that his construction is successful. Given that what God

does is invariable and a consequence of his nature, independent of any external influences, the law of sufficient reason entails that the effects of God's action are no less necessary than God's action itself. They are only contingent in one sense; they follow necessarily from an external cause rather than from their own essence independent of their cause. In other words, they are what Spinoza would later call "modes."

127 Gersonides makes a radical break from all of his predecessors, Maimonides in particular, by arguing that the more venerable entities perform their self-defining functions for the sake of the less venerable entities.

128 According to Gersonides, scripture expresses the fact that creation is an instantaneous, single act with the phrase, "and it was so."

129 The following summary is based almost entirely on *Wars* 6:1.

130 "Know that with a belief in the creation of the world in time, all the miracles become possible and the Law becomes possible, and all questions that may be asked on this subject, vanish." (*Guide* 2:25, p. 25) "In the same way, if the philosophers would succeed in demonstrating eternity as Aristotle understands it, the Law as a whole would become void, and a shift to other opinions would take place." (*Guide* 2:26, p. 330) Gersonides believed that Maimonides' fear that reason leads to the Aristotelian position on creation led him to adopt his extreme view on God's attributes and knowledge. In his words, "We say that it would appear that this view of (Maimonides) . . . did not result from speculative foundations. This is because Philosophic Thought rejects this (position), as I shall explain. Rather it would appear that the Torah put great pressure on him in this matter." *Wars*, Treatise 3, chapter 3. Pp. 183–184 of the N. Samuelson English translation (Toronto, Pontifical Institute of Mediaeval Studies, 1977). Also see footnote 202, pp. 184–185 of this same translation.

131 "The reader should not think it is the Torah that has stimulated us to verify what shall be verified in this book, [whereas in reality] the truth itself is something different. It is evident, as Maimonides (may his name be blessed) has said, that we must believe what reason has determined to be true. If the literal sense of the Torah differs from reason, it is necessary to interpret those passages in accordance with the demands of reason . . . He, therefore, maintains that if the eternity of the universe is demonstrated, it would be necessary to believe in it and to interpret the passages of the Torah that seem to be incompatible with it in

such a way that they agree with reason." *Wars*, Introductory Remarks, p. 98 of the S. Feldman English translation. (Philadelphia, Jewish Publication Society of America, 1984).
132 *Wars* 6:1:17.
133 I.e., the biblical "dry land," viz., those places on earth that are above the waters at the surface of the air.
134 Gersonides maintained that, while the element out of which the cosmos is constructed is different from the four elements of the sublunar world, all five elements themselves are created from a common stuff.
135 In the next chapter we will look at what the rabbis of this same period said about creation from the perspective of biblical text commentary.
136 Gersonides makes this claim despite the fact that he, no less than Rosenzweig, sees all concrete creatures and particular events in the world, both terrestrial and celestial, to be contingent. My judgment is that his two claims are incoherent and the one that Gersonides should have abandoned is the judgment that this is the best of all possible worlds. From this perspective the point of greatest difference between Gersonides and Rosenzweig turns on an incoherence in Gersonides' thought.
137 *Guide* 2:13.
138 *Guide* 2:14.
139 HA-HE'DER MUCHLAT.
140 MI-DAVAR.
141 Any single use of any one of these terms, depending on the particular ways that authors read earlier texts and what they wanted to say, could yield up to 18,824 distinct usages (i.e., $8 \times 12 \times 12 \times 12$)! In other words, authors familiar with the use of Greek, Latin, Hebrew and Arabic had the linguistic capability of drawing close to 19,000 distinctions in what they meant by "creation out of nothing." For example, the possible lines of linguistic association and influence of the use by a single Jewish or Muslim philosopher of the Arabic expression (A1) "MIN AL-MA'DUM" could be diagrammed as in Fig. B below.
142 CHOMER RISHON.
143 TSURAH RISHONAH and TSURAH GASHMIT. The two terms seem to function as synonyms.
144 GESHEM MESHULACH.
145 TSURAH ACHARONAH.
146 Viz., essentially something that is "MERCHAQIM HA-BILTI MUGBALIM."
147 This is the reason why Gersonides interprets the "MAYIM" of

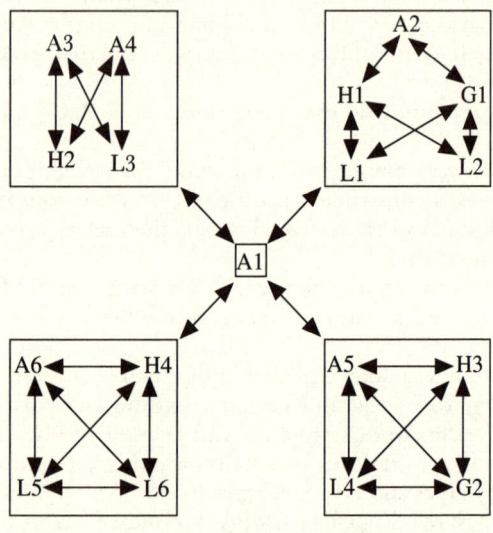

Fig. B

Genesis 1 to be the original, undifferentiated space of the entire universe. Cf. the diagram of Gersonides' interpretation of Day One above. Because MAYIM has this function of being primordial space, Gersonides uses another term, viz., "TEHOM," for the element, water. Both MAYIM and TEHOM are listed in the biblical text as entities that precede God's act of creation.

148 In terms of truth values.
149 In terms of moral values.

NOTES TO CHAPTER 4

150 An example of the kind of issue involved in the second question is the following: As discussed in the preceding chapter, the ultimate principle through which Aristotelian science schematized physics were form and matter – the former being the principle that accounted for the extent to which every individual thing in the universe is united with other individuals and is inherently intelligible, the latter being the principle that accounted for the extent to which every thing existent in the universe is singular and unique. While there was general agreement among these philosophers over the nature of form and its role in cosmology, there was considerable disagreement about the character of

matter and its role in cosmogony. As in most cases, the clearest statement of the disagreement was presented by Gersonides. In his terms, the issue was the following: The accepted view of science in the Aristotelian traditional was (in the language of contemporary philosophy of science) both foundational and reductionist, i.e., the understanding of everything that was, is and will be is an understanding of some basic or primitive entities or categories from which all complexity is constructed. Since every thing in the universe is a composite of form and matter, and the form and matter of every existent composite is itself composite, what are the most primitive forms and matter of the universe? There appears to be a general consensus that an answer to this question refers back ultimately to two entities – God as the ultimate source of form, and space as the ultimate reference for matter. Hence the foundation of science becomes divine action on passive space. However, what is the nature of this most fundamental relation? In general, the answer is that matter becomes something by God imposing upon it a certain form which renders space dimensional. In other words, at the most primitive level what is is space and what it is is dimensional. The "what" it is here is called "first form," the "that" that is is called "first matter," and the resulting foundation is called "first body." In other words, the origin of the universe is the generation of dimensional space. However, just what kind of thing is this space? Here the dispute is most precise. Ibn Sina said it was a definite potential that is not actually anything. Al-Ghazali objected and said something merely potential cannot be anything indefinite, and the foundation of everything that is must itself also be something, i.e., be something actual. In other words, from what is (actually) nothing you cannot build something (actual). On the other hand, as an actual entity the choice of space as a first entity seems arbitrary, for why would a space composed of form and matter be any more fundamental than an element in space (viz., fire, air, water and earth)? As the respective forms of the elements are what make them the elements on the kind of matter receptive to these forms, so there must be a form of space that makes it receptive to locating elements. Furthermore, just as the forms and the matter of the elements are not just generic form and matter, because they admit of opposites, so the form and matter of an actual space would not be generic form and matter, i.e., would not be a first form and a first matter. Hence, while Al-Ghazali's move solves the question raised about Ibn Sina's judgment (viz., how can something actual be composed of elements that are not actual?), it leaves us with an equally important problem (viz., how can what is fundamental be no different in nature than what is complex?). Ibn Rushd's solution was to claim that the ultimate building block of the universe is itself something actual that is indefinite. Because it is actual we overcome the objection to Ibn Sina's position, and because it is indefinite we overcome the objection to Al-Ghazali's position. As presented by Ibn Rushd this solution seems like the kind of political decision that committees come to when their participants disagree. It is a compromise which in itself could be worse than anything any of the participants believed independently. In other words, by what appeared to be a simple compromise, Ibn Rushd had in fact introduced a radically new way of conceiving of the foundations of physics, viz., that the fundamental reality to which all physical entities are reducible is something material that has no form and, as such, is itself not complex. However, this conclusion in itself entails the rejection of Aristotelian philosophy. The judgment that all things are composed of form and matter which are themselves irreducible is the most basic principle of Aristotelian physics. Hence, Gersonides' conclusion that ultimately these two kinds of entity are reducible to a single kind of entity that is real but has no form is a revolutionary break with Aristotelianism. Gersonides called this revolutionary hypothetical element "absolute body," which he claimed was the referent of the term "MAYIM" in Genesis 1. (As we say in the last chapter, the origin of the universe is absolute body [GESHEM MESHULACH], which Genesis identifies as MAYIM, from which arise, under divine direction, the first form [TSURAH RISHONAH] and the first matter [CHOMER RISHON] that Genesis calls [respectively] TOHU [whose characteristic positive feature is called "light"] and BOHU [whose characteristic negative feature is called "dark"].)

At this point Gersonides makes a major break with his predecessors in Jewish philosophy, and Rosenzweig's ontology of creation coheres with Gersonides in opposition to all of the other classical Jewish philosophers. (As we saw in the last chapter, there is a close coherence between Gersonides' understanding of the universe as a development from a primitive absolute body through individual composites asymptotically towards a unity in God and Rosenzweig's description of the world as a vector from an origin in B [*Besondere*, viz., an entirely distinctive particular nothing] to an asymptotic end in A [*Allgemeine*, viz., an absolutely general something that defines God's essence].)

Now, for our purposes in this chapter, the question is, whose view – those of the main tradition of classical Jewish philosophers, or that of Gersonides and Rosenzweig – is more coherent with the way that the classical rabbinic commentators interpreted the biblical text?

151 I.e., he dealt with the PESHAT of the text and/or he gave a PERUSH NISTAR. The different categories of rabbinic biblical commentaries given below are taken from the different terms that Nachmanides uses in his commentary.

152 I.e., the commentator (HA-MEFARESH) in this case presented a homily (AGADAH) and/or he gave a reason (YITEN TA'AM) for what the text says, and/or he revealed a secret (YIGLEH SOD) of the text. The tradition of rabbinic Judaism that concentrated on these "secrets" was the Kabbalah.

153 "Most" but not all. Marc Saperstein presents an excellent discussion of the attitudes of thirteenth-century commentators to earlier homiletic biblical interpretations in *Decoding the Rabbis: A Thirteenth-Century Commentary on the Aggadah*, Cambridge (MA), Harvard University Press, 1980.

154 I.e., PESHAT.

155 Which is another way of stating what a philosophical interpretation is.

156 What the rabbis sometimes call the text's "inner meaning" (PENIMIYAH); what I am here calling the "mystical interpretation."

157 Viz., Nachmanides.

158 "Tannaim" are rabbis contemporary with the legal discussions recorded in Judah's Mishnah. "Amoraim" are rabbis contemporary with the purported commentaries recorded in the GEMARA of both the Jerusalem and Babylonian Talmud. Henceforth these rabbis will be referred to as "sages."

159 b. 1040 C.E. in Troyes. d. July 13, 1105 C.E. Henceforth referred to as "Rashi."

160 Henceforth referred to as "Ibn Ezra." b. 1092 C.E. in Toledo. d. in 1167 C.E. in Rome. Lived in Cordova until 1140 C.E. The last twenty-seven years of his life were spent as a travelling scholar, mostly in Christian Provence, northern France and Italy.

161 Known to the Jews as "Ramban." b. 1194 C.E. in Gerona. Believed to have died in the Holy Land around 1270 C.E. Henceforth referred to as "Nachmanides."

162 b. around 1475 C.E. in Cesena. d. 1550 C.E. in Bologna. Henceforth referred to as "Sforno."
163 1095–1096 C.E.
164 See Daniel J. Silver, *Maimonidean Criticism and the Maimonidean Controversy. 1180–1240*, Leiden, E. J. Brill, 1965.
165 An earlier version of the discussion presented below appeared as "Creation in Medieval Philosophical, Rabbinic Commentaries," *From Ancient Israel to Modern Judaism: Intellect in Quest of Understanding: Essays in Honor of Marvin Fox*. Jacob Neusner, Ernest S. Frerichs and Nahum M. Sarna (eds.) Volume II, Brown Judaic Studies 173, Part 10. Atlanta, Scholars Press, 1989. Pp. 231–259.
166 The English translations of the Hebrew text presented here in this discussion of Rashi and the sages is conventional and syntactically literal. Its purpose is to help readers unfamiliar with the Hebrew to follow the summaries given below. It is not to suggest that this is the best translation of the biblical text itself or that it is how Rashi and the sages interpreted the text.
167 "R. Oshaya interprets" means that *Genesis Rabbah* asserts that someone called Rabbi Oshaya interprets. No claim is made here that in historical fact Rabbi Oshaya or anyone else gave this interpretation. The same qualification applies to every other tanna and amora mentioned in this essay.
168 KISE HA-KAVOD.
169 But apparently not the physical reality.
170 R. Ahabah B. R. Zeira adds repentance (TESHUVAH) to this list.
171 Viz., God.
172 Viz., the world.
173 Viz., TOHU, BOHU and CHOSHEKH.
174 Because the sky is mentioned before the earth in the first verse.
175 I.e., in God's case to will something is to do it, where in our case willing and doing are separate acts.
176 My sense is that something important is involved in this description, but I am not able to say what it is.
177 BOHU.
178 A xeste is slightly less than 0.5 liters.
179 Fire (ESH) is not mentioned in the text, which in itself is a problem. The assumption underlining this addition is the following: It was generally assumed that the four elements of classical science/philosophy somehow enter into Genesis' account of creation. Now, of these elements – fire, air, water and earth – only two are not mentioned by name, viz., fire and air. One way of

solving this problem is to decide what other terms in the text refer to these two elements. On the present interpretation, the consonants of the Hebrew word for sky (SHAMAYIM) are parsed to be a composite of water (MAYIM) and fire (ESH).

180 Rashi accomplishes this interpretation by assuming a point that doubles the letter in the M of the Hebrew term in the unvocalized text, so that the word for sky should be read SHAM-MAYIM.

181 In Gen. 1:10.

182 Note the implicit tension underlying this discussion between what the biblical text seems clearly to imply – viz., that what God is doing is introducing separation into what is initially unified, and what these commentators assume to be true of God – viz., that God only does what is good, unity is a virtue, and diversity is a vice.

183 In this way the commentators attempt to reconcile the apparent contradiction between their belief that the existent world is (at best) imperfect and that the deity who created it can only do what is good. Evil (or, imperfection) is introduced not by God's act of creation, but by the events that occur in Gen. 2 in the Garden of Eden. Of course this is hardly a satisfactory solution. Even if human beings themselves introduce evil, the question remains, why does God introduce into his creation an element (viz., humanity) who can make what is initially perfect imperfect?

184 In *Genesis Rabbah* X:4.

185 Some scholars think that the intended cycles of Mercury, Venus and Mars are 480 days rather than years.

186 In *Genesis Rabbah* VI:6.

187 *Ibid*.

188 What follows is based on the discussion of creation in Louis Ginzberg's *The Legends of the Jews*. Philadelphia, Jewish Publication Society of America, 1942. It should be noted that in several respects Ginzberg's cosmology contradicts the picture developed here from Rashi and our sages. For example, in Ginzberg the TEHOM is a primordial element from which BOHU and MAYIM are formed on the first day, whereas for Rashi and our sages the MAYIM is a pre-first day primary element, one of whose forms is the TEHOM. Furthermore, as we have already noted, according to our sages, TOHU, BOHU, CHOSHEKH, MAYIM and TEHOM were created prior to the first day, whereas according to Ginzberg all but the TEHOM were created on the first day. All of this says no more than it probably is a mistake, if the goal were historical accuracy, to

treat each entry in the midrash as if it were coherent. We do so here solely as a heuristic device, viz., to see to what extent a coherent picture can be formed from these homiletic/linguistic commentaries so that we will have a model by which we can compare what these earliest commentators said about the text of Genesis with what the later philosophical commentators said. This comparison will be employed in the next section of this chapter. In this sense, the sole purpose of the present summary is to set the stage for the summary in the next section, because the more conceptually sophisticated text commentators to be discussed there presupposed knowledge of the material we are summarizing here. (All of them assume familiarity with midrash, and Ibn Ezra is the only one who does not use [more accurately, did not know] Rashi's commentary.)

189 In which God subsequently will wrap himself to generate the light of the first day.
190 Whom we call "human beings."
191 Which suggests the influence of Plato's *Timaeus*.
192 Here the textual problem being addressed is the following: The grammatical form of the noun, TENINIM is plural. If it in fact names something plural, it (together with the celestial objects) is an exception, since everything else generated is singular. The text itself provides two kinds of models for speaking of the specific creatures generated in the plural, each of which is adopted by different commentators. R. Pinechas and R. Idi use the subsequent model of HA-ADAM who is created male and female. R. Mattenah and R. Huna adopt the model of the celestial objects which are truly a plural. And Rashi suggests what is (in my judgment) the best reading of the text, viz., that while the grammatical form of the noun is plural, its referent is singular. (In this sense, Rashi's model is the term for the sky.) Note also that for Rashi this is the real meaning of the text. The other interpretations are midrashic. My feel for the Rashi text (granted, it is subjective) is that the sense of this judgment is that this latter kind of interpretation is less venerable but nonetheless legitimate.
193 In this way Rashi reconciles the sense that everything that God creates is singular with the plural grammatical form of the noun.
194 After the flood.
195 The term NEFESH – which is taken in the context of Gen. 1 to mean that the creature (CHAYAH [which itself literally means something that is alive]) is living – can literally mean "soul."

Hence, a living creature is interpreted to be a creature that has a soul.
196 'ASAH.
197 LETAKEN. The term also means to mend or repair, i.e., to take something that already is there and improve it. It is the verb that is used in the idiom, "to mend the world" (LETAKEN HA-'OLAM) which expresses Israel's moral mission to contribute to the salvation of the universe.
198 TSEVAYON. While the term means character, nature or form, it also expresses desire and pleasure. The choice of word in this case suggests a direct proportionality between being natural, desirable and pleasurable.
199 QUMAH.
200 I.e., brought into existence out of nothing in an amazing way solely by God's word.
201 ZAKHAH.
202 The underlying assumption that God as legislator of the universe is subject to the laws he legislates. As work on the Sabbath is forbidden to his creatures, so God is forbidden to work on the Sabbath.
203 Viz., TOHU, BOHU, CHOSHEKH, RUACH AND TEHOM.
204 Viz., the divine robe, the element fire, the Throne of Glory, the site of the Temple, heaven, hell and Torah.
205 Viz., earth and water.
206 Most notably, the divine robe.
207 For example, in that world, unlike ours, the sun and moon have the same size, and trees are as edible as their fruit. Furthermore there exist species in the Garden of Eden that have never existed in This-World of ours, and humans are by nature vegetarians.
208 For example, sin accounts for the physical characteristics of trees and the number and size of celestial objects.
209 For example, as some things (e.g., TOHU, BOHU, CHOSHEKH, etc.) were created in a prior universe to be used to create this one, some things created in this world exist for use in the next world (e.g., primordial light and the upper waters).
210 The clearest example of this defining character of their commentaries is the identification of the 6th of Sivan as the date of both the origin of our world and the giving of the Torah at Sinai. As such their commentary is coherent with Rosenzweig's close link between creation and revelation.
211 In a sense the former were rhetoricians and the latter logicians. In this respect, Rosenzweig's study of creation is logical, his study of redemption is rhetorical, and his study of revelation

bridges the two. In other words, while Rosenzweig emerges as a logician within the context of the doctrine of creation, his total work in Jewish thought (*The Star of Redemption*) utilizes both approaches. This is yet another respect in which Rosenzweig emerges as a Jewish thinker.

212 Which is what "BERESHIT" means.
213 Which is what "ELOHIM" means.
214 Viz., the angels.
215 CHAFETS, which is what God's "RUACH" means.
216 Viz., the TOHU.
217 Which is what "BARA" means.
218 Note that the Stoics made the element air the active causal agent of all biological activity. While Ibn Ezra may be influenced in this case by Hellenistic thought, we shall see that it is not an unreasonable interpretation of scripture itself.
219 I.e., "This-World" (HA-'OLAM HA-ZEH).
220 HA'OLAM HA-BA.
221 HA-'OLAM HA-ZEH.
222 Which is what "BERESHIT" means.
223 Which is what "ET HA-SHAMAYIM VE-ET HA-ARETS" means.
224 Which is what "ELOHIM" means.
225 Which is what "BARA" means.
226 Which is what "TOHU" means.
227 Which is what "BOHU" means.
228 Which is what "CHOSHEKH" means.
229 Which is what "RUACH" means.
230 Which is what "TEHOM" means.
231 I.e., during day one.
232 The close similarity of Nachmanides and Plato's *Timaeus* cannot be ignored. In this dialogue deity differentiates space into distinct regions through whose essential motion arise the elements.
233 Which is what "BERESHIT" means.
234 Which is what "BARA" means.
235 Which is what "TOHU" means.
236 Which is what "BOHU" means.
237 Which is what "RUACH ELOHIM" means.
238 Viz., the "TEHOM."
239 Viz., the "CHOSHEKH."
240 In his commentary on Gen. 2:3 Nachmanides seems to contradict himself when he suggests that each day represents 1,000 years. The context of his discussion of the meaning of the term "day" in Genesis 1:3 is to give the simple, linguistic meaning of

the term. In contrast, the context in Gen. 2:3 is a discussion of messianism. There the interpretation is presented within the framework of Jewish mysticism, and, as we noted above, linguistic/philosophical commentaries cannot be compared with kabbalistic ones. There is no incoherence in the fact that a statement within one context literally contradicts a statement confined to the other context.

241 This is not the only parallel between Nachmanides' commentary and Plato's *Timaeus*. For example, in both cases what deity initially does is to demarcate undifferentiated regions of empty space into separate regions from which arise the elements of the physical world of generation and corruption.

242 As we shall see below, in this respect (as well as others) Sforno's commentary is closer to the implicit physics of the biblical text than any of his predecessors.

243 Desire and thought, in God's case, in consequence of His oneness, is the same thing.

244 Viz., wisdom and the Throne of Glory.

245 See days three and four below.

246 I.e., God's wind (RUACH ELOHIM).

247 Note that even those commentators who want to maintain that creation is a temporal event insist that the seven days are not seven days but are in fact differentiations of a single event. Implicit in this insistence is their understanding of what divine unity entails. That God is one says more than the claim that the set of deities contains only one member. More significantly, it says that in God's case there is an identity between God as actor and his action such that God as subject is a simple undifferentiated entity and God's action is a simple undifferentiated act. In other words, as God is not a compound with multiple parts, so God's act is not a general term for a set of multiple acts.

248 "RAQI'A."

249 "SHAMAYIM."

250 "TEHOM."

251 "'AFAR."

252 "OR" for Ibn Ezra and "CHOSHEKH" for Nachmanides.

253 "RUACH" for both Ibn Ezra and Nachmanides, and "CHOSHEKH" for Sforno.

254 "MAYIM" for Ibn Ezra and Sforno, and 'TEHOM" for Nachmanides.

255 "ERETS" for Ibn Ezra and Sforno, and "'AFAR" for Nachmanides.

256 Viz., "OR."

257 I.e., the laws of the universe are not morally neutral.
258 KOACH.
259 Viz., OR.
260 I.e., those physical objects that have souls.
261 The issue between them is the meaning of "NEFESH CHAYAH." Ibn Ezra says it means the soul of something that is alive, whereas Nachmanides and Sforno say that it means a thing that lives in virtue of having a soul.
262 According to Nachmanides, these are things that have constant motion.
263 According to Nachmanides, these are things that creep upon the earth. Ibn Ezra calls them tiny things that live on the land.
264 According to Nachmanides, these are the great sea-monsters.
265 According to Nachmanides, these are animals that eat plants. Ibn Ezra says that they are domestic animals. Sforno says that they are land animals in general.
266 According to Nachmanides, these are carnivorous animals. Ibn Ezra says that they are wildlife.
267 KAVOD.
268 Viz., HA-ADAM.
269 For a Hellenistic parallel to this homily, see Aristophanes' speech in Plato's *Symposium* (189a–194a).
270 Cf. his commentary on NA'ASEH ADAM in Gen. 1:26.
271 "BEHEMAH."
272 I.e., to become a "TSELEM ELOHIM."
273 "ELOHIM."
274 Nachmanides' commentary captures (in my judgment) the literal intent of the Genesis narrative, viz., that God acts to differentiate space into definite regions while the regions themselves (in response to divine command) generate the species-paradigms that occupy space, and the species generate the endless/infinite number of their particular/material/individual exemplifications. On Nachmanides' interpretation, HA-ADAM is the only instance of a paradigm being generated by more than a single region of space. In this case the generators who carry out God's order are the spaces, sky and earth.
275 From Nachmanides' statement that "the souls of men as well as angels are included in the host of heaven" (*Ramban [Nachmanides]: Commentary on the Torah. Genesis.* Translated into English by Charles B. Chavel, New York, Shilo, 1971, p. 59, n. 240), Chavel comments that human souls were created at the beginning of creation, and he refers readers to Nachmanides' correspondence in support of this interpretation. However, this

explanation does not follow from Nachmanides' actual words in this passage, and in fact directly contradicts what he says in his commentary on Gen. 1:26. Furthermore, even if it is the case that Nachmanides says one thing in his correspondence, it does not automatically follow that that statement can legitimately be used to interpret his published commentaries.

276 As we shall see, of all these rabbinic commentators, Nachmanides' interpretation of creation comes closest to the cosmogony of Plato's *Timaeus*. This is especially true in this case, where both writers picture the culmination of creation in the souls of living things and not their bodies. However, this is not to say that Nachmanides' doctrine of creation is identical with that of the *Timaeus*. There are notable differences, not least of which is that on Plato's account the first soul is that of a male member of the human species, and only in consequence of its imperfection do other forms of life emerge. In contrast, according to Nachmanides, the souls of the other life forms are created first, and the human soul is not differentiated by gender.

277 In this respect the distinction between philosopher and theologian is irrelevant.

278 I have chosen this specific topic because it exhibits, at first reading, the greatest degree of disparity between the rabbinic commentaries. In other words, if any part of the classical rabbis' account of creation is incoherent, it is this one.

279 Cf. Leo Strauss, *Philosophy and Law: Essays Toward the Understanding of Maimonides and His Predecessors*. Philadelphia, Jewish Publication Society of America, 1987. E.g., "Philosophy owes its authorization, its freedom, to the Law; its freedom rests on its bondage." (p. 67) This particular passage applies specifically to Averroes. However, Strauss makes the same point about the thought of Maimonides and Gersonides. For Gersonides the Torah is "not a barrier to research – since research comes to no barrier in uncovering the wisdom and mercy contained in it – but is a direction for research . . . The primacy of the Law is just as secure for Levi as for Maimonides and Averroes." (p. 78).

280 Those commentators (notably Nachmanides and Sforno) who speak of creation occurring in some time qualify the claim by adding that (1) it is a different time than the time of our world which, with respect to our time, is not time at all, and (2) the time that they mention follows the initial act of creation itself; that initial act is single, for the consquence of God's unity is that God cannot have multiple acts. Rather, what occurs during the seven days of creation is a progressive unfolding of the process

that God sets into motion at the origin of the universe. Hence, at first there is God's act, which itself is eternal, i.e., atemporal, and next is the time in which that act unfolds into a spatial model for our universe, all of which precedes both our universe and its time. Hence, with respect to the time of our world, there is consensus that creation does not occur in time.

281 In this sense this next discussion forms a transition from the philosophical reading of texts in parts 1 and 2 to the more direct inquiry about what is believable about creation in part 4.

NOTES TO CHAPTER 5

282 Notably Exodus 40:12–25, Psalm 74:12–17, Psalm 104:1–35, Job 26:7–14, and Job 38:4–39:30.

283 It is commonly held that Genesis presents two different accounts of creation, one in Gen. 1:1–2:3 and another in Gen. 2:4–25. However, my reading of the text suggests otherwise. The second text deals, not with the origin of the universe but with the origin of humanity. Humanity's beginning occupies only three verses of the first story (viz., Gen. 1:28–30); conversely the universe's beginning occupies only three verses in the second story (viz., Gen. 2:4–6). Whereas humanity appears in the first account only as one character in the story of the creation of the universe, the universe appears only as a context for the story of the beginning of human history. Many modern scholars believe that these two narratives are incompatible. My own judgment is that the classical rabbis were correct in viewing them as consistent and coherent. However, given the significantly different subject matter of the two stories, no judgment on this question need be considered in this book, i.e., it will not affect anything that has or will be said here about the Genesis account of creation.

284 The reading that results is significantly different in many respects from how the text has been interpreted by many contemporary biblical scholars. In fact the reading is sufficiently different that a mere summary of my conclusions probably will not be convincing to many of the readers of this book. I attempted to defend my reading with a careful, detailed (linguistic and logical) analysis of each word, phrase and verse of what the text itself says. However, in this instance, the length of the argument turned into a separate book, viz., *The First Seven Days: A Philosophical Commentary on the Creation of Genesis*. Atlanta, GA, Scholars Press, 1993. It is there that readers should turn for

the demonstration of the conclusions that are summarized in this present chapter.

285 I.e., the joining of the verb at the beginning of a phrase or a sentence with the consonant of conjunction.

286 In other words, a "stable act" is one that achieves its end and persists in it, while a "dynamic act" is one that moves towards (or away from) its end but has not reached it. We can picture the difference mathematically by analogy with a function. A stable act is like a function (Fx) where the value of the function for every instantiation of the variable x is the same number (n). Conversely a dynamic act is like a function (Fx) where the value of the function is different for any two instantiations of the variable x. The kind of function I have in mind is one that is end directed, i.e., there is a number (n) such that the value of every instantiation of the variable x in the equation (Fx) either approximates n or is n.

287 The verb in this case is BARA. In this text I will use its conventional translation as "creates." However, what the English verb states is not really what the Hebrew verb means. Our primary use of the English term, "to create" is as a direct action verb with which we associate what artists or craftspeople do as agents to objects like stones or pieces of cloth or paper in order to produce certain artificial products like statues, paintings or poems. However, there are perfectly good Hebrew terms to express this usage, viz., "YATSAR" (which never occurs in our narrative) or even "'ASAH" (more literally, "to do"), but not "BARA."

The verb "BARA" is a simple, active construction in the third person, singular form of the perfect tense. The verb is used in this form four times in Genesis, once more in the Pentateuch, and six more times in the Hebrew scriptures. The simple, active construction of the verb appears in other forms twenty-three times in the Bible. In every one of these cases the agent of the action is God. The object or product (as the case may be) of God's distinctive action in every instance but one is either the universe or human beings, and in most cases God's act describes or alludes to, or parallels, what is described in this first chapter of Genesis. In other words, the term "BARA" functions primarily to express the unique act, whatever it is, that God does here to or with the universe and humanity.

Hebrew verbs appear in seven different constructions. Grammar books list them as (1) the simple active, (2) the simple passive, (3) the intensive active, (4) the intensive passive,

(5) the causative active, (6) the causative passive, and (7) the reflective. The names suggest that the root verb is the same in every construction, except for the way it is used. However, more often than not, root verbs in different constructions have very different meanings that cannot be explained simply by saying one form is active or passive, intensive and/or causative and/or reflective or not. This is the case with the verb in question. Its root consonants are bet, resh, and aleph. Besides its use in the simple active construction, it also appears in the Hebrew scripture as the intensive active four times. In one case, the actor cuts people with a sword; in two cases he cuts down trees with an axe; in one case he makes a sign-post, which presumably involves carving wood with a knife. What is common to all of these examples is that someone cuts something with something else. Hence, the verb expresses a three-term relation in which an acting agent changes the nature of a recipient object by means of using some kind of tool.

This root-form also occurs thirteen times as BARIY (i.e., as bet, resh, yod, aleph). In these cases the term expresses something being fat and/or healthy. Because the root means "to cut" in the intensive active, and "to be fat and/or healthy" in a non-verbal form does not mean that "BARA" means "God cuts something (a) with something (b) to make something (c) fat and healthy." Still all of these forms share a common root, which prima facie suggests that, whatever it is that God does at first, it has some connotations of cutting and being fat/healthy. However, whatever those connotations are, they need not have any association with time.

In modern Hebrew the tense of a verb expresses time. The perfect tense is used for the past, the imperfect for the future, and the active participle for the present. However, as a general rule, the form of verbs do not themselves express time in biblical Hebrew. This role tends syntactically to be the function of adverbs. The tense of the verb itself tends instead to express the state of the action asserted, viz., whether or not it is complete. Hence, "BARA" asserts some sort of completed action directly performed by God, irrespective of any temporal reference for the action. Furthermore, that action may express a three-term relationship in which God uses some thing(s) as a means to affect some other thing(s) for the latter's well-being.

How then should this critical term be translated? The choice is somewhat arbitrary, since no English verb adequately expresses what it should stipulate here, viz., an action uniquely characteristic of a deity. The term encompasses everything that God does to the planet earth and the sky. At this general level, telling (YOMER) functions as a general synonym of BARA, since everything comes to be by God telling it to be. At the same time this one general act is expressed as three distinct kinds of actions through three separate verbs. One, telling (YOMER) some things to bring forth other things, viz., earth to bring forth vegetation and land life, and water to bring forth sea life. Two, naming (YIKRA) things, viz., day and night, sky, earth and seas. Three, making ('ASAH) things, viz., a (so-called) firmament, the sun and the moon, things that live on the planet earth, and the human being. Given the long historical association of the verb BARA with human creativity, it would be least misleading to avoid the otherwise acceptable English term, "creates." Better choices would be "produces" or "brings into being," because they are more ambiguous than "creates." Even better would "declares to be," since it highlights the text's emphasis on God's act as a kind of speech. The best choices are either to use the Hebrew term itself as an English verb or to make the noun "God" into a verb, i.e., to translate either "God baras" or "God gods." Of the two, I prefer the latter. Making "god" a verb has the advantage of indicating that there is some form of identity between what God does to the universe and what God is. It also helps convey the sense that the verb that expresses God's act is independent of any time frame.

It should be noted that there is a second verb whose subject always is God. Its root is samekh, lamed, chet. Cf., Jacob Milgrom's commentary on Numbers in *The JPS Torah Commentary*. Philadelphia/New York, The Jewish Publication Society, 1990 Excursus 32, pp. 395–396. The usual translation of this verb is "to pardon." However, as Milgrom indicates, this term no more has the ordinary sense of pardoning than "BARA" has of creating. Both terms express ways that God acts as God, viz., that God gods. They are not different because of what God does; rather, they differ because of the object that receives God's single act. In the case of "SALACH," the object is the human; in the case of "BARA," the object is the universe.

288 I.e., Gen. 1:1–2:3. Henceforth we will refer to this narrative simply as "Genesis."
289 What God's wind (RUACH ELOHIM) is is open to debate. Either it is, as Rosenzweig suggests, an expression of what God is prior to creation, or it is something distinct from God that, as such, is another element whose existence preexists the act of

creation. My own opinion agrees with the latter judgment. As we discussed above, Proverbs (and the first verse of the Gospel of John as well) identify it with Torah in the sense of divine wisdom. I tend to see it as something that resembles the ancient Stoic concept of PNEUMA, viz., the breath of the universe, that is a compound of the elements fire and air (both missing from Genesis), that functions as an active principle on the passive elements, water and earth.

290 "RAQI'A'" is a masculine noun that appears several times in the Hebrew scriptures, sometimes (but not always) with whatever meaning it has here. Its verbal root is resh, kuf, ayin. The verb occurs several times in different constructions. Jeremiah uses it once in the intensive passive, and Job uses it once in the causative active conjugation. Job uses it to state what God does to the sky, and Jeremiah uses it to describe silver that is beaten into plates. The verb is used three times in the intensive active to express stretching and/or spreading out one thing on something else. In Isaiah 40:19 a goldsmith spreads melted gold over an object; in Exodus 39:3 gold is beaten into thin, flat plates; and in Numbers 17:4 the priest Eleazar beats copper fire-pans into a cover for the altar. The verb's most frequent use is in the simple active conjugation, where it means either "to stamp upon something with feet" or "to stretch something out." Whenever it is used as Job used it, viz., with reference to God producing the universe, it means "to stretch something out." Isaiah says that "alone God NOTEH the sky, and by himself ROQA' the earth." The parallelism between the two parts of the verse suggests that, as "alone" and "by himself" are parallel, so "NOTEH" and "ROQA'" are parallel. "NOTEH" means to spread or stretch something out or to expand it. In other words, God produces the universe by expanding both the sky and the earth beyond what they are at first. In fact, Isaiah associates the three verbs – BARA, NATA, and RAQA' – together when he says that "the Lord/God (HA-EL YHVH) is BORE of the sky and their expansions (NOTEY-HEM) [and] ROQA' of the earth and its progeny."

The fact that the verb RAQA' parallels NOTEH in meaning "to expand" has led some scholars to translate "RAQI'A" as "expanse." Isaiah pictures the earth as a sphere (CHUG) with the sky spread out (NOTEH) around it like a curtain or tent (DOK), i.e., like a canopy. Similarly, Job identifies God as the one who by himself stretches out (NOTEH) the sky. On the other hand, Psalm 136:6 states that what God stretches out

(ROQA') over the water is the earth. While Isaiah/Job and the Psalmist disagree about what God expands, they agree that what is expanded is an expanse. However, this translation is misleading. It suggests that a RAQI'A is a region of space, like the domains of the earth and the water, which it is not. Rather, it is a new kind of stuff, like the light of the first day, that fills space.

The usual translation of "RAQI'A" is "firmament." However, as we have seen, this English term is misleading, since the Hebrew term's referent is not itself firm and it does not make anything else firm. (I assume that this translation is based on an interpretation like the one we encountered in the classical commentaries, viz. that the RAQI'A is formed by liquid water becoming solidified through the action of fire.) Rather, this RAQI'A, like its feminine noun counterpart, REQI'AH, is something pliable that can easily be stretched out, i.e., a ductile material. In other words, it is a kind of elastic, spread-out-able stuff, distinct from the previously listed elements, that God stretches out above the sphere earth in order to divide the encircling space of water into two distinct regions. "Stretcher" is a better translation than "firmament," because it preserves the sense that the function of this stuff, viz., stretching, defines it. However, this word has other associations in English that are not appropriate for the Hebrew. My personal preference is for "spread." (The image of something of the consistency of a cheese spread seems to me to be perfectly appropriate to what the Genesis narrative intends to communicate.) A spread is a ductile material that is precisely what its verb expresses, viz., something of indeterminate substance that someone extends over something else. In this case the material is spread throughout the interior of the sphere of elementary water that encircles the sphere of earth, dividing two distinct areas of space into three.

291 The one possible exception of this generalization is the human. In this case God makes him/her (the text says that the human is both male and female) with someone/something else. Who the "else" is is debatable. The text has God saying "let us . . . in our . . ." (Gen. 1:26). Of the candidates suggested by the rabbinic commentators for God's partner in making an ADAM, the least likely are the previous five days of creation and the souls of the righteous. In contrast, the most likely interpretations are the following three: First, there is no partner. God is simply invoking what is called the "royal we." Second, God's partners are the earth and the sky. As God's productions on the fifth day involve the space of the waters with the sky, so God's productions on the

sixth day (viz., animals, creepers and an ADAM) involve the earth with the sky. Third, God's partner is God's wind. The textual basis for this judgment is clear. The above textual analysis would suggest that God's partner differs qualitatively from anything else mentioned as a co-producer. It should be something mentioned within the narrative that, in one sense, is distinct from God (to account for the use of the plural), but, in another sense is identical with him (to account for the use of the first person and the fact that only God is named as an actor).

All three views seem plausible readings of our text. However, my own judgment comes close to the second "likely" interpretation given above. I think that God is speaking to the earth. In general, the work of the sixth day consists of God ordering the earth to generate fresh life. First he generates the animals and creepers, and here he generates the ADAM. There is no need to mention the earth again, given the context of the previous two verses. In other words, given that this verse stipulates that God once again is producing something by ordering a specific region of space to act in a certain way, the most obvious candidate for a direct originator of this particular form of life is the earth.

The model for the partnership between God and the earth is political. As a king may order his scholars (for example) to write a translation of the Bible and thereby share with them the authorship of the product (e.g., the so-called "King James" English translation), so God orders his minister, the earth, to produce a living occupant. What God joins the earth in making is ADAM. The term is a singular, masculine noun. It is not a proper name. Rather, like the sprout (DESHE) of the third day, the swarm (SHERETS) of the fifth day, and the fresh life (NEFESH CHAYAH) of the sixth day, it is a prototype, whose ontological place within the picture of our textual model functions to explain the infinite power of the earth that enables its occupants to reproduce without beginning or end. In other words, from the perspective of reproduction, there is no first nor last human being. ADAM is not a first person; rather, he is an object situated in a model that accounts for endless human procreation. This particular earth-originating fresh life is distinct from others in two ways. First, God is specified as a partner in his origin. Second, ADAM originates in ADAMAH, i.e., earth covered with vegetation that decays into the most excellent kind of soil. In other words, this life form is special in that it originates in "humus" and, hence, is called a "human."

The particular richness of his origin enables our narrative to claim that the human is something that is in both the "TSELEM" and the "KADMUT" of God and the earth. "TSELEM" is a masculine noun whose root is tsadik, lamed, mem. It occurs fifteen times in the Hebrew scriptures. The verbs formed from this root – TSILEM (to photograph) and TSULAM/HITSTALEM (to be photographed) – are creations of modern Hebrew from the biblical noun. In this and the next verse of our narrative, the noun expresses some kind of similarity between the human and God. Of particular interest is the use of the noun in its two appearances in the Psalms. Ps. 73:20 mentions the TSELEM of human beings as something that God will despise, and Ps. 39:7 says that a man (ISH) walks as a TSELEM, where the context makes it clear that this is something of no value (HEVEL). Similarly, in every other appearance, TSELEM expresses something negative, viz., the figures that idolators craft as part of their ritual. In other words, a TSELEM is a figure or image that seems to be uniquely associated with the work of sacred artists, viz., those who create physical objects to function in communal worship. In every case, to call something an "image" seems to have a negative connotation. Hence, while the human is in God's "image," the sense is that he is MERELY an image.

I suspect that the negative sense of the noun has to do with an implicit ontology in our narrative that admits of degrees of reality. In general, something that is a figure of something else is less real than its original. Hence, while a human is something rich by comparison with every other occupant that space generates, he remains, nonetheless, a mere "figure," i.e., something less perfect/excellent, and consequently, less real, than his divine original. In this ontology, the notions of real, perfect and complete entail each other. God is most real. Next are the divisions of space. Next is the human, and finally are the other varieties of fresh life.

"DMUT" is a feminine noun formed from a verb whose root is daled, mem, he. The noun often means "likeness," "form," "shape," "resemblance," "figure" and/or "character." Both the noun and the verb occur frequently in the Hebrew scriptures. Ezekiel often uses the noun in conjunction with the term "MAREH," which means something that is seen, viz., an appearance and/or a vision. Similarly, in 2 Kgs. 16:10, the term is used interchangeably with "TAVNIT" (form, model, pattern, figure, image, paradigm, and/or structure) as a model for constructing an altar. The verb can have two very different meanings. On the one hand, it often means to "be like" and/or "resemble." On the other hand, it can also mean to "cease," "stop" and/or "destroy." Clearly the term "DMUT" in our text is meant to be interchanged with the term "TSELEM," where together they share the variety of meanings of the English term, "figure." However, whereas the term "image" (TSELEM) adds the negative connotation of something of comparatively lesser value, the term "likeness" (DMUT) adds the sense of something perishable. In other words, for all of the exalted status of the human as a prototype within our model that shares common characteristics with his generators, he remains, nonetheless, inferior to them as something perishable.

Both terms – image and likeness – suggest that in some respect the human is, and in another respect is not, like both God and the earth. The analogy to the earth is clear. Like the earth, the human originates in and consists of the stuff of the earth; unlike the earth, the human is a transient occupant of the intransient space of the earth region. In the case of God, the analogy is less clear. While there is no difficulty in saying how God and the human differ, the question remains, how are they alike. In my judgment, the answer lies in the next half of this verse, viz., as God governs the space of the universe, so the human is given authority to rule the occupants of the earth.

To summarize, the general context of the passage is that God's work on the sixth day is to command the earth to produce its occupants and the human is one of its occupants. Note also that the "us" need not mean that God in this case uniquely acts directly on a spatial object. It may simply mean (which I think it does) that God commands the earth to produce the human. The earth does the actual work. God's role as a producer is as the agent whose command causes the earth to act.

292 All of the occupants created on day six are referred to (in one form or another) as NEFESH CHAYAH. The noun, "CHAYAH" is formed from a verb whose root is chet, yod, he. The verb means "to live." Hence, a CHAYAH is an object whose characteristic act is to live, viz., a life.

The term, "NEFESH" is formed from a verb whose root is nun, pe, shin. The noun often is translated as "soul," but this translation is not correct. This English term presupposes a theory of being that makes a radical separation between what is physical/material and what is mental/spiritual, and this particular dualism does not fit any possible ontology in the Hebrew scriptures. Similarly, the terms "mind" and "spirit" are common but erroneous translations of the noun, "NEFESH."

While the noun occurs frequently throughout the Hebrew scriptures, the same cannot be said for the verb. It occurs only three times, and in each of these cases, it is a third person, singular form of the imperfect tense in the simple passive conjugation, viz., YINAFESH. In 2 Samuel 16:14 we are told that David and his followers, in flight from Saul, were tired and "YINAFESH" there. Here, clearly the verb means to make a rest stop, i.e., to rest and become refreshed. In other words, the men were fresh, their labor made them cease to be fresh, and now, through rest, they once again (which is what the "re" in "refreshed" means) become fresh.

In Exod. 23:12 the verb is associated with the verbs for rest (HENIYACH) and for the characteristic form of activity (SHABAT) on the Sabbath. The seventh day is proclaimed a day on which you should SHABAT ("TISHBOT"), so that your lifestock will rest ("YANUACH") and your guests will "YINAFESH," i.e., will regain their state of being fresh. Similarly, Exod. 31:17 joins YINAFESH to SHABAT. In this case the subject of the action is God. The context is an explanation of the work prohibitions on the Sabbath, with specific reference to the end of our narrative. We are told that, since God SHABATs and NAFASHes on the seventh day, Israel should observe the Sabbath by refraining from work.

Consequently, the noun, NEFESH describes something that is NAFASH, viz., is fresh. The term, "fresh" is associated both with a living animal being active, i.e., being "lively" and with a dead animal being fit to eat, i.e., being "fresh meat." Both senses of the term fresh are appropriate here for two reasons. First, there is the immediate conjunction of the term "NEFESH" with "CHAYAH." Second, our narrative relates every living thing to man as food.

Again, the verb NAFASH in the simple active conjugation usually means to "rest," "recuperate" and "be refreshed," all of which are appropriate here, given the key role that the Sabbath plays in our narrative. However, most translations associate the noun, not with this usage of the verb, but with the way the term generally is used in the intensive active conjugation, – viz., to "animate," "breathe life into" and "enliven," – even though this verb construction never occurs in the Hebrew scriptures. They do so because of the association of the noun in this verse (and in verse 24) with the noun CHAYAH. In other words, these translations take the term NEFESH to describe what is distinctive about a CHAYAH, viz., that it is a form of animate life. However, the verb that expresses this distinctive characteristic is CHAYAH. For this purpose there is no need to add that a CHAYAH is a NEFESH. Hence, it is a better reading to say that the term NEFESH says something more about this entity than the fact that it lives, viz., that it is the kind of living thing than has a capacity to rest and to become refreshed, and, as such, it has a capacity to do whatever it means to perform the characteristic activity of the seventh day.

293 In yet another way the notion of space is primary in our narrative both ontologically and morally.

294 In our narrative only vegetation is designated as food. However, Abel sacrifices domestic animals with God's approval (Gen. 4:4), and the descendants of Noah are given every other living thing as food (Gen. 9:2–3). The rules and regulations of the Temple cult itself, given in the heart or center of the Torah (viz., at the end of Exodus, in Leviticus, and in the beginning of Numbers), specify which living things (including vegetation and fresh life of the earth and fowl of the sky, but nothing that resides in the water) God eats.

295 "And there has never again arisen a prophet in Israel like Moses whom God knew face to face" (Deut. 34:10).

296 Perhaps it is understandable that authors whose main "problem" was to explain the loss of their land would compose a story where what is most valued is space. Less understandable (and hence, more interesting) is the fact that this tale, whose central value is space, becomes the foundation text for a religious nation (viz., classical rabbinic Judaism) which makes time a central value (viz., the *raison d'être* of rabbinic Judaism is the execution of rituals that are defined by time) and who assigns minimal real worth to space (viz., the people of rabbinic Judaism live two thousand years in a nation that is defined by government that is independent of any land). In part this contradiction is overcome by rabbinic Judaism placing the possession of land into the distant future of the Messianic Age. In other words, in the World-To-Come (but not in This-World) Israel will again possess its own land and God's universe will again be defined (as it was at the beginning) by distinct territories of space.

297 Deut. 28:3–14.

298 Deut. 27:15–26; 28:16–68.

NOTES TO CHAPTER 6

299 Among the more obvious examples are the description in Plato's *Symposium* of the view that the original human being was both male and female (189e–190a), and the reference at the beginning of his *Timaeus* to a pre-historical universal flood (22c–e).

300 The following summary of Plato's *Timaeus* was researched at the Oxford Centre for Postgraduate Hebrew Studies while I was on a research leave from Temple University from January to May, 1987. I wish to express my appreciation to Temple University for giving me the time and to the Oxford Centre for providing me with the space to write this chapter.

301 The *Timaeus* is not the only work by Hellenistic philosophers that influenced rabbinic thought. Among the other authors worthy of consideration by historians of ideas are the writings of the sixth-century B.C.E. Miletians (Thales [624–546 B.C.E.], Anaximander [610–545 B.C.E.] and Anaximenes [died *c.* 525 B.C.E.]), Heracleitus (540–475 B.C.E.), Empedocles (*c.* 500–430 B.C.E.), Anaxagoras (*c.* 488–428 B.C.E.), Pythagoras and his disciples (middle of the sixth century to middle of the fourth century B.C.E.), Plato of Athens (430–347 B.C.E.), Plato's disciples (Eudoxus of Cnidus [*c.* 490–356 B.C.E.], Callippus [370–300 B.C.E.] and Aristotle of Stagira [348–322 B.C.E.]), Aristotle's student Theophrastus (372–287 B.C.E.), the Atomists (Leucippus of Miletus [332–262 B.C.E.], Democritus of Abdera [*c.* 460–370 B.C.E.], Epicurus of Samos [341–270 B.C.E.] and Lucretius [*c.* 95–55 B.C.E.]), the Stoics (Zeno of Cition, Cyprus [*c.* 332–262 B.C.E.], Chrysippus of Solí, Cilicia [*c.* 280–207 B.C.E. and Poseidonius of Apamea, Syria [*c.* 135–51 B.C.E.]), Aristarchus of Samos (310–230 B.C.E.), Archimedes (287–212 B.C.E.), Seneca (*c.* 3 B.C.E. – 65 C.E.), Plutarch (*c.* 46–120 C.E.) and Ptolemy (*c.* 150 C.E.). Of these philosophers the most influential were Plato, Aristotle, the Stoics, Archimedes and Ptolemy. Furthermore, there are multiple references to creation in the corpus of Plato's writings. Of these works the most important are found in the *Meno*, *Republic*, *Parmenides*, *Theaetetus*, *Timaeus*, *Critias*, *Philebus*, *Sophist* and *Laws*. However, of these dialogues the most elaborate account is given in the *Timaeus*, which (for the reasons discussed above) will be our sole textual focus in the discussion that follows.

302 *Timaeus* 22e.

303 Literally, "ΕΙΚΟΤΑ ΜΕΤΗΟΝ" (*Timaeus* 29d). Cf. Cornford's discussion of this passage in F. M. Cornford, *Plato's Cosmology*.

Indianapolis, Bobbs-Merrill, 1956 (hence forth referred to as "Cornford"). pp. 28–32.
304 Viz., in *Timaeus* 51b–52c. More will be said about this particular text in the discussion below.
305 Or, "what is" (TO ON).
306 Or, "what comes to be" (TO GIGNOMENON).
307 Viz., the other principles.
308 Or, "primary entities."
309 More literally, with what requires "skill" and "craft."
310 Anaxagoras, who makes NOUS a first principle, is an important exception to this generalization.
311 Or, density.
312 Or, volume.
313 Generation.
314 Corruption.
315 Viz., the differentiated body and soul of the world.
316 Viz., number and the infinite dyad.
317 For a more complete account of the generation of these intervals that employs a greater knowledge of music than I possess, see Cornford p. 61.
318 For example, see S. Sambursky, *The Physical World of the Greeks*. Translated from Hebrew into English by Merton Dagut. London, Routledge and Kegan Paul, 1956. Chapter 3, p. 61.
319 In this connection, Aristotle's *De Caelo* could be read as an attempt to solve this problem. The only qualification of this judgment is that Aristotle's intention was that of a scientist and not a historian. His major concern was to explain the heavens, and in doing so, he used Plato's judgments. His interests in faithfully representing what Plato actually said was important to him, but its importance was secondary. Hence, Aristotle's astronomy can be read more as a construct of Plato's thought than a historical report. Their intent and their picture of the universe was to a great extent the same. However, there is one fundamentally important difference between them. It is not that Aristotle made Timaeus' mathematical model into a real picture of the heavens. Rather, it is that Aristotle thought that he could give a scientific explanation of what Plato, like the author(s) of Genesis, thought could only be related through a story.
320 Which is expressed in terms of their domain.
321 Characterizing the direction of spin or rotation up an axis as "up" or "down" is standard in speaking about elementary particles in contemporary physics. To say that a spin direction is

"up" means that it is

Conversely, to say that it is "down" means that it is

Extend your right arm without bending your elbow until your arm is perpendicular to your body. Make a fist and stretch out your thumb. When the thumb is up with your other fingers curled in your right palm, the circular direction to which your finger tips point is the up direction. Conversely, when the thumb is down, the circular direction to which your finger tips point is the down direction.

322 That the earth has a motion of its own, counter to the direction of the motion of the fixed stars, raises a problem. If the earth were stationary, then it would be reasonable to say that the speed of the motion of the same is such that it completes a circuit in twenty-four hours, because every evening we see the same "fixed" star reappear in the sky. However, if the earth itself is spinning in a direction counter to the sphere of the fixed stars, then the absolute motion of the latter must be shorter than one day. Given that the motion of the same is faster than the rotation of the earth, the completion of the circuit of the fixed stars in twenty-four hours would be a relative and not an absolute circuit. I found no explanation of this apparent discrepancy in any of the commentaries listed in the bibliography. This mathematical problem could be a reason for Aristotle rejecting the claim that the earth spins on its axis.

323 The Evening Star or Lucifer.
324 Hermes or Stilbon.
325 A. E. Taylor argues that the views presented in this dialogue are not Plato's, but merely those of the leading character in this drama. Plato's own views are found in the *Laws*, chapter 10. Here he merely constructs what he thinks a late fifth-century B.C.E. Italian Pythagorean would say about creation. Cornford presents convincing arguments (in my opinion) against Taylor's position.
326 See Timaeus' description of the qualities that appear in the receptacle in 52c and Cornford's discussion of this passage on p. 194. The expression "ON PUS" itself actually occurs in the *Sophist* (240b).
327 Although slightly strained.

328 Or, sometimes, the nurse.
329 Or, bodies.
330 I.e., the receptacle.
331 Rather than eternal.
332 Rather than independent.
333 "LOGISMU TINI NOTHU" (The passage continues, "MOGIS PISTON.") (52b). This passage is the single most important text in the *Timaeus* for the epistemological status of judgments in cosmology and cosmogony. Timaeus states it after completing his discussion of the receptacle (ch. 18, 48e–51b) as a product of necessity rather than of reason. (The products of necessity are discussed in ch. 17, 48e–51b, and the products of reason are discussed in chs. 6–16, 29d–47e.) In other words it is introduced specifically to explain the puzzling epistemological status of knowledge of space, which itself has an equally puzzling ontological status. The more specific context of this sentence occurs in 51b–52c. There Timaeus uses the example of the element fire to distinguish at a more general level three kinds of entities and three ways of knowing them. The three kinds of entity (ONTA) are the rational form (EIDOS EKASTON NOETAN), appearances (FANTASMA) and space (CHORA). They are known respectively by reason (NOUS), true opinion (DOXA ALETHES) and "an understanding that is bastard [or, illegitimate] that is hardly credible [or probable or confirmable or believable]." The second are known in a dream-like state as located in the third kind of entity. Yet, properly the third kind of thing is not a thing at all because "We say that every entity of necessity is somewhere at some place occupying some space, and that which is neither on earth nor somewhere in heaven is nothing (OUDEN EINAI)" (52b). Of course, space itself is not somewhere and is no place and occupies no space in either heaven or earth. Hence, it clearly is a nothing (MEDEN).
334 Or, the receiver = the receptacle.
335 Or, reflects.
336 Or, "fate" (EIMARMENE).
337 Or, weight.
338 I.e., any number whose counterpart is.
339 My impression is that in the Pentateuch 0, 1 and (possibly) 2 are not numbers. 0 is absence, 1 is unity and 2 is duality (viz., what Timaeus refers to as nothing [MEDEN, 51d], one and pair [TAUTON KAI DUO, 52c]). Hence, 3 and 4 are the first numbers. Note that $3 \times 4 = 12$ (e.g., four matriarchs multiplied by three patriarchs produce twelve tribes) and $3 + 4 = 7$ (e.g. out

of three pre-creation elements [divine wind, water, earth] arise four divisions of space [upper water, sky, lower water, and earth] whose construction into a world is expressed as seven days). Similarly, all biblical Hebrew word roots have either three or four consonants. However, (at least in the case of Genesis), this "numerology" is totally speculative. It is not "gematria," but it has no more firm foundation than gematria.

340 I.e., divisible at an infinite number of points.
341 An octahedron.
342 An icosahedron.
343 I.e., before the cooperative activity of reason and necessity.
344 Dodecahedra.
345 Or, primary bodies.
346 The examples are Plato's (56c–57c). The mathematical explanation is Gregory Vlastos' in *Plato's Universe*. Oxford, Clarendon Press, 1975, ch. 3.
347 Viz., their appetites and practical judgment.
348 Viz., that faculty concerned with mathematics and theoretical wisdom.
349 = SHEOL?
350 Morality.
351 Physics.
352 With justice and harmony.
353 With love and conflict.
354 With reason.
355 A notable exception is Moshe Weinfeld, "God the Creator in Genesis 1 and in the Prophecy of Second Isaiah." (in Hebrew) *Tarbiz* vol. 37, no. 2 (Jan., 1968) pp. 105–132.
356 Or, receptacle.
357 TOHU and BOHU.
358 TEHOM.
359 But not absolute.
360 Viz., from the ideal of creation to the ideal of redemption.

NOTES TO CHAPTER 7

361 I wish to express my thanks to Don Lichtenberg (Indiana University) who was kind enough to read the first draft of this chapter, and to make many thoughtful, critical suggestions for its improvement. The final draft was written while on a study leave from Temple University in the spring semester of 1992 at the Chicago Center for Religion and Science (in affiliation with the Lutheran School of Theology at Chicago). I wish to express

my appreciation to Temple University for giving me the time to write and to the CCRS, together with the LSTC, for providing me with an ideal space for working and sharing ideas. In this connection, I am especially appreciative of the time I had to discuss my ideas on science and religion with Philip Hefner and Thomas Gilbert.

362 Philadelphia, Jewish Publication Society of America, 1966.
363 Cambridge, Harvard University Press, 1927.
364 With papers by Seymour Feldman, Bernard R. Goldstein, Alfred L. Ivry, Barry S. Kogan, Jonathan W. Malino and Jacob J. Staub.
365 With papers by Eric J. Chaisson, Margaret J. Geller, J. Richard Gott, John P. Huchra and Harry L. Shipman.
366 *Creation and the End of Days: Judaism and Scientific Cosmology.* Edited by David Novak and Norbert M. Samuelson. Lanham, University Press of America, 1986.
367 Stephen W. Hawking, *A Brief History of Time: From the Big Bang to Black Holes.* New York, Bantam Books, 1988. P. 162.
368 Viz., length, width, depth and time. Usually this kind of space is called "space-time."
369 $\Delta x \Delta p_x \geq (h/2\pi)$, i.e., the position times the momentum of a particle at any particular instant is equal to or greater than a specific number, viz., Planck's constant (6.626 times 10^{-34} Joule-seconds), divided by two times π. "10^{-34}" means $1/(1000 \ldots 0)$ with 34 0's, i.e. ".00 ... 06626" with 31 $(34-3)$ zeros in the decimal.
370 The judgment that the basic building blocks of the physical universe are inherently indeterminate obviously correlates with the claim in Jewish philosophy that the individuals that constitute the world are at the origin indefinite. However, it remains difficult to picture what either statement could mean in an atomist view of the universe (viz., one that reduces all complex entities and states to quantitative and formal relations between simpler entities). Now the Jewish dogma of creation need not assume any such atomism. (In fact, a process view would be more likely from this perspective.) But most contemporary philosophers of science assume (dogmatically and naively in my judgment) that atomism is realism, and they are committed to preserve as much of (what they call) realism as scientific data can permit.
371 According to Einstein's Special Theory of Relativity, energy (E) is the product of mass (m) and the speed of light squared (c^2), i.e., $E = mc^2$. ("c" is approximately 2.998×10^8 meters per second.) At this initial stage of the universe, the temperature is

sufficiently high that the mere collision of elementary particles without any rest mass (viz., photons) has sufficient energy to produce other particles. Each collision between two photons annihilates them and produces either two more photons or other kinds of particles (e.g., electrons). The photons are the constituents of radiation. That the radiation produced by the collisions is in equilibrium means that at each moment the number of particles generated equals the number destroyed.

372 I.e., $1/(10\ldots0)$ of a second with 41 zeroes in the denominator.

373 I.e., $10\ldots0$ with 32 zeroes. "K" is "kelvins." A given number, n, of kelvins (viz., n K) is the same as that number plus 273.15 degrees Celsius (viz., $n + 273.15\,°C$). The significance of the number "273.15" is that the pressure of a great many gases becomes zero at $-273.15\,°C$, i.e., at 0 K. A given number, n, degrees Celsius minus thirty two, multiplied by five and divided by nine is the number of degrees Fahrenheit. [I.e., $\{(\frac{5}{9})(T_F - 32)\} = T_C$]. Consequently, "$10^{32}$ K" is $[\{(\frac{9}{5})(10^{32} - 273.15)\} + 32]$ about $1.8 \times 10^{32}\,°F$.

374 The weak force or interaction is responsible for the decay of some unstable fundamental particles as well as the emission of electrons (also called beta particles) from radioactive nuclei.

375 The electromagnetic force or interaction includes both electric and magnetic forces. It is used to account for why some particles are attracted to each other while other particles repulse each other. As explained below, the designations "positive charge" and "negative charge" are associated with different kinds of particles in order to say that particles with opposite charges attract while other particles with like charges repel each other.

376 In general, given any particle, n, there is an anti-n that has the same mass as n, but it has an opposite charge. When a particle and an antiparticle collide, they can annihilate each other, and, in so doing, produce radiation.

377 The most basic particles of which we currently know are leptons, quarks and their antiparticles. The lightest charged leptons are electrons. Quarks are the basic constituents of the protons and neutrons. Protons and neutrons are the constituents of atomic nuclei, which, in association with electrons, constitute atoms.

378 That one set is positive and the other is negative is an arbitrary designation. The source of the terminology in this case is Benjamin Franklin, who called the kind of electric charge produced when a plastic rod is rubbed with fur "negative," and the kind of charge produced when a glass rod is rubbed with silk "positive." Since the first kind of charge is associated with electrons, they

are said to have a negative charge. Given that particles with unlike charges attract and those with like charges repulse each other, whatever elementary particle an electron attracts has a positive charge, whatever elementary particle it repulses has a negative charge and whatever is not affected by an electron or any other charged particle is said to have no charge.

379 The strong force or interaction is responsible for holding the nuclei of atoms together. Since the particles in a nucleus are all neutrally or positively charged, an additional force is needed to bind these particles together, in opposition to the influence of the electromagnetic force that inclines particles with like charges to repulse each other.

380 Viz., a particle subject to the strong force as well as gravity and the weak and electromagnetic forces that affect the earlier elementary particles.

381 I.e., material, i.e., a universe that contains matter with mass.

382 These figures are taken from James S. Trefil, *From Atoms to Quarks*. New York, Charles Scribner's Sons, 1980. pp. 20–21.

383 The term "empty" needs some qualification. Between the nucleus of an atom and its associated electron(s) there is an electric field. Contemporary physicists use the "concept" (Cf. F. W. Sears, M. W. Zemansky and H. D. Young. *University Physics*, Reading, MA, Menlo Park, CA, *et alibi*, Addison-Wesley Publishing Co., 1987. P. 546.) of this field to account for the fact that a force can be transmitted across a region of space that in every other relevant respect is empty. That an electric field is said to be at a given point in space that is not occupied by any particle means that, when some particle is placed at that point, the particle will experience the electric force. Now, the field is in some sense something. However, it is not a "thing" in the same way that particles are things. In other words, it is a kind of no-thing. It is in this sense that the claim is made that most of the space occupied by an atom is empty, i.e., most of the space occupied by an atom on this standard model is not occupied by any-thing.

384 *Ibid.*

385 Given that the gravitational force that applies to everything of which we know within the universe applies to our universe as a whole, and given that the present rate of speed at which everything is moving away from everything else (i.e., the current rate of expansion) remains constant, either the expansion will continue for ever, or at some point in time the process will reverse and everything will retract (i.e., condense) in the direction of its

origin. Which alternative is more likely depends on the total mass of the universe itself. To understand what is at stake in these alternative scenarios for the future of our physical universe, consider a rocket that is fired from the surface of the earth. The force directing the missile is resisted by the force of the gravitational attraction of our planet. The strength of the earth's gravitational force is determined by the planet's mass. The greater the mass, the greater the force of gravity. Since we know the total mass of the earth, we also know the velocity necessary for an object of a given mass to break free from our planet's attraction. Any velocity greater than that critical number for an object will propel that object away from the earth. However, if the force is less than that critical number, then eventually the object will return to earth. Give our initial assumptions, the same thing can be said of the universe as a whole. There is a critical number such that if galaxies are moving away from each other with a velocity greater than that number, then they will move away forever. If, on the other hand, the velocity is less, then at some point the expansion will come to an end and these inhabitants of the universe will reverse their direction. What that critical number is depends on what is the mass of the universe, and that quantity has not as yet been determined.

386 I.e., something infinitesimal.
387 Viz., 1, 2, 3, . . .
388 No matter how large a number you reach, you can always form another number by adding any lower number to the number in question. I.e., given any two integers, m and n, you can always form a new, higher integer, p, that is equal to m + n, and this new integer, p, can be added to any other integer, such as m or n, to form a new even higher integer.
389 Mathematicians use the term "asymptote" to designate the line towards which the points on a curve extend as they recede indefinitely from the origin. I will use the expression "asymptotic function" (as I have earlier in the book) to designate any equation whose geometric expression is such a curve. In other words, an asymptotic function is a function that has an asymptote. When diagrammed in two-dimensional geometric space, the curve can be seen to move indefinitely towards a line. The line is the asymptote. The rational number that defines this asymptote, viz., "n" in "y = n," is the function's limit. For example, consider a pump attached to a closed container where each use of the pump eliminates half of the air that is contained. At the end of the first use of the pump the container is half full, at

the end of the second it is ($\frac{1}{4}$) full, at the end of the third it is ($\frac{1}{8}$) full, etc. The function that expresses this relation between the amount of air in the container and the number of times that the pump is used is an asymptotic function. The limit of the function is zero, and the asymptote is the line that "y = 0" expresses. Another way to express this function is to say that with each use of the pump the amount of air that is contained approximates zero, so that if the pump were used an infinite number of times, then there would be no air in the container.

390 Light is sent through a screen (as in Fig. C) that contains a single slit (A), on through a second screen that contains two slits (B and C), on to a third screen where the intensity of the light is measured. The light passes through slit A as a single wave that divides into two different waves as it passes through slit B or C on to the third screen. In this case the intensity of the light (viz., the number of photons that strike a given space [comparable to the impact of a wave of water being the quantity of drops of water that impact against a given object]) exhibits an interference pattern.

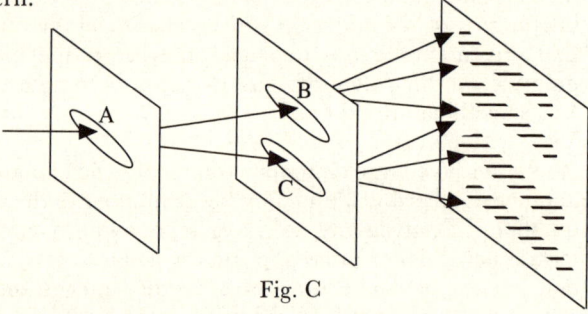

Fig. C

391 In this case light is directed at a piece of metal in a vacuum. The impacts of the light-quanta release electrons from the metal plate whose energy is measured.

392 I.e.,

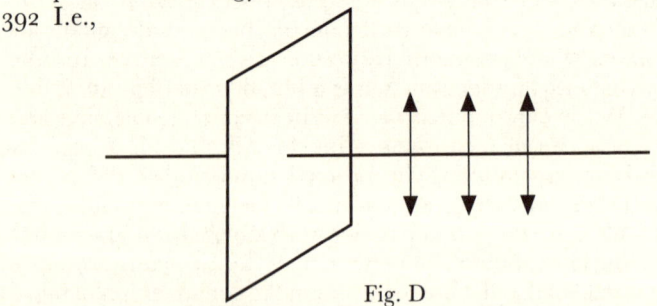

Fig. D

393 The above line of reasoning about the paradox of Schrödinger's cat assumes (as do most physicists and philosophers of science) an atomistic view of reality, i.e., it takes it for granted that what ultimately exists is a plurality of individual simple entities from which more complex entities are formed. However, as we have seen above, this is not the only way to view reality, and it is not even the most coherent way to interpret the Jewish dogma of creation. Given instead that reality consists ultimately of events in which individuals are more or less defined (depending on the character of the event), the paradoxical nature of Schrödinger's thought experiment dissolves, for on this interpretation no incoherency arises from claiming that some individual parts of an event are indefinite or indeterminate. What this example shows is not that time is bi-directional or that measurement as such changes reality. Rather it shows that, on final analysis, reality is not made of substances (be they sub-atomic particles or cats), but is constituted out of events whose nature is expressed (in probability equations) as correlations between groups or sets of individuals. In other words, the collectives to which individuals belong have ontological priority over the individuals themselves. However, this solution to Schrödinger's cat-in-the-box raises far more serious problems of its own about the nature of political ethics. These problems will be considered in the next chapter.

394 This is a third sense of the term, "universe."

395 I.e., not just with respect to the limits of human knowledge.

396 I.e., each state of affairs has a probability of $(\frac{1}{8})$.

397 A possible solution is to exclude the application of this law from the origin of the universe. It has been argued, for example, that there was a fluction of disorder at the beginning of our universe that decreases the then existent order.

398 I have adapted this example for my own use from Paul Davies, *God and the New Physics*. Harmondsworth, Penguin, 1983. p. 53.

399 This argument ignores the existence of fields between particles. While a field is no-thing, it is not absolutely nothing. Given its presence, the probabilities for a particle occupying any particular space in relation to other particles are not equal. Attraction and repulsion between particles make some locations more likely than others. However, it is not altogether clear just what kind of entities fields are. They may be no more than conceptual devices to account for the very fact in question, viz., that particles tend to form into ordered, non-homogeneous patterns rather than being randomly scattered over spatial regions. Furthermore, even after this attraction/repulsion would be

400 E.g.,

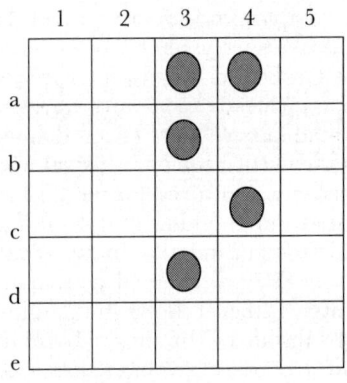

Fig. E

becomes

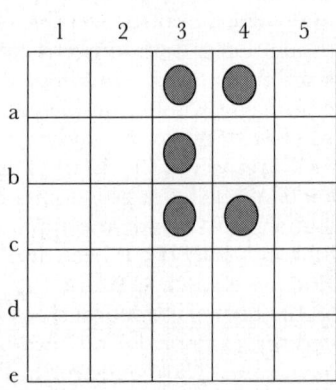

Fig. F

In Figs. E and F the life forms in 3a, 4a, 3b and 4c persist from one generation to the next, while a new life form is born in 3c and an old life form dies in 3d.

NOTES TO CHAPTER 8

401 In other words, religious claims are not as superficial as most Anglo-American philosophers of religion make them out to be when they subject them to philosophical analysis. I think this is

true of all important religious claims; it certainly is true (as I hope this book has demonstrated) of creation.
402 In *The Star* Part 2, Book 1, "Der Schöpfer, Willkür und Muss."
403 Compare *The Star*, Part 2, Book 1, "Die Kreatur, Vorsehung und Dasein" with Maimonides' *The Guide of the Perplexed*, Book 3, chapter 18.
404 *The Exalted Faith*, Part 1, Abstract, 4b19–5b6; Part 1, chapter 2, 26b17–27b14; and Part 1, chapter 6, 63b10–15.
405 *The Guide Book* 2: chapters 13, 16 and (most significantly) 19.
406 *The Star*, Part 1, Übergang, "Das herrschende 'Wer weiss'."
407 This argument will constitute all that I have to say directly in this book (on creation) about a more general concern with the relation of science and religion from the perspective of modern Jewish thought. In particular I will argue here against any radical separation between the two. The position I will critique is specifically that of Rosenzweig. However, it should be noted Rosenzweig's view has many important disciples in modern Jewish philosophy, notably those so-called "post-modern" Jewish thinkers (such as Emil L. Fackenheim and Emmanuel Levinas) whose view of reason comes from the neo-Kantian tradition of modern philosophy, viz., from Hermann Cohen to Ernst Cassirer and Steven Schwarzschild.
408 *The Star*, Part 1, Introduction, "Mathematik und Zeichen, Der Ursprung," pp. 22–24 in the 1921 German edition and pp. 20–21 in William Hallo's English translation. (Subsequently these dual page references will be indicated as follows: pp. 22–24/20–21, viz., listing the German edition first and then the English translation.)
409 *The Star*, Part 1, Introduction, "Mathematik und Zeichen."
410 *The Star*, Part 2, Book 1, "Idealistische Logik, Die Logik der Schöpfung gegen die Logik der Idee." p. 154/139.
411 What it is is the Koch curve.

412 Viz.,

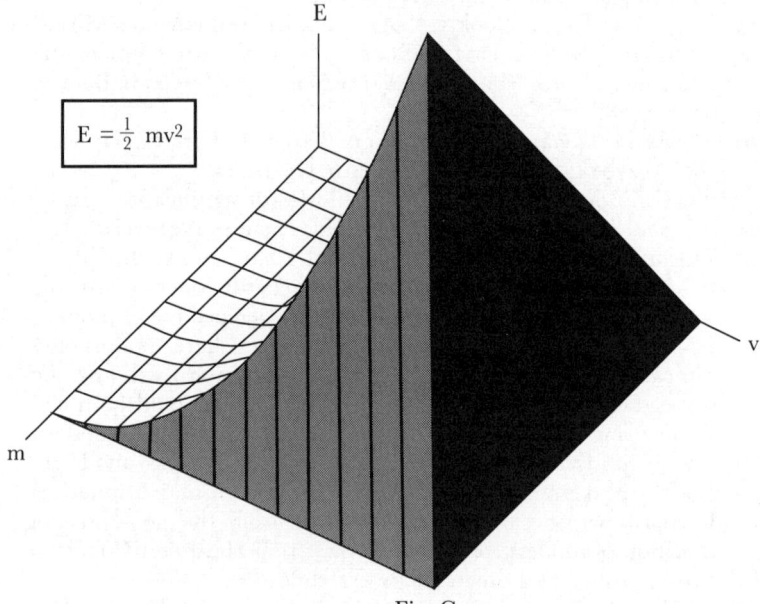

Fig. G

413 The ordinary language philosophy of Wittgenstein is but one example of an alternative approach to philosophy that Rosenzweig's critique ignores. In this connection it is of interest to note that Rosenzweig exhibited some slight uneasiness with his critique of Hegelian philosophy as philosophy when he observed in passing the confidence of English philosophy in language. (Zeweiter Teil, Erstes Buch, Idealistische Ästhetik, p. 162/146.)
414 *The Star*, Part 1, Übergang, "Rückblick: Das Chaos der Elemente."
415 In *The Star*, Part 2, Book 1, "Grammatik des Logos (Die Sprache der Erkenntnis)."
416 He doesn't provide a picture, but it is not too difficult to construct one. For example, Fig. H.

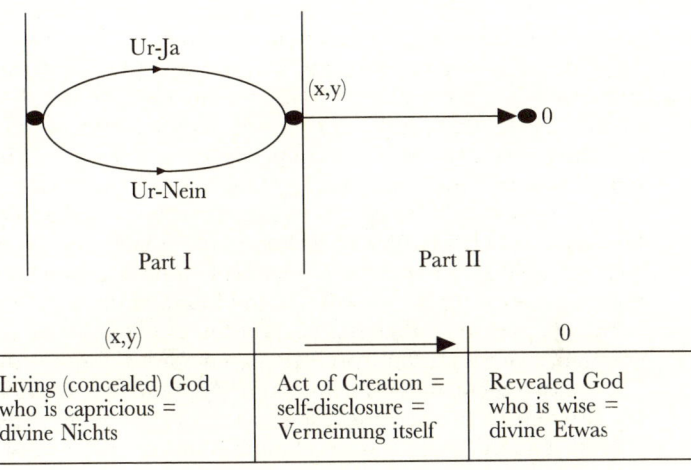

Fig. H

417 *The Star*, Part 2, Book 3, Schwelle Rückblick: Die Ordnung der Bahn, "Die neue Beziehung" and "Der neue Zusammenhang." pp. 284–285/255–256.

418 In fact, in the passage under discussion, Rosenzweig says that the new kind of relation (*Beziehung*) should be pictured as a *Bahnverlauf* that extends between two particular (*einzelne*) points, i.e., between the elements God, man and/or world. Now, it is reasonable in this context to translate the term *Bahnverlauf* into English as "vector."

419 Rosenzweig seems to contradict himself in this respect when he argues in the next section (*Die neue Ordnung*) that position in Euclidean geometry is contingent whereas in his "new" geometry it is necessary. Rosenzweig argues as follows: The initial protocosmic triangle of elements determines the consequent cosmic triangle of relations, because the points of the cosmic triangle are symbols (*Symbole*) for the lines of the protocosmic triangle. This characteristic entails that the distances between each element/point is equal. Hence, the consequent triangles must be equilateral. Furthermore, the point/element God must be located at the top of the picture, since God is the origin (*Ursprung*) of both relations (creation and revelation), which in turn determine the final product (redemption) to be at the bottom. Hence, each triangle of Rosenzweig's star must have

a fixed position (*Stellung*) in space (*Raum*). In other words, each "point" and "line" that forms each triangle in Rosenzweig's model must have absolute location and absolute direction. (The absolute points are the elements, and the absolute directions are from God to man [revelation] and from God to world [creation].) In contrast, in Euclidean geometry position in space is arbitrary. In other words, the properties of vectors (viz., "direction" [*Richtung*] and "quantity" [*Größe*]) have universal relativity. I have not discussed this section in the body of the chapter because it is only marginally related to the subject, viz., Rosenzweig's understanding of mathematical thinking. However, it does illustrate a point that will be emphasized in the next section of this chapter (on ontology), viz., that in the Jewish concept of creation, space is logically prior to spatial objects, for the critical point that Rosenzweig is making in this argument is the following: Euclidean geometry is a logic of laws that govern spatial objects, but that logic says nothing about the space in which these objects are located. In contrast, Rosenzweig's "new" geometry is a logic in which the characteristics of spatial forms is determined by their spatial location. Now, what we know (but he didn't) is that the kind of modern mathematical equations to which Rosenzweig here alludes are field equations, viz., the kind of equations that are dominant in the formalism of both quantum mechanics and relativity physics.

420 Literally, they occur or happen "unterim Zeichen des Vielleicht." Note in this context that field equations in quantum mechanics deal with probability values.

421 Literally, they are grounded in a historical-genesis (*Entstehungsgeschichte*).

422 Cf. Ivar Ekeland, *Mathematics and the Unexpected*. Chicago, University of Chicago Press, 1988; James Gleick, *Chaos: Making a New Science*. New York, Penguin, 1988; J. M. J. Thompson and H. B. Stewart, *Nonlinear Dynamics and Chaos*. New York, John Wiley and Sons, 1986.

423 For example, Rosenzweig's understanding of the relationship of creation can be diagrammed as in Fig. I.

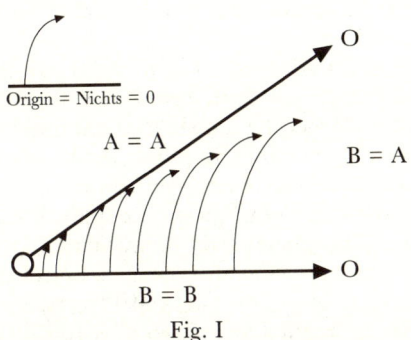

Fig. I

Both revelation and redemption would be modeled in the same way. "A = A" is Rosenzweig's symbolism for the element God, "B = B" for man, and "B = A" for the world.

424 For example,

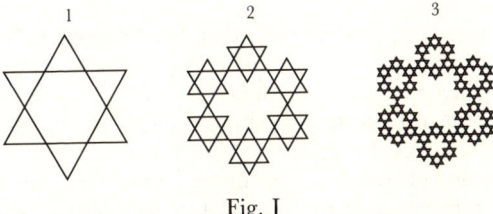

Fig. J

425 *Bahnstreck*.
426 The exception to this generalization is Hermann Cohen.
427 Possible exceptions are Whitehead and Cassirer. It is ironic that philosophers like Rosenzweig who were most opposed to mathematical reasoning operated with a more sophisticated intuitive understanding of mathematics (viz., with calculus and fractals) than philosophers like Russell and Frege who were most committed to mathematical thinking but who limited their mathematical model to simple arithmetic and linear algebra.
428 For example, Emil Fackenheim (in *Encounters Between Judaism and Modern Philosophy: A Preface to Future Jewish Thought*. New York, Basic Books, 1973) and Paul Van Buren (in *Discerning the Way: A Theology of the Jewish Christian Reality*. New York, Seabury, 1980).
429 For example, science cannot inform us why the Sabbath should be on Saturday rather than some other day of the week, and

religion has little to say about the usefulness of planting marigolds next to tomato plants to eliminate the threat of Japanese beetles.
430 The only alternative would be to claim that there exists only one individual and everything else that would otherwise qualify as one is simply a modification of that one individual. This is in fact the alternative taken by Spinoza. As modifications of substance there can be relations between individuals, and the nature of these relations become the philosophical foundation for Spinoza's ethics. In other words, on this reading, Spinoza's rejection of Cartesianism – which leads him to reject the "Cogito" as the starting point of philosophy and to return to God as the only substance as his first definition – is rooted in what Hermann Cohen would later identify as a distinctive "Jewish" impulse in philosophy to give priority to ethics over ontology. (Cohen himself never saw this Jewish, i.e. ethical, dimension in Spinoza's thought. Neither did Cohen's spiritual teacher [Moses Mendelssohn] nor his disciples [Martin Buber and Franz Rosenzweig].)
431 Lev. 19:2 and 20:26 are the primary references. Also relevant is Lev. 11:44–45.
432 This generalization applies no less universally to galaxies and to the world as a whole than it does to physical particulars within the galaxies of the universe.
433 I am told (probably apocryphally) that someone, on hearing on the radio that a certain number of people were expected to die on Memorial weekend on the highway, called the station to find out if they were among them. While the station could predict (with reasonable accuracy) life and death about groups of people (in this case motorists), it could say nothing at all about individual members of the groups.

Select bibliography

FOR PART 1: MODERN JEWISH PHILOSOPHY

PRIMARY

Borowitz, Eugene B. *A New Jewish Theology in the Making*. Philadelphia, Westminster, 1968.
Cohen, Hermann. *Briefe*. Edited by B. Strauss. Berlin, Schocken, 1939.
Das Prinzip der Infinitesimal-Method. Frankfurt a.M., Suhrkamp, 1968.
Jüdische Schriften. Edited by Franz Rosenzweig. Berlin, C. A. Schwetschke, 1924.
Reason and Hope. New York, W. W. Norton, 1971.
Religion d. Vernunft aus den Quellen d. Judentums. Frankfurt a.M., 1929. Translated into English by Simon Kaplan. *Religion of Reason*. New York, Ungar, 1972.
Schriften zur Philosophie und Zeitgeschichte. Berlin, 1928.
Fackenheim, Emil L. *The Religious Dimension in Hegel's Thought*. Bloomington, Indiana University Press, 1967.
To Mend the World: Foundations of Future Jewish Thought. New York, Schocken, 1982.
Krochmal, Nachman. *The Guide to the Perplexed of the Time*. Edited by Simon Rawidowicz. Waltham, Ararat Press, 1961.
Rosenstock-Huessy, Eugen and Rosenzweig, Franz. *Judaism Despite Christianity*. New York, Schocken, 1969.
Rosenzweig, Franz. *Das Beuchlein vom Gesunden und Kranken Menschenverstand*. Translated into English by Nahum N. Glatzer. *Understanding the Sick and the Healthy: A View of the World, Man, and God*. New York, Noonday Press, 1953.
Der Stern der Erlösung. Frankfurt a.M., J. Kaufmann, 1921. Translated into English by William W. Hallo. *The Star of Redemption*. Boston, Beacon Press, 1971. Translated into Hebrew by Yehoshua Amir. כוכב הגאוה. Jerusalem, Bialik Institute, 1970.

Franz Rosenzweig: Der Mensch und Sein Werk, Gessamelte Schriften. The Hague, Martinus Nijhoff, 1976–1979.
Hegel und der Staat. Berlin, 1920.
Kleinere Schriften. Berlin, Schocken, 1937.
נהריים. Jerusalem, Bialik Institute, 1977.
On Jewish Learning. Edited by Nahum N. Glatzer. New York, Schocken, 1955.
Rotenstreich, Nathan. *Tradition and Reality*. New York, Random House, 1972.
Spinoza, Baruch. *The Chief Works of Benedict De Spinoza*. Translated into English by R. H. M. Elwes. New York, Dover, 1955.
The Collected Works of Spinoza. Edited by Edwin M. Curley. Princeton, Princeton University Press, 1985.
The Correspondence of Spinoza. Translated into English by Abraham Wolf. New York, Russell and Russell, 1966.
Ethics. Translated into English by W. H. White and revised by A. H. Stirling. 1949.
The Ethics and Selected Letters. Translated into English by Samuel Shirley. Indianapolis, Hackett, 1982.
The Ethics of Spinoza and De Intellectus Emendatione. Translated into English by A. Boyle. London and New York, 1955.
On the Improvement of the Understanding. Translated into English by Joseph Katz. Indianapolis, Bobbs-Merrill, 1958.
Opera. Edited by Carl Gebhardt. Heidelberg, C. Winter, 1972.
Principles of Cartesian Philosophy. Translated into English by Harry E. Wedeck. New York, Philosophical Library, 1961.
Spinoza, Earlier Philosophical Writings: The Cartesian Principles and Thoughts on Metaphysics. Translated into English by F. A. Hayes. Indianapolis, Bobbs-Merrill, 1963.
Spinoza's Short Treatise on God, Man and His Well-Being. Translated into English by Abraham Wolf. New York, Russell and Russell, 1963.
The Tractatus Theologico-Politicus. Translated into English by A. G. Wernham. Oxford, Oxford University Press, 1965.

SECONDARY

Agus, Jacob B. *Modern Philosophies of Judaism: A Study in Recent Jewish Philosophies of Religion*. New York, 1941.
Altmann, Alexander. *Moses Mendelssohn*. Philadelphia, Jewish Publication Society of America, 1973.
Belaief, G. *Spinoza's Philosophy of Law*. The Hague, 1971.
Bend, J. G. van der (ed.). *Spinoza on Knowing, Being and Freedom*.

Proceedings of the Spinoza Symposium at the International School of Philosophy in the Netherlands. Assen, The Netherlands, Van Gorcum, 1974.

Bergman, Samuel Hugo. *Faith and Reason: An Introduction to Modern Jewish Thought.* Edited by Alfred Jospe. New York, Schocken, 1963.

Berkovits, Eliezer. *Major Themes in Modern Philosophies of Judaism.* New York, Ktav, 1974.

Bidney, David. *The Psychology and Ethics of Spinoza: A Study in the History and Logic of Ideas.* New York, Russell and Russell, 1962.

Borowitz, Eugene B. *Choices in Modern Jewish Thought: A Partisan Guide.* New York, Behrman House, 1983.

Bubner, R. *Modern German Philosophy.* Cambridge, Cambridge University Press, 1981.

Caird, John. *Spinoza.* New York, Arno, 1981.

Caspar, Bernard. *Das dialogische Denken: Eine Untersuchung der Religionphilosophischen Bedeutung Franz Rosenzweigs, Ferdinand Ebners und Martin Bubers.* Freiburg, Herder Verlag, 1967.

Crenshaw, James L. and Sandmel, Samuel. *The Divine Helmsman.* New York, Ktav, 1980.

Curley, Edwin M. *Spinoza's Metaphysics: An Essay in Interpretation.* Cambridge (MA), Harvard University Press, 1969.

De Deugd, Cornelius. *The Significance of Spinoza's First Kind of Knowledge.* Assen, Van Gorcum, 1966.

Dunin-Borkowski, S. *Spinoza.* Munster, Aschendorff, 1910.

Efros, Israel. "Rosenzweig's Star of Redemption." *Journal of Religion* 27 (1963): 93–100.

Fox, Everett. "Technical Aspects of the Translation of Genesis of Martin Buber and Franz Rosenzweig." Unpublished Ph.D. dissertation, Brandeis University, 1975.

Freeman, E. and Mandelbaum, M. (eds.) *Spinoza: Essays in Interpretation.* La Salle, Open Court, 1973.

Freudenthal, J. *Die Lebensgeschichter Spinozas in Quellenschriften, Urkenden und nichtamtlichen Nachrichten.* Leipzig, Verlag von veit, 1899.
Spinoza, sein Leben und seine Lehre. Stuttgart, Frommanns Verlag, 1904.

Glatzer, Nahum. *Franz Rosenzweig: His Life and Thought.* Philadelphia, Jewish Publication Society of America, 1953.

Grene, Marjorie (ed.). *Spinoza: A Collection of Critical Essays.* Garden City, Anchor, 1973.

Gueroult, Martial. *Spinoza.* Paris, Aubier Montaigne, 1968.

Hallett, Harold F. *Benedict De Spinoza.* London, Athlone Press, 1957.

Creation, Emanation and Salvation: A Spinozistic Study. Hague, Martinus Nijhoff, 1962.
Hampshire, Stuart. *Spinoza.* Harmondsworth, Penguin, 1951.
Harris, Jay M. *Nachman Krochmal: Guiding the Perplexed in the Modern Age.* New York and London, New York University Press, 1991.
Haserot, Francis F. "Spinoza's Definition of Attribute." *Philosophical Review* 62 (1953).
Heckelei, Hermann-Josef. *Erfahrung und Denken: Franz Rosenzweigs theologisch-philosophischer Entwurf eines 'neuen Denkens'.* Bad Honnef, Bock und Herschen, 1980.
Hesse, Mary B. *Models and Analogies in Science.* Notre Dame, University of Notre Dame Press, 1966.
Hessing, S. *Speculum Spinozanum: A Kaleidoscopic Homage 1677–1977.* London, 1977.
Hessing, S. (ed.). *Spinoza: Dreihundert Jahre Ewigkeit: Spinoza-Festschrift 1632–1932.* The Hague, 1962.
Horwitz, Gertrude Rivka. "Gnosticism and Creation in Buber's Philosophy." *Immanuel* 12 (Summer, 1981): 135–151.
Horwitz, Gertrude Rivka. "Speech and Time in the Philosophy of Franz Rosenzweig." Unpublished Ph.D. dissertation, Bryn Mawr College, 1963.
Hubbeling, H. G. *Spinoza's Methodology.* Anselm, Van Gaeum, 1967.
Jarrett, Charles. "The Concept of Substance and Mode in Spinoza." *Philosophia* 7 (1977): 83–105.
Jaspers, Karl. *Spinoza.* Translated into English by Ralph Manheim. New York, Harvest Books, 1974.
Joachim, Harold H. *A Study of the Ethics of Spinoza.* Oxford, Oxford University Press, 1901.
Spinoza's Tractatus de Intellectus Emendatione: A Commentary. Oxford, The Clarendon Press, 1940.
Kashap, S. P. (ed.). *Studies in Spinoza: Critical and Interpretative Essays.* Berkeley, University of California Press, 1972.
Katz, Steven T. *Jewish Ideas and Concepts.* New York, Schocken, 1977.
Jewish Philosophers. New York, Bloch, 1975.
Kaufman, William E. *Contemporary Jewish Philosophies.* New York, Reconstructionist Press & Behrman, 1976.
Kellner, Menachem. *Dogma in Medieval Jewish Thought.* Oxford, Oxford University Press, 1986.
Kennington, R. (ed.). *The Philosophy of Baruch Spinoza.* Washington, DC, Catholic University of America, 1980.
Klatzkin, J. *Hermann Cohen.* Berlin, 1921.
Kluback, William. *Hermann Cohen: The Challenge of a Religion of Reason.* Chico, Scholars Press, 1984.

The Idea of Humanity: Hermann Cohen's Legacy of Philosophy and Theology. Lanham, University Press of America, 1987.
Kogan, Barry S. (ed.). *Spinoza: A Tercentenary Perspective.* Cincinnati, Hebrew Union College–Jewish Institute of Religions, 1979.
Loewith, Karl. "Martin Heidegger and Franz Rosenzweig on Temporality and Eternity." *Philosophy and Phenomenological Research* 3 (1942): 53–77.
"Martin Heidegger und Franz Rosenzweig: Ein Nachtrag zu 'Sein und Zeit'." *Gesammelte abhandlungen zur Kritik der geschichtlichen Existenz.* Stuttgart, Kohlhammer, 1960. pp. 68–92.
Luby, Barry J. *Maimonides and Spinoza: Their Sources, Cosmological Metaphysics and Impact on Modern Thought and Literature.* New York, Las Americas, 1973.
Mark, T. C. *Spinoza's Theory of Truth.* New York, Columbia University Press, 1972.
Martin, Bernard (ed.). *Great Twentieth Century Jewish Philosophers: Sheshtov, Rosenzweig, Buber, with Selections from their Writings.* New York, Macmillan, 1970.
Mayer, Reinhold. *Franz Rosenzweig Eine Philosophie der dialogischen Erfahrung.* Munich, Chr. Kaiser, 1973.
"Franz Rosenzweigs Stern der Erlösung und seine bedeutung für die Gegenwart." *Fifth World Congress of Jewish Studies* 3 (1972): 27–36.
Melber, J. *Hermann Cohen's Philosophy of Judaism.* New York, Jonathan David, 1968.
Mendes-Flohr, Paul (ed.). *The Philosophy of Franz Rosenzweig.* Hanover, University Press of New England, 1988.
Myers, Henry Alonzo. *The Spinoza–Hegel Paradox: A Study of the Choice Between Traditional Idealism and Systematic Pluralism.* Ithaca, Cornell University Press, 1944.
Naess, A. *Freedom, Emotion and Self-Subsistence: The Structure of the Central Part of Spinoza's Ethics.* Oslo, 1975.
Neu, Jerome. *Emotion, Thought and Therapy: A Study of Hume and Spinoza and the Relationship of Philosophical Theories of Therapy.* Berkeley, University of California Press, 1977.
Novak, David and Samuelson, Norbert M. (eds.). *Creation and the End of Days: Judaism and Scientific Cosmology. Proceedings of the 1984 Meeting of the Academy for Jewish Philosophy.* Lanham, University Press of America, 1986.
Odebrecht, R. *Hermann Cohens Philosophie d. Mathematik.* Berlin, 1906.
Oko, Adolph. *The Spinoza Bibliography.* Boston, G. K. Hall, 1964.
Oppenheim, Michael David. "Taking Time Seriously: An Inquiry into the Methods of Communication of Sören Kierkegaard and Franz Rosenzweig." *Studies in Religion* 7 (1978): 53–60.

What Does Revelation Mean for the Modern Jew? Rosenzweig, Buber and Fackenheim. Lewiston, Edwin Mellon Press, 1985.

Parkinson, G. H. P. *Spinoza's Theory of Knowledge.* Oxford, Clarendon Press, 1954.

Pines, Shlomo. "Spinoza's Tractatus Theologico-Politicus, Maimonides and Kant." *Scripta Hierosolymitana* 20 (1968).

Preposiet, Jean. *Bibliographie Spinoziste.* Besançon, Université de Besançon, 1974.

Rahel-Freund, Else. *Die Existenzphilosophie Franz Rosenzweigs.* Leipzig, Meiner, 1933. Translated into English by Stephen L. Weinstein and Robert Israel. *Franz Rosenzweig's Philosophy of Existence: An Analysis of the Star of Redemption.* The Hague, Martinus Nijhoff, 1979.

Robinson, L. *Kommentar zu Spinozas Ethik.* Leipzig, F. Meiner, 1928.

Rotenstreich, Nathan. *Jewish Philosophy in Modern Times.* New York, Holt, Reinhart, and Winston, 1968.

Roth, Leon. *Spinoza.* Boston, Little Brown, 1929.

Spinoza, Descartes and Maimonides. London, Clarendon Press, 1924.

Schaeffler, Richard, Kasper, Bernhard, Talmon, Shemaryahu, and Amir, Yehoshua. *Öffenbarung im Denken Franz Rosenzweigs.* Essen, Ludgerus, 1979.

Shahan, R. and Biro, J. (eds.). *Spinoza: New Perspectives.* Norman, University of Oklahoma Press, 1978.

Soloweijczyk, J. *Das Reine Denken u.d. seinskonstituierung bei Hermann Cohen.* Berlin, 1932.

Sorabji, Richard. *Time, Creation and the Continuum.* London, 1983.

Strauss, Leo. *Spinoza's Critique of Religion.* Translated into English by Elsa M. Sinclair. New York, Schocken, 1965.

Tal, Uriel. *Christians and Jews in Germany: Religion, Politics and Ideology in the Second Reich, 1870–1914.* Ithaca, Cornell University Press, 1975.

Tewes, Jacob. *Zum Existenzbegriff Franz Rosenzweigs.* Meisenheim a. Glan, Anton Hain, 1970.

Twersky, Isadore and Septimus, Bernard. *Jewish Thought in the Seventeenth Century.* Cambridge (MA), Harvard University Press.

Wetlesen, Jon. *A Spinoza Bibliography, 1840–1940.* Oslo, Universitets-Forlaget, 1971.

Internal Guide to the Ethics of Spinoza. Index to Spinoza's Cross References in the Ethics, Rearranged so as to Refer from Earlier to Later Statements. Oslo, University of Oslo, 1974.

The Sage and the Way: Spinoza's Ethics of Freedom. Assen, Van Corcum, 1979.

Spinoza's Philosophy of Man: Proceedings of the Scandinavian Spinoza Symposium 1977. Norway, Universitets-Forlaget, 1978.

Wienpahl, Paul. *The Radical Spinoza*. New York, New York University Press, 1979.
Wilbur, J. B. *Spinoza's Metaphysics: Essays in Critical Appreciation*. Assen, Van Gorcum, 1976.
Williamson, Raymond Keith. *Introduction to Hegel's Philosophy*. Albany, State University of New York Press, 1984.
Wolfson, Harry Austryn. *The Philosophy of Spinoza: Unfolding the Latent Processes of His Reasoning*. Cambridge (MA), Harvard University Press, 1934.
Zac, Sylvain. *Le Moral de Spinoza*. Paris, Presses Universitaires de France, 1959.

FOR PART 2: CLASSICAL JEWISH PHILOSOPHY

PRIMARY

Abraham Ben David Ha-Levi (Ibn Daud). *The Exalted Faith* (האמונה הרמה). Edited by Norbert M. Samuelson and Gershon Weiss. Translated into English by Norbert M. Samuelson. Cranbury, NJ, Associated University Presses, 1986.
Chasdai Abraham Crescas. ספר אור יהוה. Tel Aviv, Esther, 1962/1963.
David Ibn Merwan Al-Mukammas. ספר יצירה (*The Book of Creation*). Edited by Judah Ben Barzilai. Berlin, H. Itzkowitz, 1885.
Levi Ben Gershon (Gersonides). *Commentary on the Book of Job*. Translated into English by A. I. Lassen. New York, Bloch, 1946.
מלחמות יהוה (*The Wars of the Lord*). Riva di Trento, s.n., 1560 and Leipzig, K. B. Lark, 1866. Translated into German by B. Kellerman. *Die Kampfe Gott's von Lewi Ben Gerson*. Berlin, Mayer and Muller, 1914. Book I translated into English by Seymour Feldman. *The Wars of the Lord: Book One: Immortality of the Soul*. Philadelphia, Jewish Publication Society of America, 1985. Book III translated into English by Norbert Samuelson. *Gersonides on God's Knowledge*. Toronto, Pontifical Institute of Mediaeval Studies, 1977. Book IV translated into English by J. David Bleich. *Providence in the Philosophy of Gersonides*. New York, Yeshiva University Press, 1973. Books III and IV translated into French by Charles Touati. *Les Guerres du Seigneur, Livres 3 et 4*. Paris, Mouton, 1968. Book V, Part 1, Chapters 1–20. Critical edition with English translation and commentary by Bernard R. Goldstein. *The Astronomy of Levi Ben Gershon*. New York, Springer Verlag, 1985.
פירוש התורה. Venice, Daniel Bomberg, 1547.
Moses ibn Maimon (Maimonides). *The Eight Chapters of Maimonides on Ethics*. Edited by J. I. Gorfinke. New York, AMS Press, 1966.

"Letter on Astrology (Letter to the Rabbis of Marseilles)." Translated into English by R. Lerner. *Medieval Political Philosophy*. Edited by R. Lerner and M. Mahdi. New York, The Free Press of Glencoe, 1963.

דלאלה אלחאירין (*The Guide of the Perplexed*) [מורה הנבוכים]. Translated into Hebrew by Judh Ibn Tibbon. Wilna, I. Funk, 1904. Translated into Hebrew by Joseph Bahir David Kapach. Jerusalem, מוסד הרב קוק, 1972. Translated into French by Solomon Munk. Paris, A. Franck, 1856–1866. Translated into English by Shlomo Pines. Chicago, University of Chicago Press, 1963.

משנה תורה. Tel Aviv, עם עולם, 1959.

עגנות (*Correspondence*). Edited by M. D. Rabinowitz. Jerusalem, מוסד הרב קוק, 1960.

פירוש למשנה (*Commentary on the Mishnah*). Translated into English by Fred Rosner. *Commentary on the Mishnah*. New York, Feldheim, 1975.

פירוש למשנה אבות (*Commentary on the Mishnah Fathers*). Edited by M. D. Rabinowitz. Jerusalem, מוסד הרב קוק, 1960.

צנאעה אלמנחיק (*The Book of Logic*). Translated into Hebrew as ספר הגיון. Jerusalem, מוסד הרב קוק, 1956.

Philo, "Selections." Edited by Hans Lewy. In *Three Jewish Philosophers*. New York, Atheneum, 1969.

Two Treatises on Philo of Alexandria: A Commentary on De Gigantibus and Quod Deus Sit Immutabilis. Edited and translated into English by David Winston and John Dillon. Brown Judaic Studies, no. 25, Chico, Scholars Press, 1983.

Saadia Ibn Joseph Al-Fayyuymi. כתב אמנת ולאעחקאדאת (*The Book of Beliefs and Opinions*) [ספר אמנות ודעות]. Leiden, Brill, 1880. Translated into Hebrew by Judah Ibn Tibbon. Leipzig, C. W. Vollrath, 1864. Translated into English by Samuel Rosenblatt. New Haven, Yale University Press, 1948. Selections translated into English by Alexander Altmann. *The Book of Doctrines and Beliefs*. New York, Atheneum, 1969.

SECONDARY

Adlerblum, Nina. *A Study of Gersonides in His Perspective*. New York, Columbia University Press, 1926.

Baumgardt, David. *Maimonides, the Conciliator of Eastern and Western Thought*. Bangalore, The Indian Institute of Culture, 1955.

Bekker, J. סודו של מורה נבוכים. Tel Aviv, s.n., 1957.

Böhner, Philotheus. *Medieval Logic: An Outline of Its Development from*

1250 to c. 1400. Manchester, University of Manchester Press, 1966.
Burrell, David B. and McGinn, Bernard (eds.). *God and Creation: An Ecumenical Symposium*. Notre Dame, Notre Dame University Press, 1990.
Copleston, Frederick. *A History of Philosophy*. Garden City, Image, 1962.
Davidson, Herbert. "Alfarabi and Avicenna on the Active Intellect," *Viator*, 3 (1972): 175–177.
"Arguments from the Concept of Particularization in Arabic Philosophy," *Philosophy East and West*, 18 (1968): 305–314.
"John Philoponus as a Source of Medieval and Jewish Proofs of Creation," *Journal of the American Oriental Society*, 89 (1969): 364.
"Maimonides' Secret Position on Creation." In I. Twersky (ed.). *Studies in Medieval Jewish History and Literature*. Vol. 1. Cambridge (MA), Harvard University Press, 1979. pp. 16–40.
Diesendruck, Zevi. "The Philosophy of Maimonides." *Central Conference of American Rabbis Yearbook* LXV (1935): pp. 355–368.
Efros, Israel. *Philosophical Terms in the Moreh Nebukim*. New York, Columbia University Press, 1924.
Studies in Medieval Jewish Philosophy. New York, Columbia University Press, 1974.
Fackenheim, Emil A. "The Possibility of the Universe in Al-Farabi, Ibn Sina and Maimonides." *Proceedings of the American Academy for Jewish Research* XVI (1946/1947).
Fakhry, Majjid. *A History of Islamic Philosophy*. New York, Columbia University Press, 1970.
"The Antimony of the Eternity of the World in Averroes, Maimonides and Aquinas." *Museon* 66 (1935): 139–155.
Feldman, Seymour. "The Doctrine of Eternal Creation in Hasdai Crescas and Some of His Predecessors." *Viator* 2 (1980): 289–320.
"Gersonides' Proof for the Creation of the Universe." *Proceedings of the American Academy for Jewish Research* 35 (1967): 113–137.
"Platonic Themes in Gersonides' Cosmology." In *Salo W. Baron Jubilee Volume*. Jerusalem, American Academy for Jewish Research, 1975. pp. 383–405.
Fontaine, T. M. *In Defense of Judaism: Abraham Ibn Daud, Sources and Structures of Emunah Ramah*. Assen, Van Gorcum, 1990.
Fox, Marvin. *Interpreting Maimonides: Studies in Methodology, Metaphysics, and Moral Philosophy*. Chicago and London, The University of Chicago Press, 1990.
Freudenthal, Gad. "Cosmogonie et Physique Chez Gersonide." *Revue des Études Juives*. 145, 3–4 (July–Dec, 1986): 295–314.

Gilson, Etienne. *History of Christian Philosophy in the Middle Ages.* New York, Random House, 1955.

Glicker, J. "עיון 10 "הבעית המודלית בפלוסופיה של הרמבם (1959): 177–191.

Goldstein, Bernard R. *The Astronomical Tables of Levi Ben Gerson.* Hamden, Archon Books, 1974.

"Levi Ben Gerson: On Instrumental Errors and the Transversal Scale." *Journal for the History of Astronomy* 8 (1977): 102–122.

"Levi Ben Gerson: On the Relationship between Physics and Astronomy." *Proceedings of the International Conference on the Interrelationship between Physics, Cosmology, and Astronomy: 1300–1700.* Tel Aviv/Jerusalem, s.n., 1984.

"Medieval Observations of Solar and Lunar Eclipses." *Archives Internationales d'Histoire des Sciences* 29 (1979): 101–156.

"The Origins of the Doctrine of Creation *Ex Nihilo*." *Journal of Jewish Studies* 35 (1984): 127–135.

"Preliminary Remarks on Levi Ben Gerson's Contributions to Astronomy." *The Israel Academy of Sciences and Humanities* Proceedings 3, 9. Jerusalem, s.n., 1969.

"The Status of Models in Ancient and Medieval Astronomy." *Centaurus* 24 (1980): 132–147.

Theory and Observation in Ancient and Medieval Astronomy. London, Variorum Reprints, 1985.

Grant, Edward (ed.). *A Source Book in Medieval Science.* Cambridge (MA), Harvard University Press, 1974.

Harvey, W. Z. "A Third Approach to Maimonides' Cosmogony-Prophetology Puzzle." *Harvard Theological Review* 74, 3 (1981): 287–301.

Harvey, W. Z. "Albo's Discussion of Time." *Jewish Quarterly Review* (1981): 220–221.

"The Term *Hitdabbekut* in Crescas' Definition of Time." *Jewish Quarterly Review* 71 (1980): 44–47.

Henry, Desmond P. *Medieval Logic and Metaphysics: A Modern Introduction.* London, Hutchinson, 1972.

Hepburn, Ronald W. "The Religious Doctrine of Creation." In Paul Edwards (ed.). *The Encyclopedia of Philosophy.* Vol. II. New York, Macmillan, 1967.

Herst, R. E. "Where God and Man Can Touch: An Inquiry into Maimonides' Doctrine of Divine Overflow." *Central Conference of American Rabbis Journal* 23 (Autumn, 1976): 16–21.

Husik, Isaac. *A History of Mediaeval Jewish Philosophy.* Philadelphia, Jewish Publication Society of America, 1958.

Hyman, Arthur. "Some Aspects of Maimonides' Philosophy of

Nature." *La Filosofia della Natura Nel Medioevo.* Milan, Societa Etrice Vita e Pensiero, 1966.

Ivry, Alfred. "Maimonides on Creation," in *Creation and the End of Days: Judaism and Scientific Cosmology*, David Novak and Norbert Samuelson (eds.), Lanham/New York/London, University Press of America, 1986. pp. 185–214.

Kaplan, Lawrence. "Maimonides on the Miraculous Element in Prophecy," *Harvard Theological Review* 70 (1977) no. 3–4: 233–256.

Klein-Braslavy, S. "Interpretation of Maimonides of the Term 'Create' and the Question of the Creation of the Universe," (in Hebrew) *Da'at* 16 (1986): 39–55.

Maimonides' Interpretation of the Story of Creation (in Hebrew). Jerusalem, s.n., 1978.

Kogan, Barry S. *Averroes and the Metaphysics of Causation.* Albany, State University of New York Press, 1985.

"Averroes and the Theory of Emanation." *Mediaeval Studies* 43 (1981): 384–404.

"Eternity and Origination: Averroes' Discourse on the Manner of the World's Existence," in *Islamic Theology and Philosophy*, M. E. Marmaura (ed.) Albany, State University of New York Press, 1984. pp. 203–235.

Lerner, R. "Maimonides' Letter on Astrology." *History of Religions* 8 (1968): 143–158.

Lipschitz, A. "The Theory of Creation of Rabbi Abraham Ibn Ezra." *Sinai* 84 (1979): 105–125.

Malino, J. "Maimonides' Guide to the Perplexities of Creation." Unpublished rabbinic dissertation, Hebrew Union College–Jewish Institute of Religion, 1979.

Malter, Henry. *Saadia Gaon: His Life and Works.* Philadelphia, Jewish Publication Society of America, 1921.

Manekin, Charles, "Preliminary Observation on Gersonides' Logical Writings." *Proceedings of the American Academy for Jewish Research* LII (1985): 85–113.

"Problems of 'Plenitude' in Maimonides and Gersonides," in *A Straight Path: Studies in Medieval Philosophy and Culture*, R. Link-Salinger (ed.) Washington, DC, The Catholic University of America Press, n.d. pp. 183–194.

Marx, Alexander. "The Correspondence between the Rabbis of Southern France and Maimonides about Astrology." *Hebrew Union College Annual* III (1926): 311–358.

Moody, Ernst A. *Truth and Consequence in Medieval Logic.* Amsterdam, North-Holland, 1953.

Neugebauer, O. "The Astronomy of Maimonides and Its Sources." *Hebrew Union College Annual* XXII (1949): 322–364.

Novak, David and Samuelson, Norbert (eds.). *Creation and the End of Days: Judaism and Scientific Cosmology*. Lanham/New York/London, University Press of America, 1986.

Nuriel, A. "Maimonides on Chance in the World of Generation and Passing Away," (in Hebrew) *Jerusalem Studies in Jewish Thought* 2, 1 (1982/1983): 33–42.

"חדוש העולם וקדמותו על פי הרמבם" תרביץ 33 (1964): 372–387.

Ravitsky, E. "חדוש או קדמות העולם בתורת הרמבם" תרביץ 35 (1966): 333–348.

Rescher, Nicholas. *The Development of Arabic Logic*. Pittsburgh, University of Pittsburgh Press, 1964.

— *Studies in the History of Arabic Logic*. Pittsburgh, University of Pittsburgh Press, 1963.

— *Temporal Modalities in Arabic Logic*. New York, Supplementary Series of Foundations of Languages, 1967.

Rosenthal, Erwin I. J. *Saadya Studies*. Manchester, Manchester University Press, 1943.

Roth, Leon. *The Guide of the Perplexed: Moses Maimonides*. London, Hutchinson's University Library, 1948.

Sambursky, Samuel and Pines, Shlomo. *The Concept of Time in Late Neo-Platonism*. Jerusalem, Israel Academy of Sciences and Humanities, 1971.

Samuelson, Norbert M. and Weiss, Gershom (eds.). *The Exalted Faith of Abraham Ibn Daud*. Cranbury, London and Mississauga, Associated University Presses, 1986.

Samuelson, Norbert M. "Comments on Maimonides' Concept of Mosaic Prophecy," *Central Conference of American Rabbis Journal* (January, 1971), pp. 9–25.

— "Halevi and Rosenzweig on Miracles," in *Approaches to Judaism in Medieval Times*, David R. Blumenthal (ed.), Brown Judaic Studies #54, Chico, CA, Scholars Press, 1984. pp. 157–172.

Sarfatti, G. B. *Mathematical Terminology in Hebrew Scientific Literature of the Middle Ages* (in Hebrew). Jerusalem, Magnes Press, 1968.

Schwarzschild, Steven S. "Moral Radicalism and 'Middlingness' in the Ethics of Maimonides" *Studies in Medieval Culture* 11 (1977) 65–94, reprinted in Menachem Kellner (ed.), *The Pursuit of the Ideal: Jewish Writings of Steven Schwarzschild*. Albany, State University of New York Press, 1990.

Shatzmiller, J. "Gersonides and the Community of Orange in the Middle Ages," (in Hebrew) in *Research on the History of Israel and Erets Israel*. Vol. II. Haifa, s.n., 1972. pp. 111–126.

Sirat, Colette. *A History of Jewish Philosophy in the Middle Ages.* New York, Cambridge University Press, 1985.
Sorabjis, Richard. *Time, Creation and the Continuum.* Ithaca, Cornell University Press, 1983.
Staub, Jacob J. *The Creation of the World According to Gersonides.* Brown Judaic Studies #24. Chico, Scholars Press, 1982.
  "Gersonides and Contemporary Theories of the Beginning of the Universe," in *Creation and the End of Days: Judaism and Scientific Cosmology*, Novak, David and Samuelson, Norbert (eds.), Lanham/New York/London, University Press of America, 1986. pp. 245–260.
Steinschneider, Moritz. *Die Arabische Literatur der Juden.* Frankfurt a.M., J. Kauffmann, 1902.
  Die Hebraischen Übersetzungen des Mittelalters. Berlin, Bibliographisches Bureau, 1893.
  Gesammelte Schriften. Berlin, M. Poppelauer, 1925.
  Jewish Literature from the Eighth to the Eighteenth Century. London, Longman, 1857.
Strauss, Leo. "The Literary Character of the Guide of the Perplexed." In *Persecution and the Art of Writing.* Glencoe, The Free Press, 1952. pp. 38–94.
  Philosophy and Law: Essays Toward the Understanding of Maimonides and His Predecessors. Translated into English by Fred Baumann. Philadelphia, Jewish Publication Society of America, 1987.
Touati, Charles. *La Pensée Philosophique et Théologique de Gersonides.* Paris, Minuit, 1973.
Twersky, Isadore. *Introduction to the Code of Maimonides (Mishneh Torah).* New Haven and London, Yale University Press, 1980.
Vajda, G. *Introduction à la Pensée Juive du Moyen Age.* Paris, s.n., 1947.
Waxman, Meyer. *The Philosophy of Don Hasdai Crescas.* New York, AMS Press, 1966.
Winston, David. *Philo of Alexandria.* Ramsey, Paulist Press, 1981.
Wolfson, Harry Austryn. *Crescas' Critique of Aristotle: Problems of Aristotle's Physics in Jewish and Arabic Philosophy.* Cambridge (MA), Harvard University Press, 1929.
  "Emanation and Creation Ex-Nihilo," (in Hebrew) in *Studies in the History of Philosophy and Religion.* Vol. II. Cambridge (MA), Harvard University Press, 1973. pp. 623–629.
  "Halevi and Maimonides on Design, Chance and Necessity," in *Studies in the History of Philosophy and Religion.* Vol. II. Cambridge (MA), Harvard University Press, 1973. pp. 1–59.
  "The Meaning of Ex Nihilo in the Church Fathers, Arabic and Hebrew Philosophy and St. Thomas." In *Medieval Studies in*

    Honor of J. D. M. Ford. Cambridge (MA), Harvard University Press, 1948.
    "The Meaning of Ex-Nihilo in Isaac Israel," in *Studies in the History of Philosophy and Religion.* Vol. 1. Cambridge (MA), Harvard University Press, 1973. pp. 222–233.
    Philo: Foundations of Religious Philosophy in Judaism, Christianity and Islam. Cambridge (MA), Harvard University Press, 1962.
    The Philosophy of the Kalam. Cambridge (MA)/London, Harvard University Press, 1976.
    "The Platonic, Aristotelian and Stoic Theories of Creation in Halevi and Maimonides," in *Studies in the History of Philosophy and Religion.* Vol. 1. Cambridge (MA), Harvard University Press, 1973. pp. 234–249.
    Repercussions of the Kalam in Jewish Philosophy. Cambridge (MA)/London, Harvard University Press, 1979.
    Studies in the History of Philosophy and Religion. Cambridge (MA), Harvard University Press, 1973.

FOR PART 2: CLASSICAL RABBINIC COMMENTARIES

PRIMARY

מקראות גדולות. Part 1. New York, Pardes Publishing House, 1951.
מדרש הגדול. Margulies, Mordecai (ed.) Jerusalem, מוסד הרב קוק, 1947.
מדרש רבה. Warsaw, 1913. Translated into English by H. Freedman and Maurice Simon. London, Soncinco, 1939. Translated into English by Jacob Neusner. *Genesis Rabbah: The Judaic Commentary of the Book of Genesis, A New American Translation.* Chico, Scholars Press, 1985.
פירושי התורה לרבינו משה בן נחמן (רמב״ן). משה בן נחמן. Edited by Charles B. Chavel. Jerusalem, מוסד הרב קוק, 1959. Translated into English by Charles B. Chavel. *Rambam (Nachmanides): Commentary on the Torah.* New York, Shilo, 1971.
ספר האגדה. Edited by Chaim Nachman Bialik. Tel Aviv, דביר, 1956.
ספרא. Edited by Weiss, Berlin, ספרים, 1925.
ספרי. Edited by Louis Finkelstein. Berlin, ספרים, 1925.
פסקתא רבתי. Edited by S. Buber. Lyck, מקצי נרדמים, 1868.
פרקי דרבי אליעזר. Translated into English by G. Friedlander, New York, Hermann Press, 1965.
תנחומא. Edited by S. Buber. Vilna, 1885.

Doron, Pinchas. *The Mystery of Creation According to Rashi: A New English Translation and Interpretation of Rashi on Genesis I–VI.* מוזנים, 1984.
Ibn Ezra, Abraham. *Ibn Ezra's Commentary on the Pentateuch.* English translation by H. Norman Strickman and Arthur M. Silver. New York, Menorah, 1988.
Midrash on Psalms. Translated into English by William Braude. New Haven, Yale University Press, 1959.
Oles, M. Arthur. "A Translation of the Commentary of Abraham Ibn Ezra on Genesis with a Critical Introduction." Unpublished Ph.D. dissertation. Cincinnati, Hebrew Union College–Jewish Institute of Religion, 1958.
Rashi. *Pentateuch and Rashi's Commentary.* Edited by Abraham Ben Isaiah and Benjamin Sharfman. Brooklyn, S.S.&R. Publishing Co., 1949.

SECONDARY

Altmann, Alexander. "A Note on the Rabbinic Doctrine of Creation." *Journal of Jewish Studies* 7 (1956): 195–206.
Baron, Salo Wittmayer. *A Social and Religious History of the Jews.* Philadelphia, Jewish Publication Society of America, 1942.
Baron, Salo Wittmayer and Blau, Joseph L. *Judaism: Postbiblical and Talmudic Period.* Indianapolis and New York, Bobbs-Merrill, 1954.
Braude, William G. "Maimonides' Attitude to Midrash." *Studies in Jewish Bibliography, History and Literature in Honor of I. Edward Kiev.* New York, Ktav, 1971. pp. 75–82.
Casper, Bernard. *Introduction to Jewish Bible Commentary.* New York, Thomas Yoseloff, 1960.
Corré, Alan (ed.). *Understanding the Talmud.* New York, Ktav, 1975.
Eichler, Barry and Tigay, Jeffrey. *Studies in Midrash and Related Literature.* Philadelphia, Jewish Publication Society, 1990.
Feldman, William M. *Rabbinical Mathematics and Astronomy.* London, M. L. Cailingold, 1931.
Ginzberg, Louis. *Legends of the Bible.* Philadelphia, Jewish Publication Society, 1966.
Goldin, Judah. "Midrash and Aggadah." *The Encyclopedia of Religion.* Vol. ix. New York, Macmillan, 1987. pp. 509–515.
Hartman, G. H. and Budick, S. (eds.). *Midrash and Literature.* New York, Yale University Press, 1986.
Heinemann, I. דרכי האגדה. Jerusalem, Hebrew University Press, 1954.

Jacobs, Louis. *Jewish Biblical Exegesis*. New York, Behrman House, 1973.
: *Studies in Talmudic Logic and Method*. London, Vallentine Mitchell, 1951.
: *The Talmudic Argument*. New York, Cambridge University Press, 1984.
Kadushin, Max. *A Conceptual Approach to the Mekilta: How the Spiritual Values of Talmud and Midrash Arise from the Bible*. New York, The Jewish Theological Seminary of America, 1969.
: *The Rabbinic Mind*. New York, The Jewish Theological Seminary of America, 1952.
Kasher, M. M. תורה שלמה. Jerusalem, 1927.
: *Encyclopedia of Biblical Interpretation*. Vol. I. English translation by H. Freedman. New York, American Biblical Encyclopedia Society, 1953.
Kasowsky, Chaim. אוזר לשון התלמוד. Jerusalem, מדרש החינוך והתרבות של ממשלת ישראל, 1954–1982.
Lauterbach, Jacob Z. "Midrash," in *The Jewish Encyclopedia*, Isidore Singer (ed.), New York, Ktav, n.d. VIII, pp. 549–572.
: *Rabbinic Essays*. Cincinnati, Hebrew Union College Press, 1951.
Lehrman, Simon M. *The World of Midrash*. London, Thomas Yoseloff, 1961.
Moore, George Foote. *Judaism in the First Centuries of the Christian Era*. Cambridge (MA), Harvard University Press, 1927.
Neusner, Jacob. *Canon and Connection: Intertextuality in Judaism*. Lanham, University Press of America, 1987.
: *Comparative Midrash: The Plan and Program of Genesis and Leviticus Rabbah*. Chico, Scholars Press, 1986.
: *Form-Analysis and Exegesis: A Fresh Approach to the Interpretation of the Mishnah*. Minnesota, University of Minnesota Press, 1980.
: *Genesis and Judaism: The Perspective of Genesis Rabbah, An Analytical Anthology*. Chico, Scholars Press, 1985.
: *A History of the Jews in Babylonia*. Leiden, E. J. Brill, 1970.
: *Invitation to the Talmud*. New York, Harper and Row, 1973.
: *Midrash as Literature: The Primacy of Documentary Discourse*. Lanham, University Press of America, 1987.
: *The Modern Study of the Mishnah*. Leiden, E. J. Brill, 1973.
: *There We Sat Down*. New York, Ktav, 1976.
: *The Way of Torah: An Introduction to Judaism*. North Scituate, MA, Duxbury Press, 1979.
: *Why No Science in Judaism*. New Orleans, Jewish Study Program: Tulane University, 1987.
Neusner, Jacob (ed.). *The Formation of the Babylonian Talmud*. Leiden, E. J. Brill, 1970.

The Glory of God is Intelligence. Salt Lake City, Brigham Young University Press, 1978.
Prijs, Leo. *Abraham Ibn Ezra's Kommentar zu Genesis.* Weisbaden, F. Steiner, 1973.
Abraham Ibn Ezra's Commentary on Genesis Chapters 1–3: Creation and Paradise. London, 1990.
Rabinovitch, Nachum L. *Probability and Statistical Inference in Ancient and Medieval Jewish Literature.* Toronto, University of Toronto Press, 1973.
Rosenthal, Erwin I. J. "Medieval Jewish Exegesis: Its Character and Significance" *Journal of Jewish Studies* 9 (1964), pp. 265–281.
Saperstein, Marc. *Decoding the Rabbis: A Thirteenth-Century Commentary on the Aggadah.* Cambridge (MA), Harvard University Press, 1980.
Silver, Daniel J. *Maimonidean Criticism and the Maimonidean Controversy, 1180–1240.* Leiden, E. J. Brill, 1965.
Steinsaltz, Adin. *The Essential Talmud.* New York, Basic Books, 1976.
Strack, Hermann L. *Introduction to Talmud and Midrash.* Philadelphia, Jewish Publication Society of America, 1945.
Unterman, Isaac. *The Talmud.* New York, Bloch, 1973.
Urbach, Efraim E. חז"ל: פרקי אמונות ודעות. Jerusalem, Magnes Press, 1969. Translated into English by Israel Abrahams. *The Sages: Their Concepts and Beliefs.* Jerusalem, Magnes Press, 1975.
Vermes, Geza. *Scripture and Tradition in Judaism.* Leiden, E. J. Brill, 1961.
Writing, Addison G. *The Literary Genre Midrash.* New York, Alba House, 1967.

FOR PART 3: GENESIS

Albrecht, Alt. *Essays on Old Testament History and Religion.* Garden City, Anchor Books, 1968.
Alter, Robert. *The Art of Biblical Narrative.* New York, Basic Books, 1981.
Alter, Robert and Kermode, Frank (eds.). *The Literary Guide to the Bible.* Cambridge (MA), Harvard University Press, 1987.
Anderson, Bernard W. *Understanding the Old Testament.* Englewood Cliffs, Prentice-Hall, 1957.
Anderson, Bernhard W. (ed.) *Creation in the Old Testament.* Philadelphia, Fortress Press, 1984.
Baab, Otto J. *The Theology of the Old Testament.* New York and Nashville, Abingdon Press, 1949.
Barr, James. *Biblical Words for Time.* London, SCM Press, 1962.

The Semantics of Biblical Language. New York, Oxford University Press, 1961.
Barthes, R., Boven, F., Leenhardt, F. J. *Structural Analysis and Biblical Exegesis.* Pittsburgh, The Pickwick Press, 1974.
Beebe, H. Keith. *The Old Testament: An Introduction to Its Literary, Historical, and Religious Traditions.* Belmont, CA, Dickenson, 1970.
Bentzen, A. *Introduction to the Old Testament.* Copenhagen, G. E. C. Gad, 1962.
Bright, John. *A History of Israel.* Philadelphia, Westminster Press, 1981.
Buss, M. J. (ed.) *Encounter with the Text: Form and History in the Hebrew Bible.* Philadelphia, Fortress Press, 1979.
Buttrick, George A., Bowie, W. R., Knox, J. et al. (eds.) *The Interpreter's Bible.* New York and Nashville, Abingdon Press, 1952.
Cassuto, Umberto. *A Commentary on the Book of Genesis.* Translated into English by Israel Abrahams. Jerusalem, Magnes, 1961–1964.
Chase, Ellen. *The Bible and the Common Reader.* New York, Macmillan, 1952.
Life and Language in the Old Testament. New York, Gramercy, 1955.
Culley, R. C. *Studies in the Structure of Hebrew Narrative.* Philadelphia, Fortress Press, 1976.
Doukhan, Jacques B. *The Genesis Creation Story.* Berrien Springs, Michigan, Andrews University Press, 1978.
Driver, S. R., Plummer, A., Briggs, C. A. (eds.) *The International Critical Commentary.* Edinburgh, T. and T. Clark, 1910.
Eaken, Frank E. Jr. *The Religion and Culture of Israel: An Introduction to Old Testament Thought.* Washington, DC, University Press of America, 1977.
Eissfeldt, O. *The Old Testament: An Introduction.* New York, Harper and Row, 1965.
Fishbane, Michael. *Biblical Interpretation in Ancient Israel.* Oxford, Oxford University Press, 1985.
Text and Texture: Close Readings of Selected Biblical Text. New York, Schocken, 1979.
Fox, Everett. *Genesis and Exodus: A New English Rendition with Commentary and Notes.* New York, Schocken, 1983.
Gaster, T. H. Thespis: *Myth, Legend and Custom in the Old Testament.* New York, Harper and Row, 1969.
Gunkel, Hermann. *Schöpfung und Chaos in Urzeit und Endzeit.* s.l., s.n., 1895.
The Legends of Genesis, the Biblical Saga and History. Translated into English by W. H. Carruth. New York, Schocken, 1964.

Hahn, Robert F. *The Old Testament in Modern Research*. Philadelphia, Fortress Press, 1966.
Hastings, James (ed.). *Dictionary of the Bible*. New York, Charles Scribner's Sons, 1898.
Heidel, Alexander. *The Babylonian Genesis*. Chicago, University of Chicago Press, 1963.
Kaufmann, Yehezkiel. *The Religion of Israel: From Its Beginnings to the Babylonian Exile*. Translated into English by Moshe Greenberg. Chicago, University of Chicago Press, 1960.
König, E. *Theologie des Alten Testaments*. Stuttgart, s.n., 1923.
Kramer, Samuel Noah. *The Sumerians*. Chicago, University of Chicago Press, 1963.
Kramer, Samuel Noah (ed.). *Mythologies of the Ancient World*. Garden City, NY, Doubleday and Co., 1961.
Kuntz, J. Kenneth. *The People of Ancient Israel: An Introduction to Old Testament Literature, History and Thought*. New York, Harper and Row, 1974.
Lambert, W. G. "A New Look at the Babylonian Background of Genesis," in *The Bible in its Literary Milieu*. John Maier and Vincent Tollers (eds.). Grand Rapids, Eerdmans, 1979. pp. 285–297.
Leibowitz, Nehama. *Studies in the Book of Genesis*. Jerusalem, World Zionist Organization, 1972.
Levenson, Jon D. *Creation and the Persistence of Evil: The Jewish Drama of Divine Omnipotence*. San Francisco, Harper and Row, 1988.
Sinai and Zion: An Entry into the Jewish Bible. San Francisco, Harper and Row, 1987.
Long, Charles H. Alpha. *The Myths of Creation*. New York, Collier, 1969.
Neusner, Jacob. *Ancient Israel after Catastrophe: The Religions World View of the Mishnah*. Charlottesville, University Press of Virginia, 1983.
Christian Faith and the Bible of Judaism: The Judaic Encounter with Scripture. Grand Rapids, MI; Eerdmans, 1987.
Judaism and its Social Metaphors: Israel in the History of Jewish Thought. Cambridge and New York, Cambridge University Press, 1989.
Judaism as Philosophy: The Method and Message of the Mishnah. Columbia, SC, University of South Carolina Press, 1991.
The Philosophical Mishnah. Atlanta, Scholars Press, 1988.
Neusner, Jacob, and Greeley, Andrew M. *The Bible and Us*. New York, Warner Books, 1990.
Neusner, Jacob, Levine, Baruch A., and Frerichs, Ernest S. (eds.). *Judaic Perspectives on Ancient Israel*. Philadelphia, Fortress Press, 1987.

Niditch, Susan. *Chaos to Cosmos: Studies in Biblical Patterns of Creation.* Chico, CA, Scholars Press, 1985.
Noth, Martin. *A History of Pentateuchal Traditions.* Englewood Cliffs, Prentice-Hall, 1979
The History of Israel. New York, Harper and Row, 1958.
Pederson, Johannes. *Israel: Its Life and Culture.* London, Oxford University Press, 1926.
Pfeiffer, R. F. *Introduction to the Old Testament.* New York, Harper and Brothers, 1941.
Polzin, Robert. *Biblical Structuralism: Method and Subjectivity in the Study of Ancient Texts.* Philadelphia, Fortress Press, 1977.
Pritchard, James B. *The Ancient Near East.* Princeton, Princeton University Press, 1950.
Rad, Gerhard von. *Genesis, A Commentary.* Translated into English by John H. Marks. Philadelphia, Westminster Press, 1972.
Rosenberg, David. *Congregation.* New York, Harcourt, Grace, Jovanovich, 1987.
Rowley, H. H. *The Old Testament and Modern Study.* Oxford, Clarendon University Press, 1951.
Samuelson, Norbert M. *The First Seven Days: A Philosophical Commentary on the Creation of Genesis.* Atlanta, GA, Scholars Press, 1993.
Sarna, Nahum M. *The JPS Torah Commentary: Genesis.* Translation and commentary by Nahum M. Sarna. Philadelphia, Jewish Publication Society, 1989.
Understanding Genesis. New York, Jewish Theological Seminary of America, 1966.
Schneidau, Herbert N. *Sacred Discontent: The Bible and Western Tradition.* Berkeley, University of California Press, 1977.
Segal, Moses Hirsh. *The Pentateuch: Its Composition and Its Authorship and Other Biblical Studies.* Jerusalem, Magnes Press, 1967.
Sellin, E. and Fohrer, G. *Introduction to the Old Testament.* Nashville, Abingdon Press, 1965.
Speiser, E. A. *Genesis.* Garden City, Doubleday, 1977.
Stadelmann, Luis I. J. *The Hebrew Conception of the World.* Rome, Analecta Biblica, 39, 1970.
Vawter, Bruce. *On Genesis.* Garden City, Doubleday, 1977.
Wakeman, Mary K. *God's Battle with the Monster.* Leiden, Brill, 1973.
Weinfeld, Moshe. "God the Creator in Genesis 1 and in the Prophecy of Second Isaiah" (in Hebrew). *Tarbiz* Vol. 37, No. 2 (Jan., 1968) pp. 105–132.
"Sabbath, Sanctuary and the Kingdom of the Lord" (in Hebrew). *Bet Mikre* vol. 69 (1976/1977) pp. 188–193.

West, James King. *Introduction to the Old Testament.* New York, Macmillan, 1981.
Wright, G. Ernst. *The Old Testament Against Its Environment.* London, SCM Press, 1950.

FOR PART 3: PLATO'S *TIMAEUS*

PRIMARY

Archer-Hind, Richard Dakre. *The Timaeus of Plato.* New York, Arno Press, 1973.
Cornford, Francis MacDonald. *Plato's Cosmology.* London, Routledge & Kegan Paul, 1966. (First published by Kegan Paul, Trench, Trubner & Co., 1937.)

SECONDARY

Claggett, Marshall. *Greek Science in Antiquity.* London, Collier Macmillan, 1955.
Farrington, Benjamin. *Greek Science.* Harmondsworth, Penguin, 1944–1949.
Pederson, Olaf and Mogens, Phil. *Early Physics and Astronomy.* New York, American Elsevier, 1974.
Sarton, George. *Introduction to the History of Science.* Baltimore, Carnegie Institute of Washington #376, 1927.
Solmsen, Friedrich. *Aristotle's System of the Physical World: A Comparison with His Predecessors.* Ithaca, Cornell University Press, 1960.
Vlastos, Gregory. *Plato's Universe.* Seattle, University of Washington Press, 1975.

FOR PART 4: ASTROPHYSICS

Aitchison, I. J. R. and Hey, A. J. G. *Gauge Theories in Particle Physics.* Bristol, England, Adam Hilger, 1982.
Barnett, Lincoln. *The Universe and Dr. Einstein.* New York, Bantam Books, 1948.
Barrow, John D. and Silk, Joe. *The Left Hand of Creation.* New York, Basic Books, 1983.
Barrow, John D. and Tipler, Frank J. *The Anthropic Cosmological Principle.* New York, Oxford University Press, 1986.
Bohm, David. *Causality and Chance in Modern Physics.* New York, Van Nostrand, 1957.
Quantum Theory. New York, Prentice-Hall, 1951.

Bohr, Niels. *Atomic Theory and the Description of Nature.* Cambridge, Cambridge University Press, 1934.
The Physical Principles of Quantum Theory. Chicago, University of Chicago Press, 1930.
Physics and Philosophy. New York, Harper and Row, 1958.
Chaisson, Eric J. and Field, George. *The Invisible Universe.* Boston, Birkhauser, 1985.
Chaisson, Eric J. and Stengers, I. *Order Out of Chaos.* New York, Bantam, 1984.
Cornell, James and Leightman, Alan P. *Revealing the Universe: Prediction and Proof in Astronomy.* Cambridge (MA), Massachusetts Institute of Technology Press, 1982.
Davies, Paul C. W. *The Accidental Universe.* New York, Cambridge University Press, 1982.
God and the New Physics. New York, Viking Penguin, 1983.
Other Worlds. London, Dent, 1980.
Space and Time in the Modern Universe. Cambridge, Cambridge University Press, 1977.
Davies, Paul C. W. and Brown, J. R. *The Ghost in the Atom.* Cambridge, Cambridge University Press, 1986.
DeWitt, B. S. and Graham, N. *The Many Worlds Interpretation of Quantum Mechanics.* Princeton, Princeton University Press, 1973.
Duff, M. J. and Ishon, C. J. (eds.). *The Quantum Theory of Space and Time.* Cambridge, Cambridge University Press, 1982.
Edward, Le Roy. *Religious Beliefs of American Scientists.* Westport, Greenwood Press, 1978.
Fano, G. *Mathematical Methods of Quantum Mechanics.* New York, McGraw-Hill, 1971.
Ferris, T. *Galaxies.* San Francisco, Sierra Club Books, 1980.
Fine, Arthur. *The Shaky Game – Einstein's Realism and the Quantum Theory.* Chicago, University of Chicago Press, 1986.
French, A. P. and Taylor, E. F. *An Introduction to Quantum Physics.* Middlesex, Nelson, 1978.
Gal-Or, Benjamin. *Cosmology, Physics and Philosophy.* New York, Springer Verlag, 1981.
Gamow, George. *Mr. Tompkins in Wonderland.* New York, Macmillan, 1940.
One, Two, Three, . . . Infinity. New York, Bantam Books, 1947.
Gatewood, G. "On the Astrometric Detection of Planetary Systems." *Icarus* 27 (1976): 1–25.
Geller, M. J. "Research Frontiers in Astronomy," in *Research Frontiers and the National Agenda.* Washington, DC, National Academy Press, 1984.

Gott, J. R., Gunn, J. E., Schramm, D. N., and Tinsley, B. M. "Will the Universe Expand Forever?" *Scientific American* 234:20 (March, 1976): 62–65.
Gottfried, Kurt. *Quantum Mechanics Vol. 1: Fundamentals.* Reading (MA), Benjamin/Cummings, 1966.
Groth, E. J., Peebles, P. J. E., Seldner, M., and Soneira, R. "The Clustering of Galaxies." *Scientific American* 237:5 (Nov. 1977): 76–78.
Guth, A. H. and Steinhardt, P. "The Inflationary Universe." *Scientific American* 20 (May, 1984): 116–128.
Harrison, Edward R. *Cosmology.* Cambridge, Cambridge University Press, 1981.
Hawking, Stephen W. *A Brief History of Time: From the Big Bang to Black Holes.* Toronto, Bantam Books, 1988.
Heisenberg, W. *The Physicist's Conception of Nature.* Westport, CT, Greenwood Press, 1958.
Hempel, Carl G. *Aspects of Scientific Investigation.* New York, The Free Press, 1965.
Honner, John. *The Description of Nature: Niels Bohr and the Philosophy of Quantum Mechanics.* New York, Oxford University Press, 1988.
Hubble, Edwin. *The Realm of the Nebulae.* New Haven, Yale University Press, 1936.
Kuhn, Thomas S. *The Structure of Scientific Revolutions.* Chicago, University of Chicago Press, 1957.
Lake, G. "Windows on a New Cosmology." *Science* 24 (May 8, 1984): 675–681.
Mandelbrot, B. *The Fractal Geometry of Nature.* New York, W. H. Freeman & Co., 1983.
March, Robert H. *Physics for Poets.* New York, McGraw-Hill, 1970.
Munitz, Milton K. (ed.). *Theories of the Universe.* New York, The Free Press/Collier-Macmillan, 1957.
Novak, David and Samuelson, Norbert. *Creation and the End of Days: Judaism and Scientific Cosmology. Proceedings of the 1984 Meeting of the Academy for Jewish Philosophy.* Lanham, University Press of America, 1986.
Pagels, Heinz. *Cosmic Code.* New York, Simon and Schuster, 1982.
Perfect Symmetry. New York, Simon and Schuster, 1985.
Peebles, P. J. E. *Cosmology.* Princeton, Princeton University Press, 1971.
Popper, Karl. *Objective Knowledge.* London, Oxford University Press, 1972.
Prigogine, I. *From Being to Becoming: Time and Complexity in the Physical Sciences.* San Francisco, W. H. Freeman, 1980.

Powers, J. *Philosophy and the New Physics*. London, Methuen, 1982.
Rae, Alastair I. M. *Quantum Mechanics*. Bristol, Hilger, 1986.
　Quantum Physics: Illusion or Reality? Cambridge, Cambridge University Press, 1986.
Redhead, Michael. *Incompleteness, Nonlocality and Realism: A Prolegomenon to the Philosophy of Quantum Mechanics*. Oxford, Clarendon Press, 1987.
Reeves, H. *Atoms of Silence*. Boston, Massachusetts Institute of Technology Press, 1984.
Robinson, Abraham. *Selected Papers of Abraham Robinson*. Edited by H. J. Keisler, *et al*. New Haven, Yale University Press, 1979.
Salmon, Wesley. *Space, Time and Motion*. Minneapolis, University of Minnesota Press, 1980.
Sears, F. W., Zemansky, M. W., and Young, H. D. *University Physics*. 7th edition. Reading, MA, Menlo Park, CA, *et alibi*, Addison-Wesley Publishing Co., 1987.
Seielstad, G. *Ecology*. Berkeley, University of California Press, 1983.
Simmons, G. F. *Introduction to Topology and Modern Analysis*. New York, McGraw-Hill, 1963.
Shipman, H. L. *Black Holes, Quasars, and the Universe*. Boston, Houghton Mifflin, 1980.
Shu, F. *The Physical Universe*. Mill Valley, University Science Books, 1982.
Silk, J. *The Big Bang*. San Francisco, W. H. Freeman, 1980.
Strauss, Claude Levi. *Structural Anthropology*. Garden City, Doubleday & Co., 1967.
Toulmin, Stephen. *The Return to Cosmology: Postmodern Science and the Theology of Nature*. Los Angeles, University of California Press, 1982.
Trefil, J. S. *From Atoms to Quarks: An Introduction to the Strange World of Particle Physics*. New York, Scribners, 1980.
　The Moment of Creation: Big Bang Physics. New York, Scribners, 1983.
Vilenkin, A. "Creation of the Universe from Nothing." *Physics Letters* B117, 25 (1982).
von Neumann, J. *Mathematical Foundations of Quantum Mechanics*. Princeton, Princeton University Press, 1955.
Wagoner, R. and Goldsmith, D. *Cosmic Horizons*. San Francisco, W. H. Freeman, 1983.
Waldrop, M. M. "Before the Beginning." *Science* 84: 5 (January, 1984): 44–53.
Weinberg, Steven. *The First Three Minutes*. New York, Basic Books, 1977.

Gravitation and Cosmology. New York, Wiley and Sons, 1972.
Wheeler, J. A. and Zurek, W. N. (eds.). *Quantum Theory and Measurements.* Princeton, Princeton University Press, 1983.
Whitney, Charles A. *The Discovery of Our Galaxy.* New York, A. Knopf, 1971.

Indices

FOREIGN LANGUAGES

ARABIC

BA'D AL-'ADAM 101
LA MIN SHAY 101
MIN AL-MA'DUM 101
MIN LA SHAY 101
MIN LA WUJUD 101
MIN LAYS 101

GERMAN

Allgemeine 44–46, 50
Allgemeines, ein individualisiertes 44
Allgemeinheit der Gattung 62
Allgemeinheit, eine besondere 44
Also 49
Am Anfang 57
Anfang der Schöpfung 60
Augenblick 57, 60

Bahn 51
Bedürftigen 62
Bedürftigkeit 247
Begriff der Schöpfung 53
Begründeten 254
Bejahung 39
Besondere 43–48, 50
Besonderheit 48
Bestimmtheit 58
Bestimmung 38, 44, 58
Bewegung 51
Brütend 59–61

Da 48
Dasein als Schon-Dasein 57
Denken 35, 42, 44, 53, 244
Ding 56, 60, 62

Ding-an-sich 42
Dingwelt 58
dumpfste aller Tätigkeiten 59

Eigenart 51
Eigenheit 49
Eigenschaft 58, 60
Eigensein 48
Eigentümschaft 45
Einheit 35
Eins 35
Einzeldinge 58
Einzelnen 61
Einzelnis 48
Einzige 58
Element, Begriffliche 53
Elementarwort 54
Erkenntnis 61
Es 60
Es werde 60
Etwas 34, 36–38
Etwas, das einzelnen 43

Figur 255
Finsternis 58
Freiheit 247
Fülle 43

Ganze 45, 58
Gattung 44, 51
Gattungsmerkmal 44
Geburt 51
Gegebene 43
Gegenständlichkeit 57
Gegenständlichkeit der Dinge 58
Gegenwartserlebnis 52
Geheimnis 50
Geist 59
Geist Gottes brütend 58

Index of foreign languages

Geschehenswort 58
Geschöpf 54
Gestalt 33, 63, 255
gewirktes 56
Glaube 34, 52, 244
Gott schuf 57
Gott sprach 60
Gottes Geist 59
göttlichen Bejahung des Kreatürlichen Daseins 56
Grundbegriff 53
Gut 56–57
Gut gar sehr 62

im Ebenbilde Gottes 61
Individuum 44
Irdischen 62

Ja 39–40, 43, 49, 54, 63
jedes einzelne Ding 58

Kraft, eine anzeihende 44

Lasset uns . . . machen 61
Lebendigkeit des Gottes 39
Leer 58
letzten Schöpfertat 61
Licht 60
Liebe 62

Machtat 51
Mehrkeit 44
Mensch 61
Mensch, Lebendige 49
Muss 62

Nein 39–40, 43, 49, 54
Neue Beziehung 253
Nichtnichts 39, 48–49
Nichts 34, 36–38, 43, 63

Objecktivität 61
Offenbarung 34, 51
Offenbarungswunder 54
Organon der Offenbarung 54

Persönlichkeit 61–62

Satz 57
Scheidet 61
Schicksalsmuss 51
schlechtin bejahende Bewertung 60
Schlussworte 56

Schon-vergangenseins 58
Schöpfermacht 62
Schöpfertum 247
Schöpfung 34, 51
Schöpfungswort 57
Seele 62
Sehr 62
Sein 34, 36, 38, 58
Selbst 60
Setzung 38
Sinn 247
So 48–49
So ist es 62
Sprache 54
Sprache, Wirkliche 54

Taten 62
Tatsächlichkeit 51
Tatwort 59–60
Teile 45
Theologie, historische 51
Theologie, Neue 53
Tod 62
tönender Laut, . . . Wort 61

Und 40, 43, 49, 54
unmittelbar gegenwärtig er-lebten Augen-blicks des Lebens 63
Uns 61
Unteren Grenze 58
Urja 48
Ursprung 62
Urwort 40, 54
Über-mathematischen 255

Vereinung 39
Vergangenheit 52
Vergänglichsein 48
Verhängniss 62
Vermittelten 62
Verneinung 63
Vielheit 58, 62
Vielleicht 51
Vollendet 61
Vorbedingung 53

War 58
Ward 57
Weissagung 62
Wenn 50
Wesen 42, 60, 62
Willenstrotz 51
Wirklichkeit 38

Wirklichkeit, Vorhandene 53
Wissen 34, 51–52, 54
Wissenschaft 34
Wollen 54, 61
Wunder 62
Wüste 58

Zeichen 39, 43, 48, 62
Zeugnis 62
Zusammenhang 255

GREEK

ANAGKE 59, 169–170
ANALOGIA 175
AORISTOS DUAS 170
APEIRON 192
APEIRON PLETHOS 170
ARCHE 170, 192
ARITHMOS 175

CHORA 184

DEMIOURGOS 171
DIATHIGE 174

EIDOS 175, 182
EIKOS 169
EK TOU ME EINAI 101
EK TOU ME ONTOS 101

FYSEI 172
FYSOLOGOI 173

GENE 182
GENESIS 59, 170

HAPEIRON PLETHOS 59
HEN 171

IDEA 182

KOSMOS 169

LOGOS 169

MORFE 182
MYTHOS 169, 177, 245

NOUS 59, 169–170

OKEANOS 191
ON PUS 183

ONTOS ON 183
OURANOU THEOI 175
OUSIA 37, 59

PANTELOS ME ON 184
PARADEIGMA 175
PERAS 59, 170
PNEUMA 141
POIETES 171
POIOTES 182
PROS HEN 103
PSYCHE 175

RUSMOS 174

SEFAIRA 176
SIMTHITOS EPITHMIA 185
STOICHEIA 173
SUMA 176

TA PATHEMATA AISTHETICHA 190
TECHNE 172
THEOS 169
TO AUTOMATON 172
TO MEGA KAI TO MIKRON 171
TO ON 170, 177
TO OURANOS 175
TROPA 174
TUCHE 172

UPODOCHE 171
UPODOCHEN 59

HEBREW

'AFAR 140, 144, 150
'ASAH 118, 142, 171
'ATID LAVO 117
'AVAD 164
'AVODAH 164
'ESEV 120
'ETS PERI 120
'EVED 164
ACHAR HA-HE'DER 101
ADAM 61, 71
AGADAH 117, 123
ARETS 93

BARA 57, 137–138, 141–142
BARA ELOHIM 57–58
BE-TSELEM ELOHIM 61
BEHEMAH 127, 146
BERESHIT 57, 113–114, 137–138, 141, 225

BIRQI'A HA-SHAMAYIM 123
BOHU 58, 93, 114–116, 124, 132,
    137–138, 140, 149–150, 204,
    226–227

CHAFETS 137
CHAYTO-ARETS 127, 146
CHOSHEKH 58, 93, 114–115, 121, 132,
    134, 138–139, 141, 144
CHOVOT EVARIM 15
CHOVOT HA-LEVAVOT 15

DESHE 120

ELOHIM 137–138, 140, 147

EPIKOROS 16
ERETS 140–141, 144
ESH 119
ET HA-SHAMAYIM VE-ET HA-ARETS 58,
    137
HA- 58
HA-'OLAM HA-BA 137
HA-'OLAM HA-ZEH 137
HA-ADAM 146
HA-MEFARESH 110
HA-TSADIKIM 117
HALACHAH 15
HAYETAH 58

KAVOD 146
KEDUSHAH 260
KHEN 48–49, 144–145
KI TOV 119
KIDMUTE-NU 130
KOACH 145
KOFER BE-'IKKAR 16
KOL MELAKHTO 131

LE-OTOT 121

MAGEN DAVID 28
MAYIM 93, 114, 119, 132, 140–141, 144,
    227
ME-AYIN 101
ME-LO DAVAR 101
ME-LO METSIYUT 101
MELAKHAH 56
MEOROT 94, 121–122
MERACHEFET 58, 60–61

NA'ASEH 61

NEFESH CHAYAH 127–128, 146, 175

OR 60–61, 93, 121, 134, 141, 144–145,
    204

PENIMIYAH 111
PERUSH NISTAR 110
PESHAT 110

RAQI'A 93, 118–119, 123, 125, 131–132,
    143, 145, 161
REMES 126–127, 132, 146
ROSH HA-SHANAH 121
RUACH 59–60, 93, 114, 132, 137–138,
    140–141, 144
RUACH ELOHIM 59, 93, 116, 133,
    139
RUACH ELOHIM MERACHEFET 58

SHAM 119
SHAMAYIM 93, 119, 140, 143
SHERETS 126, 146

TEHOM 93, 114, 116, 123–124, 132,
    138–139, 141, 144, 226
TENINIM 126, 146
TOHU 58, 93, 114–116, 124, 132,
    137–138, 140, 149–150, 204,
    226–227
TOV 56–57, 61, 144–145, 233
TOV MEOD 62, 99
TSELEM ELOHIM 146

VA-YEHI KHEN 62–63
VA-YOMER ELOHIM 60

WAW, *see* Grammar, Verb, Consonant of
    Conjunction 158

YAVDEL 61
YEHI 60
YIGLEH SOD 110
YITEN TA'AM 110
YOM 139

LATIN

ex nihilo 101
ex non esse 101
ex non existente 101
non ex aliquo 101
non fit ex aliquo 102
post non esse 101

BIBLE REFERENCES

Aaron 163
Abraham 124, 163
Adam 132–133
Amos 4, 114
David 165
Deuteronomy 165
Eve 132
Exodus 157
Gen. 1: 22, 56–57, 59–60, 71, 109, 117
Genesis 3–4, 20–21, 56, 91, 153, 157–169, 172, 175, 191, 193–198, 205–206, 220–221, 226–227, 232, 238, 241, 262
Isa. 06: 117
Isa. 45: 114
Isaac 124, 163
Jacob 163
Job 157
Leviticus 260
Miriam 163
Moses 15, 163
Moses, Death of 164–165
Noah 124
Prov. 8: 114, 134
Ps. 104: 117
Ps. 148: 114
Psalms 157
Qohelet 165
Rebecca 163
Sarah 163

BOOKS

Abraham Ibn Daud, *Exalted Faith* 83, 107, 220
Genesis Rabbah 21, 80, 106, 111–112, 114, 123, 136, 149, 207, 220, 225, 230
Gersonides, *Wars of the Lord* 12, 83, 91–94, 107, 207, 220
Ginsberg, *Legends of the Bible* 208
Kogan, Barry, *Reason, Revelation, and Authority in Judaism, A Reconstruction* 13
Krochmal, *Guide of the Perplexed for the Time* 31
Maimonides, *Commentary on the Mishnah* 91
Maimonides, *Guide of the Perplexed* 91, 220
MIQRAOT GEDOLOT 111–113, 207
Mishnah 77–79

Moore, *Judaism in the First Centuries of the Common Era* 208
Novak and Samuelson, *Creation and the End of Days: Judaism and Scientific Cosmology* 209
Plato, *Critias* 168, 191
Plato, *Hermocrates* 168
Plato, *Laws* 168
Plato, *Phaedo* 191
Plato, *Philebus* 168
Plato, *Republic* 168–169, 196, 232
Plato, *Timaeus* 3, 6, 13, 21, 37, 59–60, 70, 72, 100, 104–105, 134, 139, 153, 156, 167–198, 205, 207, 211, 213, 220–221, 226, 232, 238–243, 245, 248, 262
Rosenzweig, *Star of Redemption* 21, 31, 53, 107, 176, 197, 207, 220, 241, 250, 253, 255
Russell and Whitehead, *Principia Mathematica* 43, 51
Saadia, *Book of Beliefs and Opinions* 82
Spinoza, *Ethics* 177, 246–247
Spinoza, *Tractatus Theologico-Politicus* 246–247
Steinheim, *Die Offenbarung nach dem Lehrbegriffe* 30
Talmud 77–78
Talmud, Mishnah, Sanhedrin 16
Trefil, *From Atoms to Quarks* 215
Weinberg, *First Three Minutes* 209

NAMES

Abraham Ibn Daud 3, 16, 18, 21, 29, 79–81, 83–91, 98, 112, 210, 241, 248
Abravanel, Isaac 17
Albo, Joseph 18, 29, 77, 79, 107
Alghazali 102–103
Anaxagoras 193
Anaximander 173, 192
Anaximenes 173, 192
Anderson, Carl D. 210
Aristotle 12, 34, 37–38, 51, 79, 90, 99–101, 104, 178, 181, 187, 193, 244, 256
Averroes, *see* Ibn Rushd 6

Bachya 16
Baeck, Leo 77
Bar Kappara 115
Bar Sira 129

Index of names

Bibago, Abraham 18, 29
Bleich, David 9–10
Bohm, David 210
Bohr, Niels 210, 231
Booth, John Wilkes 2
Buber, Martin 12, 21, 30, 50, 77, 257

Calliphus 178, 181
Cantor, George 218
Chaisson, Eric J. 209
Christiani, Pablo 113
Cohen, Hermann 12, 19, 27, 36–37, 77, 238, 249–250
Crescas, Chasdai 12, 14, 17–18, 29, 79, 231
Critias of Athens 168

Democritus 174, 193
Descartes, Rene 42, 47, 56, 257
Dirac, Paul 210
Duran, Shimon Ben Zemach 18

Einstein, Albert 210–211, 237, 252
Empedocles 175, 192–193
Epicurus 193
Euclid 37, 254
Eudoxus 181
Eudoxus of Cnidus 178
Everett, Hugh 230

Feldman, Seymour 209

Gell-Mann, Murray 210
Geller, Margaret J. 209
Genibah 131
Gersonides, Levi Ben 3, 9–10, 12–13, 21, 79–81, 83, 89–109, 113, 149–150, 210, 225, 227, 231, 237
Ginsberg, Louis 208
Goldstein, Bernard R. 209
Gott, J. Richard 209
Guth, Alan H. 209

Halevi, Judah 16
Hawking, Stephen 209–211
Hegel, Wilhem Friedrich 25, 31, 34, 36, 38, 43, 45, 72, 105, 245–247, 252
Heisenberg, Werner 210, 212, 223
Heracleitus 173–175, 192–193
Hermocrates of Syracuse 168
Hillel 115
Hirsch, Samson Raphael 30
Homer 192

Hubble, Edwin 210
Huchra, John P. 209

Ibn Ezra, Abraham Ben Meir 21, 80, 106, 112–113, 136–152, 210, 241
Ibn Gabirol, Solomon 79, 248
Ibn Rushd 12, 91, 102–103
Ibn Sina 102–103
Ivry, Alfred L. 209

Jefferson, Thomas 263
Jesus 52
Joseph Ibn Zaddik 112
Judah Halevi 112

Kant, Immanuel 37, 42, 47–48, 52, 56, 196, 242
Kogan, Barry S. 13, 209
Krochmal, Nachman 18, 21, 31–32, 68

Leibniz, Gottfried Wilhelm von 99, 229, 232
Leucippus 174, 193
Lincoln, Abraham 2
Lucretius 193

Maimonides, Moses Ben 3, 10, 12–13, 16–18, 21, 29, 38, 47, 56, 89, 91, 95–101, 104, 153, 210, 231, 237, 239, 247–248
Malino, Jonathan W. 209
Maxwell, James Clerk 210
Mendelssohn, Moses 18, 30–31
Moore, George Foote 208

Nachmanides, Moses Ben 5, 21, 80, 106, 136–152, 112–113, 210, 226, 241
Neo-Platonists 34
Newton, Isaac 210
Nietzsche, Friedrich 36

Onkelos 112

Parmenides 34, 43
Plato 3–4, 13, 21, 34, 37, 51, 59–60, 66, 70, 72, 79–80, 90, 99–101, 104–105, 134, 139, 156, 167–198, 205, 211, 220–221, 232, 241, 245–246, 264
Prigogine, Ilya 210
Pythagoras 2, 171, 186
Pythagorean 59

Rab 115, 119

Index of subjects

Rashi 21, 80, 106, 112–113, 134–136, 149–150, 210, 226, 241
Rav 'Azariah 122
Rav Abba Ben Kahana 118–119
Rav Abbahu 114, 116, 123
Rav Acha 120
Rav Aibu 129
Rav Ammi 129
Rav Berekiah 116, 120, 122, 129
Rav Chama 131
Rav Chama Ben Rav Chanina 129, 131
Rav Chama Ben Rav Hoshaya 127
Rav Chanina 117–119, 122, 129, 131
Rav Efes 122
Rav Gamaliel 114
Rav Hila 129
Rav Hoshaya 122
Rav Hoshaya the Elder 127
Rav Huna 115, 126, 129
Rav Idi 126
Rav Isaac 119–120
Rav Jannai 123
Rav Jassi 129
Rav Jehotsadak 117
Rav Jochanan 117–118, 122, 131
Rav Jonah 115
Rav Jonathan 123, 129
Rav Jose Ben Rav Chalafta 119
Rav Joshua Ben Levi 114, 129
Rav Joshua Ben Rav Chanina 131
Rav Joshua Ben Rav Nehemiah 118
Rav Joshua of Siknin 129
Rav Judah 116, 123
Rav Judah Ben Pazzi 115
Rav Judah Ben Rav Shalom 120
Rav Judah Ben Rav Simon 114, 116, 118, 131
Rav Leazar 129
Rav Levi 115, 119, 129, 131
Rav Luliani Ben Tabri 117
Rav Mattenah 126
Rav Menachem 114
Rav Nachman 119
Rav Nathan 120–121
Rav Nehemiah 116
Rav Oshaya 114, 118
Rav Phinechas 118–119, 126
Rav Pinechas 123, 132
Rav Samuel Ben Nachman 117, 129
Rav Simeon 115
Rav Simeon Ben Lakish 117, 123
Rav Simlai 129
Rav Simon 122

Rawls, John 22
Rosenzweig, Franz 4–5, 12, 18, 21, 25–73, 77, 79–81, 98–100, 104–109, 113, 136, 148–152, 176, 197–198, 202, 210, 222, 225, 227, 238, 241–258
Russell, Bertrand 43, 249, 256

Saadia 12, 21, 77, 79, 82, 90, 107, 146
Schelling, Friedrich Wilhelm Joseph von 25, 31, 36, 43
Schrödinger, Erwin 210
Schwarz, John 210
Schweitzer, Albert 52
Sforno, Obadiah Ben Jacob 21, 80, 106, 112–113, 136–152, 210, 241
Shammai 115
Shipman, Harry L. 209
Simon Ben Zemach Duran 29
Socrates 168
Solon 168
Spinoza, Baruch 8, 12, 21, 29–30, 41, 65, 99, 109, 177, 210, 231, 246–247, 255
Staub, Jacob J. 209
Steinheim, Solomon Ludwig 30–31

Thales 173, 192
Timaeus 2–3
Timaeus of Locrus 168
Trefil, James S. 215, 238

von Neumann, John 210

Weinberg, Steven 209
Wheeler, John 210, 222
Whitehead, Alfred North 27–28, 43, 257
Wigner, Eugene 210
Wittgenstein, Ludwig 252

SUBJECTS

= 44–45
= A 43, 48, 54
= B 54
A 38, 40, 42–48, 54
A = 49, 54
A = A 38–41, 46–47, 50
A = B 45
Absolute One 83, 88
Academy for Jewish Philosophy 9–10
Accident 88, 95
Act 62

Index of subjects 351

Act of Creation 61
Act, Dynamic 158
Act, Relational 61
Act, Stable 158
Action 66–67, 73
Active 175, 197
Active and Passive 59, 171, 195
Active Intellect 87–88
Actual 142
Actuality 142, 163, 227
Adamites 124
Affection, Perceived 190
Air 93, 116, 131, 134, 137–141, 143–144, 150, 173–176, 182–184, 188–190, 217
Air-Life 125
Algebra 46, 189, 211, 249, 251–253
Algebra, Linear 211, 255
Algebra, Non-Linear 254–255
All 34–35, 44, 105
Amora 111
And 253
Angel 85, 89–90, 117, 124–125, 129, 132–134, 137, 140, 142–143, 146–147
Animal 89, 175, 191
Animal, Land 190
Annihilation 214–215
Anthropic Principle 236
Antilepton 214
Antiquark 214
Aristotelian, Jewish 108–109
Aristotelianism 82–83, 85, 91, 97, 102, 197, 248
Aristotelians 141
Aristotelians, Jewish 79
Arithmetic 182, 188
Art 4
Artist 246
Astronomy 3, 83, 89, 97, 134, 142, 167
Astrophysics 10–11, 13, 21, 143, 248
Asymptote 2, 28, 32–33, 39, 46, 65, 69, 80, 106, 149, 197, 202, 216, 219, 222, 238, 253–254
Asymptotic Series 211
Athena 168
Athens 168–169
Atlantis 168–169
Atom 100, 215–216, 235
Atom, Volume of 215
Atomism 82, 193, 197, 243–244, 262–263
Atomist 100, 174–175

Attraction 190, 215
Attribute 58, 60–61, 64, 71, 80, 96, 152, 237
Authenticity 80
Authority, Epistemic 13

B 43–44, 47–49, 54
B = 49, 54
B = A 41, 45, 47, 50
B = B 41, 47, 50, 54
Babylonia 167
Ball 131
Bastard Reasoning, *see* Reasoning, Bastard and Thinking, Bastard
Beast 132
Becoming 59, 104, 170, 183–184, 196
Beginning 221, 226
Behemoth 126
Being 34, 37–38, 59, 170, 183–184, 196
Being and Becoming 103, 170, 176, 184
Belief 30–31, 33–34, 36, 42, 45, 52–53, 55–56, 108, 151, 245, 248
Belief, Jewish 4
Belief, Root 115, 152
Believable 67
Ben Netz 124
Bible 4, 19, 54, 67, 77–78, 91, 105, 108, 151–152, 206, 232, 245
Bible, Interpretation of 70–72
Big Bang 21, 216–217, 227–228, 258
Bird 94, 125, 132, 175, 190–191
Birth 137
Blessing 126, 147, 160, 260
Body 86–89, 93–94, 105, 109, 135, 138, 176, 185, 190, 257
Body, Absolute 93, 102–104
Body, Celestial 143
Box 211

Calculation 122
Calculus 27, 36, 70, 249, 252
Calculus, Platonic 188–189
Capacity 137, 146–147
Cartesian Coordinate 251, 255
Category 45, 51, 62
Category, Conceptual 45
Cattle 132
Causal Chain of Creation 85, 88
Causation 109, 135
Cause 95, 172, 176, 203, 231, 233
Cause, Efficient 87
Cause, Formal 85, 87–88
Cause, Material 85

Index of subjects

Celestial Object 194
Certainty 96, 105
Chance 4, 100, 135, 172, 174–175, 191–193, 203, 231, 233, 240
Chaos 174–175, 225, 254
Character 48–50, 182
Child, Birth of a 183–184
Choice, Human 231
Christian 17
Christian, Protestant 245
Christianity 18
Christianity, Liberal Protestant 77
Christianity, Protestant 245–246, 249
Christianity, Roman Catholic 79
Circle 211
Circuit, Eccentric 96
Civilization, Destruction of 168
Clarity 203
Class 170
Clergy 246
Cognition 61
Coherence and Consistence 110
Cohesion 102
Collection 263
Commanding 160
Commandment 18, 60, 90, 126, 133
Commandment to Know 105
Commentary 136–152, 201, 208
Commentary, Jewish 241
Commentary, Medieval 225
Commentary, Medieval Rabbinic 79
Commentary, Rabbinic 202, 227
Commentator, Jewish 220
Commentator, Rabbinic 226
Common Sense 243, 263
Community 45
Complex Number 211
Compulsion 62
Compulsion-of-Destiny 51
Concatenation 174
Concept 45
Conclusion-expression 56
Conflict 175
Conjunction and Disjunction 174
Conjunction and Separation 100
Constellation 132
Contingency 96–97, 105, 109, 172, 254
Continuity 33
Continuum 192–193
Correlation 226, 256, 262
Cosmogony 1, 5, 21, 80, 90–106, 108, 145–153, 167, 191, 197, 248

Cosmology 1, 3, 5, 21, 72, 80, 83–90, 108, 134, 142–143, 167, 191, 197–198, 248
Cosmology, Negative 42
Cosmos, see Universe 195
Course 28, 51, 63, 250
Creation 26, 28, 32–34, 37, 41, 90–91, 169, 202, 204, 250, 252
Creation and Creator, no distinction between 2
Creation, as a Basic Principle of Belief 18
Creation, Dogma of, 2, 5
Creation, Out of Nothing 101–102, 123, 133–134, 137–138, 142, 150–151, 195, 221, 226–227, 251
Creation, Single act of 2
Creation, Unity of 115, 132–133, 159–160
Creative word 57
Creature 57
Creeping-Life 126
Criminal 191
Cube 177, 187–188
Curse 121
Curve 254

Dark 71, 93, 117, 124, 135, 138–139, 150, 157, 160, 227
Darkness 58–60, 64, 69
Day 92, 98, 121, 130, 132–133, 139, 157–160, 195, 204
Day and Night 93
Day, Zero 160
Day, First 57–61, 93, 113–119, 121, 124, 128, 132, 136–137, 140, 142–143, 160, 162, 227
Day, Second 117–119, 124–125, 132, 137, 143–145, 160–162, 227
Day, Third 93, 119–121, 125, 132, 137, 144–146, 157, 160, 162
Day, Fourth 93, 121–125, 132, 137, 143, 145–146, 157, 160, 162
Day, Fifth 94, 117, 125, 127, 132–133, 137, 146–147, 160, 162
Day, Sixth 61–63, 94, 99, 115, 125, 127, 130, 132–133, 137, 146–147, 160, 162
Day, Seventh 130–136, 147–148, 160, 204, 227
Death 33–34, 62, 137, 190, 261
Death and Birth 165
Death and Life 260–261

Index of subjects

Deep 150, 195
Deity 3, 169–173, 175–176, 180–183, 185, 187, 190–196
Democracy 8
Demon 124, 127–128, 132
Density 26, 62, 174, 185, 218
Design 172, 175
Design, *see* Purpose 192
Design, Argument from 235–236
Determinism 88
Difference, Intermediate 177
Different, Motion of 180–181
Differential Equation 249
Differentiation, Partial 211
Dimension 211
Dimensionality 5, 93, 102–103, 251–252
Disposition 102
Distinctive 105
Diversity 109, 194
Diversity, Source of 83–85
Dogma 2, 4, 6, 14–18, 20, 29, 105–106, 152, 198, 242, 245
Dogma, Jewish 239–240, 243
Doxis 212
Duties of the Heart and Limbs 15
Duty 151
Duty, Jewish 205
Duty, Religious 5, 15, 201–202, 242
Dynamics 27, 33, 63, 67, 69–70, 109, 205, 226, 232, 244
Dynamism 204

$E = \frac{1}{2}mv^2$ 252
Earth 3, 95, 115, 119–120, 123, 125, 129, 131–132, 137–147, 150, 157, 159–164, 171, 174, 188, 191, 194, 226
Earth, Element 173, 175–176, 182–184, 187–189
Earth, Planet 168, 176, 178, 180–181, 190, 194
Ecliptic 87
Ego 41
Egypt 78, 163
Egypt and Mesopotamia 79
Eisegesis and Exegesis 71
Electric Charge 214
Electromagnetic Force 214
Electron 215, 235, 238
Element 28, 32–33, 36–38, 40–42, 46–47, 50–51, 53, 57, 87, 89, 93–94, 102, 119, 123, 131–132, 137, 140–141, 146, 149–150, 161, 173–176, 182–185, 187–190, 192–193, 242, 250–255, 262
Element, Fifth 93, 150
Element, Static 248
Emanation 32, 45, 58, 248
Emanation and Creation 248
Emanationism 248
Emotion 190
Empirical Observation 97–98, 105, 108, 201
Empiricism 55
Empiricist 43
Emptiness 118
End 158–159, 164, 203–204, 222, 226, 250, 258–259
Energy 189, 214–216, 226–227
Entity, Real 181
Entropy 234–235
Epicureans 100
Epicycle 87, 96
Epistemology 55–56, 97, 105, 242, 244–256
Equilibrium, Thermal 214
Essence 42
Essence of the Throne 86
Ethics 34–35, 79, 81, 90, 99, 135, 152, 167, 257–264
Europe 262
Evening 114, 138
Evening and Morning 93
Event 60, 203–204, 238, 244, 257
Evil 59
Evil and Good 232
Evolution 52
Exile 164, 167
Existence 103, 177, 180, 184, 203
Existence, Intermediate 177
Existence, Sort of 183–184, 245
Existent 58
Experience 55, 254

Faith, Biblical 152
Female 191
Festival 15, 121, 133
Field 238, 257
Field, Electric 224
Figure 255
Fire 93, 118–119, 123, 125, 131–132, 134, 138–141, 144–145, 150, 168, 173–176, 182–184, 188–189, 226
Fish 94, 125–126, 132, 175, 191
Flier 160
Flood 168

Index of subjects

Flying-Life 94, 176
Food 126, 162
Force 39, 260
Form 35, 37, 103, 105, 109, 138, 182–183
Form and Matter 86–88, 96, 102–104, 109, 149
Form, Elementary 102
Form, Last 93, 102
Form, Primary 102–103, 138, 140, 142, 149
Forward and Backward 182
Foundation 16
Fourier Series 211
Fractal 254–255
Freedom 36, 50, 151
Function, Hyperbolic 211
Future 40, 52–53

Galaxy 217
Game of Life 234, 236–237
Gamma Ray 214
Garden of Eden 121, 125, 132, 134
Garment, Great 123
Gas 173, 176, 189, 217
Gehenna 119, 123, 132
Generation 184
Generation and Corruption 34
Genus and Specific difference 35
Geometry 28, 70, 175, 178, 182–190, 249, 251–255
Geometry and Algebra 253
Geometry, Euclidean 254–255
Geometry, Pythagorean 211
Germany 11, 31, 33, 37
Glory 146
God 3, 13, 15, 17–18, 26, 29–36, 38, 41–51, 55–66, 69, 82, 88, 90, 103–104, 108, 137, 140, 147, 160, 163–164, 176, 195, 203–204, 226–227, 231, 237–239, 250–252, 258
God creates 142
God says 142
God sees 144
God, Action of 64, 152
God, Caprice of 39, 57
God, Concealed 63
God, Creativity of 247
God, Deed-of-Power of 51
God, Desire of 137–138, 142, 144
God, Essence of 32, 39–40, 42, 45–46, 50, 60, 62, 149, 247

God, Eternality of 39
God, Fate of 39
God, Freedom of 39–40, 45–46, 49–50, 57, 247
God, Hidden 62
God, Image of 61–62, 94
God, Imitation of 259–261
God, Infinite Being of 48
God, Knowledge of 39, 89
God, Mind of 237
God, Mouth of 116
God, Mythical 62
God, Nature of 43, 47, 233
God, Need of 247
God, Obligation of 40
God, Perfection of 231
God, Power of 39, 41, 49, 57, 62, 247
God, Reality of 38, 203
God, Service of 164–166
God, Single act of 3
God, Speech of 64
God, Spirit of 58–59, 64, 66, 69, 71
God, Unity of 2, 18, 29, 83, 90, 92, 152, 237, 259
God, Vitality of 39
God, Volition of 100
God, Will of 4, 49, 66, 72, 92, 100, 104, 109, 135, 138, 142, 150–151, 193, 203–204, 233, 260
God, Wind of 124, 129, 150, 160
God, Word of 13, 17, 54, 57, 61, 98, 108, 116, 118, 120, 132
God, Work of 57, 260
Good 56–57, 59–62, 71, 93, 119, 133, 143, 152, 171, 194, 196, 203–204, 221, 232–239, 260
Good and Bad 171–172, 196
Good, Very 65, 93, 99, 114
Governance 146
Governing 160, 163–164
Grammar 65, 110, 251, 253
Grammar, Definite Article 58
Grammar, Imperative 60, 63, 66–67
Grammar, Past Tense 57–58, 60, 63, 67
Grammar, Present Tense 60, 63, 67
Grammar, Proposition 57
Grammar, Royal We 129
Grammar, Verb 63, 67, 158–159
Grammar, Verb and Noun 257
Grammar, Verb, Consonant of Conjunction 158–159
Great and Small 171, 175
Great Ocean 230

Index of subjects

Greece 168, 262
GUT (Grand Unified Theory) 209

Harmonics 250
Harmony 175, 177
Heaven 85–86, 93, 95–96, 115, 125, 133, 138, 142–144, 147
Heaven and Earth 221, 227–232
Hegelianism 252
Hell 123–124
Hellenism 78–79, 134
Heresy 14
History 30, 52, 167, 254
History of Ideas 213
History, Christian 261
History, Intellectual 206
History, Jewish 78–79, 261
History, Muslim 261
Holiness 260
Holy of Holies 124
Homiletics 110
Host 93
Human 88, 127, 133, 160, 162–163, 177–178
Humanity 145, 148, 164

I-Thou and I-It 257
Icosahedron 187–188
Ideal 40, 104, 202–204, 216, 222, 226, 232, 238, 259
Idealism 31, 33, 43, 45, 59, 255
Ignorance 11, 13
Ignorance, Veil of 22
Image 184
Image of God 146
Impulse, Innate 185
Indefinite 143, 203, 212, 256
Indefinite and Definite 232
Indefinite Particle Position and Velocity 216
Indeterminacy 244, 258
Indeterminate 231
Indeterminate and Determinate 259
Individual 34–35, 40, 43–47, 50, 58, 61–62, 64, 88, 105, 152, 160, 204, 238, 257, 262–263
Individual, Knowledge of 34
Individuality 35
Indivisibility and Divisibility 177
Infinite 218
Infinite Dyad 170–171, 185
Infinite Judgment 36–37, 42
Infinite Series 211

Infinitely Small 218
Infinitesimal Residuum 42
Infinitesimal Calculus 250
Infinity 204, 219, 229
Integral, Multiple 211
Intellect 85, 88–89, 93–94, 109, 134–135, 139–140, 143, 146
Intention 172
Interference 223
Interpretation 157
Interpretation, Rabbinic 110–111
Interval 177
Irrationality 2, 9, 36
Islam 16, 77, 82–83
Israel 13–14, 16, 114, 164
Israel, Ancient 262
Israel, Nation of 163–165

Jerusalem 124, 164
Jewish 1, 6, 9–10, 77, 106–107, 151, 197, 205, 241
Jewish Thought, Parameters of 148–152
Jewishness 29, 70, 107–108
Judaism 29, 77–79
Judaism, Beliefs of 15–18, 90
Judaism, Orthodox 26
Judaism, Rabbinic 9–16, 26, 79
Judgment, A Priori and A Posteriori 242
Judgment, Analytic and Synthetic 242
Jupiter 87, 122, 181
Justice 175

Kabbalah 32, 111–112
Kalam 79
Karaites 16
Kind 182
Knowledge 34, 52, 89, 96, 130, 145, 247
Knowledge and Truth 3, 8–9, 11
Knowledge, Limits of 1–4, 7–8, 91, 95–97, 108, 201–202
Knowledge, Precise and Imprecise 3–4
Koch Curve 255

Land, Dry 93, 119–120, 125, 132, 143–144, 161, 175
Land-Life 88, 93–94, 125–126, 128, 130, 176
Language 4–5, 33, 54, 98, 110, 135, 171, 210–213, 246, 251–252
Language, Biblical 257
Language, Mathematical 213
Language, Silent and Spoken 251–252
Law 4, 260

Law, Jewish 5, 15, 95, 151
Law, Jewish, Mosaic 15
Law, Rabbinic 79
Lepton 214
Leviathan 126, 132
Levite 164
Liberal 2, 9–10, 12, 22, 263
Life 45, 63, 88, 133, 146–147, 160, 175–176, 190–192
Light 60–61, 64, 69, 71, 93, 117, 119, 124, 132, 135, 137–143, 145–146, 150, 157, 160–161, 222–224, 227
Light and Dark 171–172, 196
Light, Speed of 228
Lighter 160, 162–163
Limit 59, 171, 175, 216, 226, 238–239, 250
Limit and Unlimit 171–172
Limit and Unlimited Multitude
Line 185, 254–255
Liquid 173, 176, 189
Living Thing 164, 175
Logic 5, 35–36, 110, 201, 251–252, 254
Logic, Universal operator 43
Logos 42–44
Love 62, 129, 175
Loyalty 11–12, 263

Maimonidean Controversy 112
Maimonides' Thirteen Basic Principles 91
Makes 128, 171
Male 190
Male and Female 59, 146, 171–172, 190, 196
Man 26, 30–36, 41–42, 45, 47, 54, 61, 63–66, 69, 71, 88, 94, 125, 128–130, 146–147, 176, 250–252
Man, Being of 48
Man, Essence of 48
Man, Freedom of 49
Man, Knowledge of 35
Man, Will of 36, 49
Mars 87, 122, 181
Mass 102, 226
Mathematics 6, 37, 50, 70, 110, 177–178, 205, 217–218, 233–234, 243, 248–257, 263
Mathematics, Geometry 32, 37, 46
Mathematics, Number 37
Mathophobia 27
Matter 35, 37, 93, 105, 109, 138, 142, 189, 193, 215

Matter, Conservation of 189
Matter, Primary 92–93, 102–104, 138, 140–143, 149, 192, 227
Mean 177–178
Meaning 247
Meaning, Hidden 110
Meaning, Simple 110
Measurement 224, 230
Mercury 87, 122, 129, 133, 181
Mesopotamia 78
Messiah 114
Messianic Age 117, 132
Metaethics 34–35, 42, 48, 51
Metalogic 34–35, 42–43, 45, 47–48, 51, 54
Metaphysics 34–36, 38–43, 47–49, 51
Methodology 19–20, 70–72, 151–152, 206–213
Midrash 4–6, 21, 78–79, 106, 111, 127, 134–136, 142, 150, 193, 207–208, 226, 230, 253
Mind 59, 175, 257
Mineral 87–88
Miracle 52, 54, 91
Mirror 183–184
Mist 144
Modality 229–230
Model 6, 64, 70, 175, 178, 182, 184, 203–204, 211, 216, 227, 232, 242, 244, 249–251, 253, 255, 261–262
Model, Geometric 250
Model, Mathematical 210–212
Molecule 215
Moment 58, 137
Moment, First 57–60, 63
Moment, Second 60–61
Monarchy 8
Moon 87, 122–123, 125, 132–133, 157, 176, 180–181
Moon, New 133
Moral and Amoral 248
Moral Idea 98
Morality 4, 18, 66, 80, 93, 109, 145, 151, 191, 203–204
Morning 138
Motion 88, 174, 178–183, 185, 192–193, 244
Motion of the Same 139
Mover 86, 88–89
Multiplicity 8, 59, 62
Music 177–178, 220
Mystery 6, 50, 252
Mysticism 110, 142

Index of subjects

Mysticism, Jewish 78, 80
Myth 3–4, 72, 164, 167, 173, 213, 219, 245–246

Name, Proper 61
Naming 160
Nation 264
Nature 172, 174–175
Necessity 4, 59–60, 64, 70, 89, 92, 100, 109, 135, 169–178, 182, 192–193, 195, 203, 231, 240
Negation 36–39, 42–43, 48–49, 63, 245
Negative and Positive 232, 259
Negativity 104, 136, 204, 232, 249
Neith, Priests of 168
Neo-Platonism 79, 85
Neutron 214–215
New Thinking 33, 37
Newtonian Mechanics 252
Night 138, 160
No 253
Noachite 163
Non-Being 103
Not-Otherwise 43
Nothing 31–34, 36–38, 43, 46, 48–49, 51–52, 55, 64, 69, 73, 99, 101, 103–104, 109, 150, 184, 195, 197, 203–204, 218–219, 226–227, 238, 240, 253, 258
Nothing, Absolute 138
Nucleus 215, 235, 238
Number 170–172, 175–178, 185–186, 188–190, 250
Number, Irrational 186–187
Number, π 170–171

Object 57–58, 63–65, 67, 69–70, 73, 79–80, 142, 195, 203, 238, 243, 249, 256–258
Object, Celestial 87–88, 95, 121, 125, 142, 144–145
Object, Indefinite 104
Objective 57
Objectiveness 57
Objectivity 20, 61
Occupant 132
Ocean, Great 124
Octahedron 187–188
Odd and Even 171–172, 185
One and Plurality 171–172
Ontology 54–55, 67, 71–73, 79, 81, 105, 134, 136, 144, 183, 194, 197, 203, 238, 244, 256–258, 263

Opinion 8, 89, 97
Opposites 170
Opposites, Equilibrium of 175
Opposites, Harmony of 195–196
Organism 165, 175, 192
Origin 104, 204, 226, 250, 258–259
Originality 151
Orthopraxis 26
Ought 152
Ought and Is 259
Overflow 85–86

Packing-Unpacking 253
Paganism 31
Panathenaea 168
Paradise 123–125
Parallel Universe Theory 230
Participle 59, 212, 226, 238, 257, 262
Particle, Elementary 214
Particle, Nuclear 214
Particular 34, 43–50, 58, 64, 80, 152
Particularity 104
Passive 171, 175, 197
Past 40, 52–53, 202, 204, 224
Patriarch 114
Peace 129
Peculiarity 51
Pegasus 101
Pentagon 188
Perception 182, 184, 190
Person 49, 61–62, 65, 67, 69
Personality 49–50, 61–62
Philosopher 191
Philosopher, Christian 228
Philosopher, Jewish 107, 136, 220, 228
Philosopher, Muslim 228
Philosophical 14
Philosophy 2, 6, 22, 31, 33, 37–38, 41–43, 45, 51–56, 63–64, 67, 69, 98, 105, 107, 110, 167, 252
Philosophy and Judaism 77
Philosophy and Religion 6, 108
Philosophy and Science 106
Philosophy and Theology 5, 37–38, 52–56, 68–69, 105–106, 136, 244, 252
Philosophy of Religion 6, 73, 240–264
Philosophy, Christian 7
Philosophy, Classical 70
Philosophy, German 26
Philosophy, Greek 79, 99, 173–175, 248
Philosophy, Historical 8

Philosophy, Jewish 5, 15, 78, 82–83, 97, 99, 105–107, 113, 135–136, 202, 208, 231, 237, 239, 241
Philosophy, Jewish, Liberal 6–14
Philosophy, Jewish, Medieval 4
Philosophy, Limits of 68
Philosophy, Modern 25, 231, 244
Philosophy, Muslim 231
Philosophy, Negative 245
Philosophy, New 38, 41
Philosophy, Rabbinic 70, 79, 92
Photoelectric Effect 223
Photon 214, 222–224, 227–228, 230
Physics 21, 35–36, 38, 47, 73, 97, 99, 109, 135–136, 141, 145, 167, 185, 189, 191–192, 205–240, 243–245, 257–258, 263
Physics and Ethics 191, 196
Physics, Newtonian 262–263
Pi Meson 214
Picture 176, 178, 191–192, 210, 213
Picture, see Model 253
Pious 121
Plane, Rectilinear 185
Planet 87
Plant 89
Platonism 79, 153, 250
Poetry 220, 247, 253
Point 185, 250, 252–254
Poland 11
Political theory 79
Politics 196, 204, 232, 246, 262–264
Polytheism 134
Position 174
Potential 142
Potentiality 142, 163
Power 137, 140, 145
Praxis 212
Praxis and Doxis 15
Pre-Creation Entity 227
Pre-Creation Universe 114, 123–124, 132–134
Pre-Existent 57–58, 60, 225–226
Pre-Existent Entity 123, 137–138, 150, 152, 160
Present 40, 52–53, 204
Priest 164–165, 167, 232, 261
Principle 170, 176–177, 203
Principle, Basic 169–176
Principle, Formal 103
Principle, Innate 190
Principle, Material 103
Probability 9, 216, 230, 235, 245, 263

Probability Series 211
Process 28, 33, 54–55, 63, 70, 80, 109, 149, 159, 202–203, 218, 222, 226, 240, 244, 257
Procreation, see Reproduction 147, 163
Proper Name 256
Prophecy 62, 165, 253
Prophet 247
Proton 214–215
Prototype 64, 160
Providence 90
Providence, Divine 247, 262
Psychology 190
Purpose 104–105, 109, 172, 191–193, 196
Purpose, see Design 192
Purpose, see End 166
Pursuer 260
Pyramid 187–188, 252
Pythagoreans 171, 185, 192

Quality 176, 182–185
Quantum Mechanics 209, 216, 222, 229, 231, 263
Quantum Mechanics, Copenhagen Interpretation of 231
Quark 26, 214

Radiation 26, 214
Radius 180
Rain 132
Ratio 87
Ratio of Elements 100
Rationalism 246, 250
Reality 18, 31–35, 37, 40, 50–51, 53–55, 66, 69, 73, 80–81, 105, 109, 124, 136–137, 171, 173, 176, 178, 195, 211–212, 216–218, 221–222, 230, 240, 244–246, 251–252, 256–257, 262–263
Reality, Conception of 63
Reality, Language of 54
Reality, Model of 250, 258
Reality, Picture of 173, 252
Reason 6, 11, 14, 95–97, 108, 115, 146–147, 169–173, 175–185, 187, 190–195, 201, 252
Reason and Faith 249
Reason and Revelation 248
Reason, Limit of 249
Reasonable 19
Reasoning 9
Reasoning, Bastard 3, 184, 245

Index of subjects

Reasoning, Contextual 7–8
Receptacle 59–60, 70, 169–172, 176, 182–184, 188, 195
Rectangle 211
Redemption 21, 26, 28, 30, 33, 52–54, 62–63, 66, 69, 106, 109, 135, 202–204, 222, 239–240, 250, 252, 258–259, 261, 264
Relation 54, 72–73, 79, 250–255, 257
Relation, Dynamic 248
Religion 30, 246, 256
Religion and Science 175
Religion, Greek 173, 175, 192
Reproduction 126–127, 146, 160, 163
Reptile 190
Rest 131–132, 147, 204, 227
Rest and Motion 171–172, 196
Revelation 5, 21, 26, 28–30, 33–34, 36, 41, 45, 51–56, 61–66, 69, 71, 89–91, 98, 105–106, 108, 130, 132, 135, 157, 165, 201, 245, 247–248, 250, 252, 256–257, 259, 261–262, 264
Revelation and Reason 242
Revelation, Mosaic 165
Reward and Punishment 165
Right 263–264
Right and Left 172, 182
Righteous 117, 126, 129, 133
Righteousness 129
Ritual 121, 133, 160
Robe of God 117

S is P 38
Sabbath 15, 121, 131, 133, 148, 160, 163, 204
Sacrifice 164, 232
Sages 67, 81, 83, 90, 104, 106–108, 113–136, 156, 167, 191, 193–194, 197–198, 202, 205, 221, 225, 227, 230–231, 242, 260
Sages, Commentaries of 107–113
Sais, Egypt 168
Salvation 17
Same 180
Same and Different 171–172, 177–180
Same, Motion of 180–181
Sameness, Intermediate 177
Sassanid empire 78–79
Saturn 86, 122, 181
Saying 160
Schema 108, 152, 192, 198, 205, 221, 233, 239–241, 243, 248
Scholasticism 53

Schrödinger's Cat Paradox 224, 230, 263
Science 3–6, 15, 26–27, 42, 52, 54–57, 63–64, 73, 89, 98, 105, 108, 151–152, 167, 177, 202, 206–240, 247, 256–257, 263
Science and Ethics 99, 221, 232–239, 244
Science and Judaism 239–240
Science and Philosophy 243
Science and Religion 10, 14, 82–83, 91, 95, 105, 108–109, 111, 196, 213, 220–221, 243–250, 256
Science and Theology 106
Science and Tradition 242
Science, Aristotelian 91, 104
Science, Asia Minor 193
Science, Cosmology 26
Science, Greek 192–194
Science, Obsolence of Conclusions 11
Science, Physics 3, 6, 26
Science, Platonic 104
Science, Roman 193
Science, Transient Nature of 209–210
Scientific 14
Scientist 246
Scripture, Approach 4–5
Sea 137, 143–144, 161, 191
Sea-Life 94, 125, 127, 130, 176, 190
Seas 119–120, 132
Season 121–122
Secular and Religious 245–246
Self 49–50, 60
Self-contemplation 82
Semantics 110
Sequence 239
Serpent 127
Set 218
Shape 174, 182
Sign 121
Sin 133
Sinai 65, 130, 132, 163, 165
Singularity 48–49, 209, 214, 218, 226–227
Sivan 130, 132
Size 174, 188
Sky 119–121, 123, 125, 129, 132–133, 137, 142–144, 159–162, 171, 188, 226
Slavery 14
Snow 131
So, It was 133, 143
Solid 173, 176, 185–186, 189

Something 32, 34, 36, 38–41, 43, 46–47, 55, 64, 73, 99, 101, 104, 109, 137, 150, 203–204, 227, 238, 240, 253, 258
Sophist 51
Soul 30, 62, 85–90, 93–94, 109, 127, 129, 133, 135, 137, 146–147, 175–184, 190, 194–195
Soul, Human 181–182, 185, 190
Soul, Transmigration of 190–191
Space 1, 3, 59, 64, 69, 72–73, 79–81, 87, 89, 93, 98–99, 102, 104, 108–109, 123–125, 133, 135–138, 140, 142–144, 147–148, 152, 160, 163–164, 203–204, 217–222, 226–227, 235, 238, 243–245, 251, 258–259
Space, Differentiation of 160–162
Space, Empty 215
Space, Occupant of 162
Space-Time 209, 217, 219, 244, 249
Species 44–45, 50, 140, 142, 147, 152
Species and Individual 88
Speech 62, 65, 69
Speech of Revelation 61
Speed, *see* Velocity 180
Sphere 89–90, 93–94, 96, 109, 124, 138–139, 144, 174, 176, 178, 180–181, 188, 211, 228
Sphere, All-Encompassing 86–87
Sphere, First 87
Sphere, Inclined 87
Sphere, Northern 87
Sphere, Right 86–88
Sphere, Southern 87
Spin 178, 180
Spirit 59
Spiritual 135
Spontaneity 172
Square and Oblong 172
Star 93, 122, 125, 137, 139, 143, 145, 157, 160, 176, 180–181, 216
Star of David 250, 252–253, 255
Steppes 124
Stoic 141
Stone of Temple Mount 124
Story 1, 3–4, 18, 163, 165, 168–170, 173, 175–176, 181–183, 190–191, 194–197, 205, 213, 232, 242, 245–247
Straight and Crooked 171
String Theory 209–211, 213
Strong Force 214
Structure 63, 255

Subject 63
Substance 37–38, 57, 65, 67, 69, 80, 101, 104, 178, 181, 194, 203, 244, 256
Substantiality 57–58, 67
Sun 87, 122–123, 125, 132–133, 157, 176, 180–181, 238
Swarm 145, 160
Syllogism 7, 169
Symbol 213
Symbol, Algebraic 253
Syracuse 168

Tale, *see* Story 163
Tanna 111
Teleology 82, 135, 151
Telos 232
Telos, *see* Purpose 196
Temperature 26, 214–215, 218
Temple 114, 123–124, 164, 261
Testimonial 62
Text 11–14, 19, 22
Text, Approach to 19, 21–22
Text, Selection of 12, 19–22
Theatre, Greek 173
Theology 28, 30–31, 36–38, 40, 52–53, 55–56, 63, 66–67, 69, 98, 105–106, 108, 257
Theology and Philosophy 37–38, 52–56, 68–69
Theology, Christian 53
Theology, Historical 51–52
Theology, Jewish 105, 136, 257
Theology, Negative 38, 47
Theology, New 245
Theology, Rabbinic 79, 110
Thermodynamics, Law of 239
Thermodynamics, Second Law of, *see* Entropy 234
Thing 58, 101, 170, 226
Thinking, Bastard 213
Thinking, Mathematical 33–34, 50, 56
Thinking, Philosophic 253
This-World 114, 116, 118, 121, 123, 126, 131, 135, 148
Thought 35, 114
Throne of Glory 114, 116, 123–124, 131, 138, 226
Thus let it be 65
Time 1, 2, 7, 67, 86, 92, 98, 105, 129, 133, 138–141, 143, 151–152, 157–158, 160, 169, 170, 176, 191, 195, 203–204, 218, 221–222, 225–226, 243, 259

Index of subjects

Time Zero 217–219
Torah 4, 14–17, 85, 90, 100–101, 114, 123–124, 129–130, 132–133, 163–165, 196, 232, 259–261
Torah, Limits of Interpretation 151–152
Tower of Babel 163
Tradition 12–14, 201–202
Tradition, Jewish 71, 205
Traditional 9–10, 12, 14–16
Transcendental Unity of Apperception 47–48
Transfinite Number 218
Transformation, Coordinate 211
Tree 120, 132–133
Triangle 37, 186–187, 250, 255
Triangle, Equilateral 187
Triangle, Equilateral Scalene Right 186–187
Triangle, Half Equilateral, Scalene Right 2
Triangle, Isosceles Right 2, 186–187
Triangle, Right 188–189
Truth 5–9, 14, 72–73, 96, 105, 110, 129, 151–152, 220–221, 242, 246–247
Truth and Praxis 212
Truth and Tradition 10–14
Truth, Factual 7
Truth, Moral 7
Truth, Religious 14
Truth, Scientific 11–12

Uncertainty Principle 212, 223–224
Underworld 191
Unit 185
Unity 35, 59, 171, 175, 194
Unity and Diversity 72, 80
Universal 38, 42–48, 50, 62
Universal, Individual 44
Universal-Soul 176
Universe 1–3, 5–6, 19, 36, 58–59, 65, 69, 72, 92–93, 95, 97, 99, 109, 123, 151, 163, 165, 169, 175, 195, 203–204, 229
Universe, Actual 229
Universe, Best Possible 231–232
Universe, Elemental 148–150
Universe, End of 147, 240
Universe, Eternity of 152
Universe, Goodness of 145
Universe, Inflationary 209, 213, 228–229
Universe, Model of 218–219
Universe, Multi-Heavened 121, 123–125, 133–135

Universe, Multiple 204, 221, 227–232
Universe, Origin of 141–143, 214–215, 221, 226, 239
Universe, Physical 132, 137–138, 215
Universe, Picture of 216–217
Universe, Possible 229
Universe, Purpose of 239
Unlimited 192
Unlimited Multitude 59, 171, 173, 175
Unmoved Mover 85–86
Up and Down 182

Value 4–5, 7, 9, 14, 20, 60, 64, 73, 99, 137, 151, 204, 239, 261
Value of Universe 166
Vector 28, 32, 39, 46–47, 69–70, 87, 180–183, 244, 250, 253–255
Vector Analysis 211
Vector-Segment 255
Vegetation 88, 93, 120–121, 125, 132, 137, 145, 160, 162–164
Velocity 97, 139, 178, 180
Velocity, Angular 87, 96
Venus 87, 122, 181
Virtue 134
Void 195
Volition 54, 61, 109, 231
Vortex 100, 139, 174

Waste 69
Waste and Empty 58–61, 64, 71
Water 93, 104, 116–117, 119–120, 123–125, 131–132, 137–141, 143–146, 150, 161–162, 168, 173–176, 182–184, 187–191, 204, 227
Wave and Particle 222–223
Weak Force 214
Weight 173–174
Will 66
Will, Defiant 49, 51
Wind 116, 150
Wisdom 138, 146, 226
Wonderwoman 178
Word, Action- 59–60
Word, Basic 30
Word, Event- 58
Word, Root- 56
Work 131–132
World 26, 30–36, 38, 40–41, 47–51, 55–56, 63, 69, 176–177, 250–252, 261
World, Being of 48

World, Birth of 51
World, Essence of 48
World, Sublunar 142, 150
World, Supralunar 150
World-Body 194
World-Soul 43, 175, 178, 180, 183, 194
World-Spirit 43
World-Thought 42
World-To-Come 117–118, 121, 126, 132, 135, 137, 148, 163, 204, 259
Worship 13, 204, 261

Worship, Communal 63

Xestes 118

$y = x$ 38–40, 251
Year 121
Year, New 133
Year, Walking 123
Yes 253

Zero 185
Zodiac 87, 180

GENERAL THEOLOGICAL SEMINARY
NEW YORK